Oil Palm

Flows, Migrations, and Exchanges

Mart A. Stewart and Harriet Ritvo, *editors*

The Flows, Migrations, and Exchanges series publishes new works of environmental history that explore the cross-border movements of organisms and materials that have shaped the modern world, as well as the varied human attempts to understand, regulate, and manage these movements.

Oil Palm

A Global History

JONATHAN E. ROBINS

The University of North Carolina Press Chapel Hill

This book was published with the assistance of the Luther H. Hodges Jr. and Luther H. Hodges Sr. Fund of the University of North Carolina Press.

Set in Charis by Westchester Publishing Services
Manufactured in the United States of America

The University of North Carolina Press has been a member of the Green Press Initiative since 2003.

Library of Congress Cataloging-in-Publication Data
Names: Robins, Jonathan, author.
Title: Oil palm : a global history / Jonathan E. Robins.
Other titles: Flows, migrations, and exchanges.
Description: Chapel Hill : University of North Carolina Press, [2021] | Series: Flows, migrations, and exchanges | Includes bibliographical references and index.
Identifiers: LCCN 2020056875 | ISBN 9781469662886 (cloth ; alk. paper) | ISBN 9781469662893 (paperback ; alk. paper) | ISBN 9781469662909 (ebook)
Subjects: LCSH: Palm oil industry—Africa—History. | Palm oil industry—Southeast Asia—History. | Palm oil industry—Political aspects—History. | Oil palm—Industrial applications. | Capitalism.
Classification: LCC HD9490.5.P343 A357 2021 | DDC 338.1/73851096—dc23
LC record available at https://lccn.loc.gov/2020056875

Cover illustration: Palm nut harvesting in Côte d'Ivoire © Cirad; black rolled ink texture © Shutterstock.com/Itsmesimon.

An earlier version of chapter 7 appeared as "Shallow Roots: The Early Oil Palm Industry in Southeast Asia, 1848–1940," *Journal of Southeast Asian Studies* 51, no. 4 (2020). Material in chapters 5 and 6 appeared in "Food and Drink: Palm Oil versus Palm Wine in Colonial Ghana," *Commodities of Empire Working Paper Series* 25 (2016); "Smallholders and Machines in the West African Palm Oil Industry," *African Economic History* 46, no. 1 (2018), and "'Imbibing the Lesson of Defiance': Oil Palms and Alcohol in Colonial Ghana, 1900–40," *Environmental History* 23, no. 2 (2018). Material in chapters 4 and 9 appeared in "Oil Boom: Agriculture, Chemistry, and the Rise of Global Plant Fat Industries, ca. 1850–1920," *Journal of World History* 29, no. 3 (2018).

Contents

Figures, Graphs, and Maps

Figures

Graphs

Maps

Abbreviations Used in the Text

CDC	Colonial Development Corporation (Commonwealth Development Corporation in 1963; CDC Group Plc. in 1999)
FAO	Food and Agriculture Organization of the United Nations
FELDA	Federal Land Development Authority
FFA	Free fatty acid
FMS	Federated Malay States
GOPDC	Ghana Oil Palm Development Company
H&C	Harrisons & Crosfield
HCB	Huileries Congo Belge
INEAC	Institut national pour l'étude agronomique du Congo belge
IRHO	Institut de recherches pour les huiles et oléagineux
MPOB	Malaysian Palm Oil Board
NEI	Netherlands East Indies
NES	Nucleus Estate-Smallholder
NGPI	National Development Corporation–Guthrie Philippines
NIFOR	Nigerian Institute for Oil Palm Research
PNG	Papua New Guinea
POEM	Palm Oil Estates Managers, Ltd.
PORIM	Palm Oil Research Institute of Malaysia
RBD	Refined, bleached, deodorized
RSPO	Roundtable for Sustainable Palm Oil
SOCFIN	Société Financière des Caoutchoucs
UAC	United Africa Company
UFC	United Fruit Company
WAIFOR	West African Institute for Oil Palm Research

Oil Palm

Introduction

In 1841, a Liberian colonist delivered a piece of advice to African Americans contemplating a new start in the fledgling West African state: plant oil palms. Besides being "the most beautiful" trees, their fruit yielded an oil that was "altogether preferred by the colonists to lard." The young stem of the palm could be eaten as a vegetable, and one could tap "a very pleasant wine" from older trees. Featured prominently on Liberia's official seal, the oil palm seemed to be a living symbol of Africa's tropical bounty. One writer predicted that palm oil would soon be "one of the heaviest articles of traffic in the commercial world."[1]

Palm oil has indeed become one of the world's most important agricultural commodities. Extracted from the flesh of the oil palm's fruit, palm oil is the most widely consumed fat on the planet. It's in cooking oil, peanut butter, cookies and crackers, soaps, cosmetics, plastics, and biodiesel. Palm kernel oil, extracted from the nut at the center of the fruit, serves in nearly as many roles. What the prognosticators of the 1840s got wrong was where and how this would happen. Farmers and settlers across western Africa launched palm oil as a global commodity in the nineteenth century. Yet today, the vast majority of the world's palm oil comes from Southeast Asia. Africa now imports ten times more palm oil than it exports.

This book traces the history of that remarkable reversal, following the oil palm's journeys out of Africa and the development of palm oil as a commodity. Not a natural history of *Elaeis guineensis* (the Linnaean name for the oil palm), this is rather the story of how humans used and lived with oil palms—a history that has been anything but natural.[2] Humans shaped where and how oil palms were cultivated and what ends they served. The dominant role of palm oil in today's food system owes something to the plant's innate biology but much more to political and economic power. The people who grew and harvested oil palms, rather the tree itself, are what made palm oil cheap and plentiful on the world market. Historically, many of the people doing that labor weren't free to choose how they worked with this plant. Elites, states, and corporations shaped systems of production, making and remaking oil palm landscapes to suit their interests.

It's a story shot through with the stark realities of power and injustice, of slavery, colonialism, and violence. Yet it's also a story of opportunity: of farmers taking up oil palm on their own initiative or joining pioneering settlements, seeing the promise of a better life in neat rows of palms.

Consumers were important in the oil palm story, too. Like most commodities, palm oil became part of an expansive commodity "web," connecting individual producers, consumers, and middlemen.[3] Decisions made in a nineteenth-century British candle factory could reverberate through palm groves in Nigeria, making or breaking the fortunes of oil producers. The reverse wasn't always true: chemical technologies allowed manufacturers to transform and interchange palm oil with other fats, weakening the market power of oil palm farmers. Cheapness, versatility, and invisibility to consumers ultimately became palm oil's main selling points.

This book isn't encyclopedic. I've left some historic uses for palm oil unexplored (ski wax, to take one example). I haven't discussed some of the places where palm oil was historically produced and traded (Lusophone Africa is conspicuously absent). In an effort to tell a global story, I have resorted to broad labels like "Africans" and "Europeans" that run the risk of overgeneralizing. The stories I've assembled here highlight the biggest groups of producers and consumers of palm oil in the past, and I include particular cases that highlight key trends—or exceptions to them.

I tell most of this story through written archives: letters, travelogues and memoirs; records left by corporations; and memos and reports filed away in government archives. These are the documents that survived time, bankruptcies and acquisitions, decolonization, and the whims of record keepers. The people who are at the core of this story—farmers and workers in Africa, Asia, and Latin America—left relatively few written records. When they do appear in the archives, their voices are often filtered through European observers complaining about their behavior. Wherever possible, I've supplemented archival sources with oral histories, ethnographies, and journalistic accounts recorded across the twentieth and twenty-first centuries. The endnotes show my deep debt to the researchers who wrote down—and are still recording—the stories and experiences of people living with oil palms.

The Story Ahead

Part I: Africa and the Atlantic World

The book begins with an introduction to the oil palm tree and its deep history in Africa. Beyond getting the reader acquainted with *E. guineensis* as a physical thing, chapter 1 highlights the role of humans in *making* oil palm landscapes. Humans helped bring the oil palm out of a marginal existence in swamps and forest edges. Together, they colonized grassland and forest regions across western Africa. Wherever there were oil palms, there were humans. And oil palms provided much more than oil, offering wine, medicine, fuel, timber and thatching, and fiber. Contrary to what you'll read in most accounts, palm oil wasn't traded across vast distances in antiquity, winding up in Egyptian tombs. Bad science and Eurocentric thinking created a false past for the oil palm, disregarding the ways Africans used oil palms on their own terms.

Europe's "discovery" of the oil palm in the fifteenth century launches chapter 2. Though European ships carried palm oil from port to port in western Africa and brought some to Europe, that trade was soon eclipsed by the traffic in captives across the Atlantic. Palm oil sustained and lubricated the Atlantic slave trade, and enslaved Africans used their knowledge of oil palms to survive and resist slavery in the Americas. Oil palms traveled, too, taking root in a new continent. Meanwhile, Europeans appropriated African knowledge about the medicinal uses of palm oil for their own benefit. By the seventeenth century, palm oil was a commonly available drug in England, and by the late eighteenth century, it was beginning to find new roles in soap and other substances.

That new demand supported a growing trade in palm oil, which in many textbook accounts neatly replaced the slave trade. The story isn't entirely wrong: as European powers moved to abolish the Atlantic slave trade, palm oil did become an important export commodity from Africa. Chapter 3 makes a case that the transition was more complicated than often imagined. The core institutions of the slave trade—and slavery itself—adapted to the palm oil trade. Enslaved Africans made much of the palm oil that Europeans celebrated as a "legitimate" product. Large plantations, new tools, and a new product appearing around 1840, inedible "hard" palm oil, showed how Africans living under diverse political and environmental conditions could produce commodities for sale to the world. The fortunes of Africa's palm oil exporters changed dramatically at the end of the century, however. Egged on by missionaries and merchants, Britain seized more and more territory along the African coast, helping to spark the "scramble for Africa" in the 1880s. Palm oil was hardly the main reason for the "scramble," but its particular significance to Britain helps explain why that country, rather than France or Germany, wound up controlling the heart of the oil palm belt.

What did the British want with palm oil? Chapter 4 traces the growing importance of palm oil in Britain and the industrializing world more broadly. First used to add color to soap, palm oil became a staple ingredient in soap and candles during the Industrial Revolution. New technologies to refine and modify fats made palm oil more versatile, replacing fats harvested from terrestrial and marine animals. Canny marketers built on the idea that palm oil was a "legitimate product," suggesting that buying a candle could "snuff out" slavery. And events entirely unrelated to Africa's palm groves had great significance for palm oil producers: war in Russia, cowboys raising cattle on the Pampas of South America, whalers prowling the Pacific, and drillers striking petroleum in Pennsylvania all pushed world demand for palm oil up and down. By the 1850s, palm kernels joined palm oil in the manufacturer's arsenal of fatty materials. Chemists tinkering with palm oil and kernels found new ways to strip out the color, taste, and smells that Europeans now found unappealing. As soon as they could, manufacturers slipped palm fats into margarine and other food products—illicitly at first, but later boldly advertising them as superior, plant-based foods that would replace unhygienic lard and butter.

Part II: Oil Palms and Empire

For European soap and margarine manufacturers hard-pressed to find raw materials, the "scramble for Africa" came just in time. Chapter 5 recounts their attempts to capture what many believed were Africa's vast, untapped palm groves. Their goal was to transform palm oil from a handmade thing found in a dazzling array of colors, odors, and qualities into a single, edible commodity. The commodity form abhors context and nuance, but as Europeans discovered, context was everything. Palm oil meant different things in different places, and Africans weren't enthusiastic about selling their palms or their labor to Europeans. At times, imperial governments took land and labor at gunpoint so that Europeans could enjoy cheap margarine on toast. Yet a string of dismal failures with mechanized production eventually forced Europeans to acknowledge the importance of people in the ecology and economics of the oil palm. By the 1930s, it had become clear that machines *could* make good palm oil, but it was just as clear that Africans would only feed those machines under crushing political and economic pressure. Anywhere they had a choice, Africans continued making oil by hand or found new lines of work.

The failure of industrialists in Africa didn't mean the failure of Africa's palm oil industry, however. Chapter 6 shows how "traditional" methods, aided by simple machines and new infrastructure, produced huge quantities of oil and kernels for the world market. British and French officials came to sympathize with these "smallholder"[4] producers, seeing them as viable competitors to steam-powered oil mills and plantations. Colonizers offered new tools and new palm varieties to boost output, but they also imposed harsh taxes and strict regulations to squeeze oil and kernels out of Africa. In the long run, colonial officials were forced to admit that "traditional" methods were well adapted to the conditions of western Africa.

Denied access to land and cheap labor in Africa, capitalists turned halfway around the world to Sumatra and the Malay Peninsula to create the first major oil palm plantations. Chapter 7 shows how they built on the earlier success of the rubber industry in Southeast Asia, clearing tropical forests with unfree "coolie" labor to make oil palm monocultures. These plantations were undoubtedly efficient, turning out top-quality oil from their factories. Yet the planters found that oil palms were no different than any other crop in monoculture, and they were soon battling nutritional deficiencies and diseases that rarely appeared in Africa. By 1939, Sumatra had overtaken Nigeria as the world's largest exporter of palm oil, but the future of the plantation sector was far from certain.

The Second World War shattered the colonial structures that had provided plantations with cheap land and labor. As chapter 8 shows, imperial governments redoubled their efforts to extract palm oil and kernels from their colonies after the war, while nationalist politicians saw the value of palm exports in financing their own dreams for development. New models emerged at places like Kulai in southern Malaysia, where the "Nucleus Estate-Smallholder" (NES) model sought to meld the efficiencies of the plantation system with the social virtues of smallholder ownership. Newly independent African states also embraced oil palm experiments, but civil war wrecked the industries of Africa's leading producers, Nigeria and Congo. Backed by state subsidies and loans from the World Bank and other organizations, Malaysia (and later, Indonesia) shot ahead, becoming the world's dominant palm oil producers.

Part III: Expanding the Oil Palm Frontier

The last section of the book follows the oil palm's career through the end of empire and the Cold War, concluding with the "neoliberal turn" of the 1990s. Chapter 9 returns to the consumer's side of the story, showing how a practically unlimited market for palm oil emerged in food and chemical industries. Whether used whole or broken up into its molecular constituents and reassembled into new chemicals, palm oil appeared in everything from instant noodles to lipstick. But palm oil's place in the world market wasn't inevitable. America's soybean lobby waged a fierce "oil war" against palm oil producers in the 1980s, raising the profile of an "invisible" commodity. By the end of the twentieth century, concerns over the health of the global environment brought palm oil into the biofuel business, creating a highly controversial new market. Coming on top of existing food and chemical uses, biofuel turned oil palm into the world's premier "flex crop," a commodity insulated from overproduction and cyclical crises because of the many different markets it serves.[5]

Chapter 10 traces the changes that this vastly expanded market for palm oil brought to the tropics. While development organizations promoted oil palm and the NES model as a means of liberating farmers from poverty, the oil palm became an instrument of oppression and a symbol of environmental catastrophe in many places. In the Philippines, a plantation manager hired paramilitaries to terrorize workers and indigenous landowners. In Indonesia, oil palm fueled violent conflicts over land in Sumatra, Kalimantan, and Irian Jaya. In Latin America, oil palm promised to revitalize ruined banana plantations, but bloody conflicts over land and labor ensued.

The final chapter, chapter 11, follows the oil palm story into the era of "neoliberal globalization" in the 1990s and 2000s. Unleashed by deregulation and liberalization, the oil palm plantation machine marched across Asia, Africa, and Latin America at unprecedented rates. The tree offers the world an amazing bounty of cheap, versatile raw materials for our food and fuel needs, but its growth has come at a significant cost to the world's tropical forests and the people who live in them.

Conclusions

Today's palm oil industry faces a number of contentious debates, most notably about the effects of oil palm cultivation on tropical ecosystems. As one headline recently warned, "Palm Oil Is in Everything—and It's Destroying Southeast Asia's Forests."[6] This book argues that the oil palm itself is hardly to blame. A conjunction of economic, political, cultural, and environmental factors propelled the oil palm industry into a position where it, more than any other industry, poses "the single most immediate threat to the greatest number of species" across the planet.[7] For many consumers, the simple message is that "palm-oil products carry a curse" into our homes.[8]

Defenders of palm oil make the case that it is, first and foremost, necessary. While Europeans fret over whether they should use palm oil in biodiesel, roughly two billion people in developing countries use palm oil to meet their basic caloric needs. Wealthy consumers in Europe and North America are eating more palm oil too, thanks to regulations that forced trans fats out of the food system. Even people who never touch palm oil—and few can honestly make this claim—benefit indirectly from the influence it has on the price of other fats we consume. The positive case for oil palm stresses the plant's high yield. It makes more fat per hectare than any other domesticated plant or animal can manage.[9] That high yield comes from the oil palm's fantastic efficiency in converting sunlight and CO_2 to plant matter, and the tree's potential role in carbon-neutral or carbon-negative fuel cycles has attracted a great deal of interest from researchers and policymakers focused on climate change. There's also a human dimension to the oil palm story that boycott campaigns rarely address: millions of smallholders rely on palm oil for their livelihoods, though big plantations do produce a majority of the oil on the world market.

Should you use palm oil or not? Unless you prepare all of your food, soaps, and cosmetics from scratch, avoid diesel vehicles and (in a few countries) unplug from the electric grid, you don't have much of a choice. Understanding how and why the oil palm became such an important tree in the global agricultural system is the key to understanding what's happening now. We need to see that history from a broad range of perspectives—not just the triumphal tales of plantation companies selling oil and a particular narrative about oil palm development to consumers, or the dire warnings of habitat destruction offered by environmental campaigners. We also need to consider the stories of Africans, Asians, and Latin Americans who grow and use palm oil in many different ways. Their stories, I argue, make a compelling case for supporting alternative ways of living with oil palms—not alternatives to palm oil itself.

Note on Translations and Figures

When a non-English language source is cited, the translation is my own unless otherwise noted. All figures given in this book, especially those before 1950, should be taken with a grain of salt. Even for recent periods, the Food and Agriculture Organization (FAO) data I use varies considerably compared with data from the United States Department of Agriculture (USDA) and other organizations. Historically, palm oil was measured—often haphazardly—in casks, barrels, pipes, puncheons, hundredweights, tuns (a whale oil measure), tons (weight), gallons, and many other measures. Unless otherwise noted, assume that a ton is 2,240 pounds, and that a gallon is an *imperial* gallon, about 1.2 U.S. gallons. Historical sources reported widely-varying weights for palm oil, from 8 to 9.5 pounds per gallon. These variations reflected inaccuracies in measurement, adulterants, confusion over gallon measures, temperature variations, and changes in the chemical structure of palm oil due to different production methods. An imperial gallon of pure oil should weigh about 9.2 pounds at 25° C. I have converted some figures given in gallons to tons at the rate of 9 pounds per gallon, the most commonly-cited historic figure.[10]

1 The Oil Palm in Africa

Getting to Know the Oil Palm

Elaeis guineensis is, as a nineteenth-century writer put it, "one of the most useful inhabitants of the forest, as well as one of its greatest ornaments."[1] A healthy adult specimen can reach thirty meters into the sky, supported by an impossibly slender trunk. Topped with a shock of dark green fronds, it looks like many of its relatives in the palm family, the *Arecaceae*. Oil palms tower over grassland and "palm bush" scrub in western Africa (figure 1.1); in wetter regions, they form dense groves draped with ferns and orchids. These palms have given Africans an abundant supply of oil for thousands of years. Freshly extracted palm oil is a rich reddish-orange color, free-flowing in tropical heat but thickening in cooler climates. While much depends on how the oil is prepared, palm oil can taste nutty, vegetal, or even smoky, with aromas reminiscent of violets, pumpkins, or carrots.

No other plant supplies as much fat today as the oil palm. Palm oil is usually the cheapest material available for food, soap, and a host of chemical industries. Nearly all of this oil comes from trees in plantations. No longer growing to towering heights, these stubby oil palms stand in neat formations nine meters apart. The long galleries formed by mature trunks and the shady fronds overhead have a pleasing symmetry (figure 1.2). There is a real sense of the sublime in these leafy cathedrals, but it is eerily unlike the tropical forest it replaced.[2] There's little chatter from birds, mammals, or insects, and few splashes of colorful plant life on the plantation.

The beauty of the oil palm fades the closer you get to it. The stems of those green fronds are lined with needle-sharp spines (figure 1.3). The leaflet edges and tips are sharp, too. The palm's trunk is covered in rough, sharp stubs left by pruning. Most plantation palms succumb to the chainsaw or bulldozer before reaching twenty-five years old, often not long enough for these stubs to drop off. The oil palm's fruit is rather menacing, too. It grows in hive-like bunches, packed between sharp spikes (figure 1.4). It's a plant best dealt with at arm's length. Even so, puncture wounds from spines are a regular occurrence on oil palm plantations.[3]

148. HALF-WAY UP 149. AT THE TOP

FIGURE 1.1
Man climbing a tall oil palm in Liberia using a single-rope technique. The palm stands roughly eighty feet high amid cleared scrub. H. H. Johnston, *Liberia*, vol. 1, plates 148–49.

People tolerate those spines because, by some evolutionary accident, oil palms produce an incredible amount of fat in their fruit. Unlike fat in a nut, this material doesn't do much to nourish a seedling. Instead it attracts animals that spread the seeds.[4] Most other plants use sugars for this trick. The size of a small plum, palm fruit has firm flesh that is full of stringy fibers. Palm fruit can be black, red, yellow, orange, green, or even white on the outside, depending on the variety and state of ripeness. The flesh inside is typically reddish orange. When this flesh (or pericarp) is smashed, it gives up its oily juice, the source of palm oil. The pericarp surrounds an oil-rich nut, too. Covered in a woody shell, the palm kernel looks like a tiny coconut. When crushed, it yields palm kernel oil, a white-colored fat similar to coconut oil.

In 1763 the Dutch botanist Nikolaus Jacquin dubbed the African oil palm *Elaeis guineensis*, combining a Greek word for oil with what he presumed

FIGURE 1.2 Mature oil palms in a Malaysian plantation. Author photo, 2017.

was its homeland: Guinea, the western coast of Africa. Jacquin encountered the plant in Martinique, but as any transatlantic traveler could have told him, it was ubiquitous in western Africa. Fossil evidence shows that the oil palm grew in Africa long before it ever appeared in Brazil, which hosts the only significant population of wild oil palms outside Africa.[5] The current consensus holds that the African oil palm and its American cousin, *Elaeis oleifera*, shared an ancestor deep in the past, before Africa and Brazil drifted apart to form the southern Atlantic.[6]

Yet many botanists in the last century puzzled over the fact that oil palms didn't appear to have a "natural" home in Africa. Foreign writers often insisted that oil palms were forest trees (see chapter 5), but it was clear that oil palms and humans lived together. Wild specimens spring up in forest

FIGURE 1.3 A cross-section of *tenera* palm fruit impaled on the sharp spines of a palm frond. The white kernel is visible in the center, surrounded by a thin kernel shell and fibrous, oil-rich flesh. Author photo, 2017.

openings and in wet lowlands, but the tree is not at home in the mature tropical rainforest. In these settings, deciduous giants tower over oil palms, shading them into oblivion. "However tall the palm-tree may be," a Fante saying tells us, "it stands within the buttress of the silk cotton tree [*Ceiba* spp.]."[7] It may take a century, but palm groves in human-made forest clearings inevitably vanish.[8] Oil palms also grow as opportunistic pioneers in drier grassland, but seasonal fires cull most seedlings before their crowns climb high enough to escape damage.

Today scientists point to swamps or floodplains as the original home of the oil palm. Palms in these environments "look poor, having a thin stem and a small crown and producing small inflorescences." Their fruit flesh is thin, with thick kernel shells. If the palm came from the swamp, it has moved on to better things with the help of humans.[9] Clearly the oil palm emerged *somewhere* in Africa, and it survived millions of years of change, including the arrival of *Homo sapiens*. Yet C.W.S. Hartley—the dean of twentieth-century oil palm science—concluded it was "speculative and un-

FIGURE 1.4
A cluster of ripening *tenera* palm fruit ("fresh fruit bunch" in industry parlance). Author photo, 2017.

rewarding" to think about the oil palm's distribution in Africa in terms of "wild" palms and "natural" habitats. As far as he was concerned, it was obvious that people were the most important force in the oil palm's life cycle and had been for centuries.[10]

Africans and the Oil Palm

Among the Ngwa Igbo of southeastern Nigeria, one tradition attests, "People came into the world and saw the oil palm."[11] A Guro tradition from Ivory Coast similarly puts oil palms before humans: Bali, the creator, made oil palms ahead of people to sustain Earth's new inhabitants.[12] According to a Bassari source from northern Togo, "there were no other trees but one, a [oil] palm" when God created man (along with an antelope and a snake).[13] In Igbo tradition, yams and cocoyams emerged when Chukwu, the creator, ordered Nri to sacrifice his son and daughter. A second sacrifice of a male

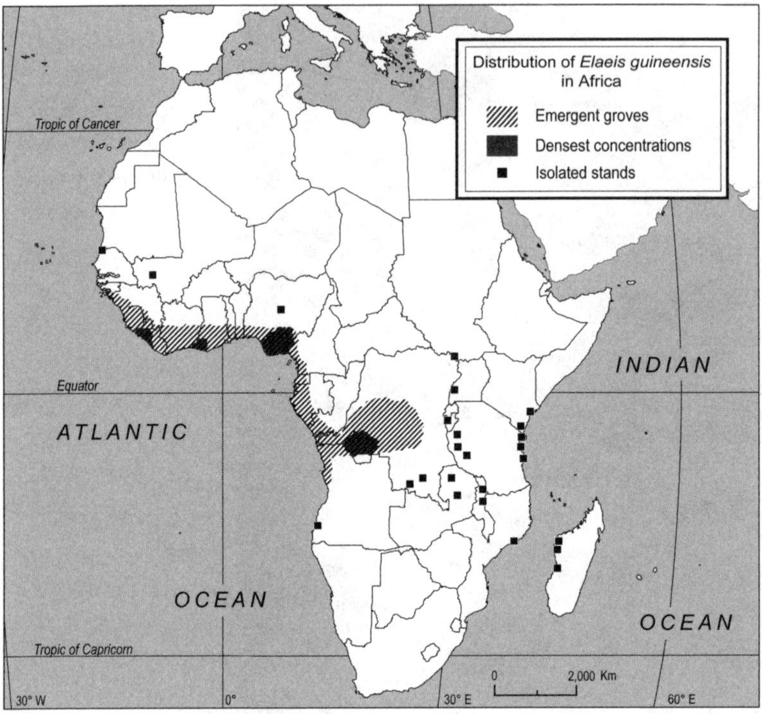

MAP 1.1 Oil palm distribution in Africa, with historical areas of dense "emergent" groves indicated. Adapted from Watkins, *Palm Oil Diaspora*.

and a female slave yielded the oil palm and breadfruit tree, respectively. The story links prestigious staples—yams and palm oil—with men, but it puts the oil palm in a decidedly inferior category.[14] Whether African traditions gave oil palms precedence over humans or not, they usually regarded the trees as something apart from agriculture and cultivated crops.

Yet palm fruit was more than a wild food gathered to supplement the agricultural diet in antiquity. "Is it a coincidence," one scientist asked, "that the densest populations in sub-Saharan Africa are in southern Nigeria where the combination of yam cultivation and the exploitation of the oil palm . . . has been most highly developed?"[15] Root crops provide most of the calories people consume across the oil palm's range (map 1.1), but palm fruit and oil form the base for stews and sauces that make yams and other starches palatable. Palm oil provides critical supplies of vitamin A (which the human body does not produce) and the fat needed to absorb it.[16]

Archeological materials gathered from Bosumpra Cave in Ghana suggest that humans have been eating palm fruit and kernels for at least 5,000 years.[17] Early agriculturalists used fire to clear the landscape but kept oil palms and other useful species.[18] They probably didn't plant oil palms, but rather exploited existing groves and encouraged their spread. A surge in *E. guineensis* pollen grains found in underwater sediments dating back 3,000 years coincided with the expansion of agricultural settlements across the West African forest zone. Palm kernel shells mixed in with pottery fragments and stone tools in archeological sites attest to the widespread use of oil palms in daily life.[19]

For many years, scientists argued that Africa's first agriculturalists hacked and burned their way through a primeval "Guineo-Congolian rainforest" stretching from Sierra Leone to Congo and beyond. In this telling, oil palms were the survivors of forests destroyed by African farmers, leaving "derived savannah" behind.[20] New research has overturned that interpretation, however. An "aridification event" about 4,000–5,000 years ago wiped out forests and encouraged the spread of grassland across western Africa. Oil palms probably expanded into these gaps ahead of human settlers, the seeds spread by animals.[21]

Humans helped the palm along, though, protecting it from grassland fires and voracious elephants.[22] Linguistic evidence shows a close link between oil palm dispersion and the arrival of Bantu-speaking agriculturalists in the Congo basin beginning around 1,000 B.C.E. Few central and southern African languages use non-Bantu terms for the oil palm, suggesting that the tree came with migrants, either carried by them or sharing the same ecological openings in the forest.[23] As a tradition among Mfumte-speakers of northern Cameroon tells us, oil palms "follow men," growing in the wake of human activity.[24] The interplay of climate and agriculture pushed the oil palm's frontier to the south and east, but progress was slow. Nineteenth-century travelers reported only scattered groves around Lakes Kivu and Tanganyika, despite amenable environmental conditions. Tanzanians interviewed in the twentieth century clearly indicated that oil palms were recent arrivals, brought by people rather than by animals.[25]

Rather than serving as agents of deforestation—with oil palms the evidence of ecological vandalism—African farmers may in fact be responsible for afforestation in many places. Ethnographic research, coupled with historic aerial photography, showed that forests grew out of the moist, nutrient-rich soils left behind in the shade of abandoned village palm groves. Rejecting earlier classifications like "semi-wild" or "sub-spontaneous," geographer Case Watkins describes these palm groves as "emergent" phenomena. They are not purely human creations, but rather develop out of human interactions with a complex set of natural forces.[26] These emergent groves often give way to other tree species, creating true forest where none had existed.

As early as the 1920s, elders in Congo told a missionary that they and their ancestors weren't "shifting cultivators" cutting out clearings in a forest: they had *built* the forest with their farming practices.[27] At the time, few Europeans cared to listen. One colonial forester recalled how blinded he had been by stereotypes: "What I had in my inexperience looked upon as glorious virgin [forest] growth, dating from the Flood, quickly revealed itself to my better experienced and disappointed eye as nothing more than secondary growth of moderately good quality." With the help of local guides, seeing a landscape was "like reading a book," revealing human history in the environment.[28] Across much of western and central Africa, forests have probably been advancing rather than retreating for the past 1,000 years or so, and this despite bouts of low rainfall.[29] Far from marking humanity's destructive impact on forests, oil palms stand across Africa as a testament to the versatility, ingenuity, and sustainability of local farming practices.

Oil Palms in African Material Cultures

Archaeology, ethnography, and oral tradition from across western Africa show the value of the oil palm in food systems and its place in daily life. Oil and kernels from the fruit, cabbage from the tree stem, wine from sap, and fronds and fibers for building and handicrafts: the tree was a bountiful resource for the settlers of western Africa and their descendants. The age and environment of an oil palm shapes how humans use it, however. Young palms can be easily harvested from the ground: walk up, hack away spiny fronds, and cut off the fruit bunch at the stem. But young palms only fruit reliably when they are growing in direct sunlight. Palms in a shady grove may spend a decade or more without producing fruit as they climb through the canopy. A palm can reach intimidating heights before it starts bearing fruit under these conditions. But humans often helped oil palms by transplanting and protecting young palms in fire-prone fields, and by hacking away competing plants in the bush.

Ripe fruit eventually drops from oil palms, but it bruises and quickly turns rancid. Cutting fruit allowed humans to avoid spoilage and harvest on their own schedule. On tall palms, this required difficult feats of climbing, often performed using a one- or two-rope system.[30] A Nupe tale credits a founding hero with proving his valor by climbing an exceptionally tall oil palm, gathering special fruit to cure a sick king.[31] Climbers had to cut away spiny fronds as they worked, and venomous snakes sometimes lurked in palm crowns. If the rope broke or the climber slipped, "the result is always horribly painful, if not fatal." As the falling climber grasps the trunk, stumps of old fronds "tear the climber's flesh in a particularly agonising manner."[32]

Palm climbing was always men's work. Among the Sherbro of Sierra Leone, to take one example, men take on "dangerous tasks," including clearing bush, ocean fishing, and palm climbing.[33] A rare exception in southeastern Nigeria proved the rule: in one town, unmarried or widowed women could climb palms, but without ropes. Even then, they could only cut fronds, not fruit.[34] While some communities associated tree-climbing with masculine strength, elite men often forced youths or slaves to do the hard work. By the twentieth century, specialist tree-climbers did much of the harvesting in tall groves.[35] The weight of palm bunches—up to 100 pounds or more—meant that palms far away from village production sites went unharvested.[36]

Accounts from the nineteenth century state that oil palms were typically harvested once or twice a year; each tree could produce about a gallon of oil annually.[37] Compared with contemporary yields, these figures are quite low. Climate was one constraint. During the rainy season—roughly April to October in West Africa, and September to May in the Congo basin—tropical downpours can be daily events. Dry periods in the rest of the year can stretch on for weeks at a time, limiting fruit growth. These rainfall patterns set a natural rhythm for oil palm harvesting. Farmers often harvested at the end of the dry season, when fruit was at its ripest. Men refused to climb rain-soaked trunks, fearing a deadly fall. However, oil-making required plenty of water, forcing producers in dry areas to wait for the return of the rains. As rain softened the ground, farmers shifted their attention back to yams and other crops.[38]

Once collected, bunches of palm fruit usually sat for a few days to ripen. Beating or chopping up the bunch releases fruit from the stem. Women typically cooked fruit in water before pounding it, though the order of operations could be reversed. Men might also pound fruit, especially in large-scale production (see figure 3.1). The cooked fruit pulp is squeezed by hand or twisted in nets to wring out the oil, which is then boiled and skimmed to separate oil from residual fruit juices.[39] Kernels are removed from the pulp and dried, after which they might be cracked with a pair of rocks to remove the kernel shell. In southeastern Nigeria, a family specializing in oil-making in the early twentieth century might produce 180 gallons of palm oil and

56 bushels of palm kernels a year.[40] Making a ton of good-quality oil might take as few as 120 hours or as many as 420 workdays, depending on which source we trust. One of the more detailed estimates shows at least 315 workdays per ton of oil: 140 for harvest and transport, 90 for stripping and pounding fruit, and 85 for cooking and purifying oil.[41]

Making palm oil was hard work, and the product was highly valued. As urban centers developed across western Africa, they provided attractive markets for farmers who might send their palm oil ten, twenty, or even a hundred miles away when water transport was available. By the time Europeans arrived in the fourteenth century, even relatively rural areas were enmeshed in regional markets, trading foodstuffs, salt, and other goods across long distances.[42] The trade in palm oil extended outside the palm belt as well: it was a popular food in the savannah zone of West Africa, where palms were scarce. In central Africa, traders carried pots of palm oil long distances into drier regions.[43]

Palm kernels were also an important source of food: they yield an edible oil and can be eaten as nuts or ground into flour. One seventeenth-century account describes a sort of bread made from palm kernels in the vicinity of the Congo River.[44] Livestock also ate kernels, and kernel shells made a particularly good charcoal favored by blacksmiths.[45] All these uses are well documented in the historic record, and it's baffling that so many historians and scientists of the past century insisted that Africans didn't use kernels at all before the export trade developed in the nineteenth century.[46]

Beyond its uses as food, palm oil was important in health and hygiene. A German visiting in 1603–1604 reported that people washed themselves three times a day, "after which they anoint themselves with tallow or with palm oil, which is an excellent medicine."[47] Palm oil protected the skin and hair, and it had cosmetic value in many cultures. Women (and sometimes men) spread palm oil on their skin to "shine the whole day."[48] Palm oil was also a useful way of applying decorative color and perfumes, like powdered camwood.[49] Many Africans considered palm oil to be a medicine in its own right, and it served as a medium for delivering other curative substances. Historical sources recount healers mixing herbs with palm oil to treat skin conditions or ease headaches.[50] A seventeenth-century Portuguese source describes palm oil as a "popular cure" in Angola, while the "leaves, roots, bark and fruit" of the oil palm were used to treat conditions ranging from arthritis to snake and insect bites.[51]

Foreign visitors praised the quality of soap made from palm and palm kernel oils, mixed with ashes from palm fronds. One writer attested that "the *Negroes* Cloathes are very clean" as a result.[52] The roasting method often used to extract kernel oil produced the characteristic color of the famous "black soap" made by West African artisans.[53] Unlike a typical bar of soap today, palm and palm kernel soaps were "soft" soaps with a short shelf life, but they were still traded extensively in regional markets.

Palm and kernel oil served as illuminants, too. Akan gold miners used palm oil lamps as they tunneled through the earth, working narrow shafts up to fifty feet deep.[54] Clay oil lamps provided a dim light inside houses across the oil palm belt, though many people abandoned them for candles and kerosene in the nineteenth century. Outside the home, torches coated in palm oil provided light for travel and night fishing.[55]

Of course, the oil palm offered many other useful things to people beyond oil. "The usefulness of the palm tree cannot adequately be described," wrote one visitor.[56] The palms provided "fibre [stripped from fronds] for the manufacture of stuffs, rafters and laths for partitions, fronds for thatching the roofs, beds and seats, planks for palisades, materials for bows and arrows, and the innumerable household utensils and apparatus for hunting and fishing; finally, oil and palm wine."[57] While wine and fiber could be gotten from other palms (such as the raffia), the widespread distribution of the oil palm and its close association with human farming meant that it was often the preferred plant. Most of a house and its fittings could be made from the oil palm, down to concrete-like floors made by mixing kernel shells with sand and lime.[58] One late twentieth-century survey of villages in Guinea found that nearly half of all mats, baskets, and nets were made with *E. guineensis* materials.[59]

Hunters and warriors could use the stiff midribs of palm leaves as arrows.[60] Tiv warriors in eastern Nigeria turned oil palm fronds into particularly deadly spears. An account recorded in the 1930s described the process: "A man who was preparing to do deeds of valour broke off a number of branches of the oil palm. He pared them down to a sharp point, and put them in the fire till they were so hard that if you speared a beast in the flank with them they would break its ribs into small pieces."[61] Across the palm belt, men fought with shields made from plaited palm fronds. The same fibers that made strong fishing line and sturdy baskets served equally well in war, though the arrival of firearms led many to abandon shields altogether.[62] But warriors armed with guns still used oil palms, employing palm fiber as wadding and even making gunpowder with kernel charcoal in desperate times.[63] One unusual military use for oil palms came when Krobo warriors poured palm oil over rocky pathways to defend their mountain stronghold from an Asante invasion in the 1770s. Asante troops slipped back down the mountain under a hail of spears and arrows.[64]

For some African communities, palm wine was even more important than palm oil in daily life. Palm wine comes from the sap of the tree, which naturally ferments into a mild wine in a matter of hours, becoming stronger and stronger before turning into vinegar. Wine was at the center of an oil palm origin story recorded from the Kuba people of Congo around 1910. When a woman dared enter a lake full of palm wine, the lake vanished, leaving a ravine with four trees standing in it. A man learned to tap wine from the palm's inflorescences in a dream, and when the community learned the secret, they planted oil palms across the country to ensure an ample supply of wine for the future.[65] An early European visitor to the region described palms "planted like vines," regularly pruned and tapped with a "little pipe or pot inserted into the tree" (figure 1.5). He found palm wine "agreeable," writing, "It makes people merry and strong, and does not bring any headache like other wines."[66]

Tapping a live tree was the most common way of extracting wine, but some communities also practiced "slaughter-tapping." After cutting the trunk or roots of a palm and toppling it, the tappers chopped a hole into the center of the tree. Sap flows into the opening for as long as a month, sometimes aided by a fire built under the trunk. This method produces a lot of wine—reputed to be stronger than tapped wine—without the work of climbing taller palms. A 1625 text described it as one of many barbaric practices of the "Gagas" (Jaga or Imbangala) in Angola: "When they settle themselves in any Countrie, they cut downe as many Palmes as will serve them Wine for a moneth: and then as many more. So that in a little time they spoile the Countrie." When the wine was gone, the warriors moved on to the next set of victims.[67] But slaughter-tapping wasn't incompatible with permanent agriculture. In the Gold Coast, wine tappers cut down palms because there were so many to spare. In Dahomey, tappers managed specialized "vineyards" of palms. Everywhere slaughter-tapping was practiced, it was a way of getting value out of palms that were too numerous or too tall to be worth climbing.

The ubiquity of the oil palm in African material culture was matched by its place in religious practice. The diversity of African religious traditions defies any attempts at generalization, but oil palms play important roles in cosmologies and rituals. Healers, diviners, and priests across the palm belt used practically every part of the tree. Fruit, palm oil, kernels, kernel oil, flowers, unfermented sap, palm wine, roots, fronds, and ashes all served different functions as offerings to ancestors or supernatural actors, or as

FIGURE 1.5
A man tapping wine from a standing oil palm in Cameroon. His knife points toward a wound cut into the tree to allow sap to drain. Bücher and Fickendey, *Die Ölpalme*, plate IX.

treatments for human bodies plagued by disease or supernatural forces.[68] Yoruba practitioners treated the red pock marks of smallpox with red palm oil, and by offering the deity Sapona—who was blamed for inflicting the disease—white palm wine and palm kernel oil.[69] Palm fronds across a path often served to ward off evil or mark the closure of a field or road. People working with corpses or facing deadly threats might carry a palm shoot, associated with life, as protection.[70] Diviners held trials by ordeal with boiling palm oil; those who could drink it or stick their hand in without burning were innocent.[71] Diviners still use polished, unshelled kernels from the uncommon "fetish" palm (formerly *E. guineensis* var. *idolatrica*) to offer spiritual guidance. In this variety of oil palm, the leaflets fuse together and sometimes grow from multiple "heads," creating an unusual appearance. Many communities reserved these palms for priests or diviners. For Ibibio religious practitioners, this variety's bright-red oil guarded against witchcraft. The oil was reputed to violently boil out of cooking pots, making it unfit for daily use.[72]

In some cultures, people anoint newborn children with palm oil, or bury a child's umbilical cord at the foot of a palm seedling.[73] Palms might receive honorary titles or serve as objects of veneration in other communities, receiving gifts and offerings.[74] Catholic priests in Angola cut down fetish palms in a futile effort to suppress local religious traditions, but Christianity wasn't incompatible with continued spiritual uses of oil palms.[75] Christian converts simply incorporated oil palms into new practices. Villagers might carry fronds on Palm Sunday, while also using them to "ward off the evil in village entrances."[76] Christian converts (and no small number of European missionaries) read the oil palm into the Bible, conflating date palms with their own familiar species. For these believers, oil palms proved the veracity of the Scriptures, letting Africans materially engage with Biblical teachings and spiritual metaphors employing (date) palm fronds and (olive or mustard) oil.[77]

Palm Oil in Global History

In 1897, a French chemist named Charles Friedel opened a window into the oil palm's history outside western Africa. He examined several jars found in tombs at Abydos in southern Egypt, dating to the first dynasty (over 5,000 years ago). After running chemical tests, he concluded that the solidified mass in one of the jars probably contained palm oil.[78] The Abydos find linked western Africa with the ancient Mediterranean, and palm oil entered world history as one of many exotic commodities gathered from afar by a great civilization. The Palmolive soap company (whose eponymous product was made from olive and palm oils) drew on this idea in its advertising in the 1910s–1930s, promising consumers the ancient beauty secrets of Egypt.

It was a good story, but it wasn't true. Chemists armed with better tools and techniques in the 1930s showed that Friedel's identification of palm oil was "demonstrably wrong."[79] The Abydos funerary jars were filled with animal fats from cattle and pigs.[80] This was consistent with other evidence documenting uses of animal fats in Egyptian foods, cosmetics, and in mummification.[81] Palm trees were a popular decorative motif in Egyptian art, but they were date palms, not oil palms.[82] Many historians and scientists failed to notice that chemists had updated Friedel's pioneering tests. C. W. S. Hartley cautiously noted in 1967 that palm oil might have been used in ancient Egypt, drawing on a chain of citations going back to Friedel's 1897 paper.[83] A more recent account in the *Cambridge World History of*

Food takes Friedel's single jar as evidence "that palm oil was most likely available in ancient Egypt." The authors state that a "rather large quantity" was found, suggesting that it was "for dietary purposes rather than as an unguent."[84] Embellishing on this account, other writers attest that Egyptians were "buried with casks of palm oil, reflecting the high societal value attributed to the product." Palm oil was thus "one of the earliest traded commodities."[85] Some writers let their imaginations run wild: one recent book asserts that palm oil was "prized by the pharaohs of ancient Egypt as a sacred food."[86]

In the absence of written evidence from African sources, writers latched on to fragmentary clues and conjured up a history for the palm. As noted above, Christian authors often asserted that oil palms featured in the Bible, confusing them with date palms. After quoting two Biblical passages involving oil, one English minister declared: "The inhabitants of Africa, where Palm Oil-trees abundantly grow, are living Comments upon these Two Parts of Scripture."[87] Another writer argued that the oil palm was proof that the Greek writer Herodotus was an accurate chronicler rather than a fanciful storyteller. Herodotus described "Ethiopians" bathing in a fountain of magical water and emerging with an oily sheen. Ignoring the rest of the account, the author decided that Herodotus must have been describing palm oil, based on the fact that many Africans put palm oil on their skin to clean and protect it.[88]

More recently, some scholars have claimed that Herodotus's *Histories* give evidence of palm oil used in Egyptian mummification. A frequently cited mid-twentieth-century English translation of the Greek text states that Egyptian mortuary specialists used palm oil to rinse the body cavities of the dead.[89] Herodotus clearly states that palm *wine*, not oil, was used for this purpose, and the substitution here is inexplicable. Another Greek account by Diodorus Siculus confirms that palm wine, not oil, was used in mummification.[90] Mildly acidic and alcoholic, palm wine would have helped clean and disinfect the body. This wine came from date palms (*Phoenix dactylifera*), or perhaps *doum* palms (*Hyphaena thebaica*) or the nearly extinct Argun palm (*Medemia argun*).[91] There is no chance this highly perishable liquid was imported from Africa's oil palm belt.

The first plausible trace of palm oil and the oil palm tree outside Africa's oral traditions and archeological record comes from an Arabic source. In the fourteenth century, the geographer Al Dimashqi described a tree growing in Samaqanda, reputed to be a province of the ancient kingdom of Ghana. When peeled open, the tree's fruit revealed an "extremely oily and sweet" interior. "For that reason they press oil out of it."[92] Like other geographers of the era, his work combined new accounts from travelers with bits from older publications, some now lost to history, and it's not clear where this description originated. If Al-Dimashqi was indeed describing an oil palm, this passage offers no evidence that palm oil was traded in North Africa or beyond. Although writers in the medieval Islamic world described many types of oil, no other reference to palm oil has been identified in the voluminous literature on exotic commodities and medicinal items.[93]

• • • • • •

The search for palm oil and the oil palm tree in ancient history reflected a Eurocentric worldview that saw no significance in a plant, its products, or the people who lived with it until it entered the Western narrative. But this should not diminish the significance of the oil palm or the peoples who used it in antiquity. Africans transformed the trunks, fronds, roots, fruit, and kernels of oil palms into an impressive variety of products. African pioneers brought oil palms with them as they carved out new settlements across western Africa. While there is still much to learn about the earliest urban centers in the forest zone of western Africa, we know from early accounts that the region hosted large cities and complex societies. Sixteenth-century Benin City, with a population numbering in the tens of thousands, invited favorable comparisons to Lisbon, Amsterdam, or Madrid. Benin sat atop a "massive surplus-extracting apparatus" that drew in labor and resources from a wide hinterland, and similar states could be found across the forest zone from modern-day Senegal to Angola.[94] The urban centers at their cores consumed vast quantities of palm oil, palm wine, and other foods imported from hinterland regions, reflecting sophisticated economic and political systems.

As chapter 2 shows, Europeans adapted to the economies of western Africa, rather than vice versa. Their ships initially carried the cargo of Africa's regional trade, before becoming more and more committed to the transatlantic slave trade. The slave trade profoundly transformed states, societies, and even landscapes across western Africa as it reached a horrific crescendo in the late eighteenth century. This traffic also introduced Europeans to the oil palm and its myriad products, creating the first markets outside Africa for palm oil and kernels and laying the groundwork for an immense industrial boom in the nineteenth century.

2 Early Encounters and Exchanges

Atlantic Encounters

In 1455–1456, the Venetian Alvise Cà da Mosto (or Cadamosto) joined two Portuguese expeditions to what is now Senegal. During a month spent enjoying the hospitality of the ruler of Cayor, he observed the customs, dress, and food of the people. He described a method for tapping palm wine, calling the drink "as sweet as the sweetest wine in the world." Cadamosto "preferred it to our own," marveling that the palms were "all in the forest, common to everyone to tap." He also described "a certain oil in their food." It had the "scent of violets, the taste of our olive oil, and a colour which tinges the food like saffron, but is more attractive." The palm wine might have come from any number of species, but the description of the oil fits *Elaeis guineensis* perfectly. Though Cadamosto admitted he never saw oil or wine made, this was likely the first European description of an oil palm and its products.[1] Cadamosto wasn't drinking palm wine and eating palm oil on a gastronomic tour, of course. He had come to trade horses for slaves.

Cadamosto wasn't the first European to seek human chattel in Africa. By the fifteenth century, European merchants buying gold, ivory, spices, and slaves were regular visitors along Africa's Atlantic coast. Duarte Pacheco Pereira's 1506 account was the first to specifically mention buying palm oil along with slaves, panther skins, cotton cloth, and exotic seashells. Most of these commodities weren't bound for Europe: according to Pereira, "they are all valuable at the castle of S. Jorze da Mina [Elmina, Ghana], where the King's factor sells them to the negro merchants for gold."[2] This pattern was in fact typical of European trading activity in Africa into the early seventeenth century. African merchants accepted European, Asian, and American goods in exchange for the captives and commodities desired by the Europeans, but they also wanted African commodities ranging from luxury items—like Pereira's panther skins—to more mundane consumer goods, like palm oil and yams. African merchants took advantage of foreign sailing ships to move both goods and captive labor up and down the coast.

It was this ocean-borne trade that launched the oil palm and its products on new paths out of Africa and into the rest of the world. Palm oil and wine sustained and lubricated commerce along Africa's Atlantic coast from the fifteenth century onward. Some palm oil entered Atlantic commerce as a novel medicinal commodity, but palm oil's main role was in service to the slave trade. This chapter explores the importance of the oil palm in the transatlantic slave trade, and the ways in which palm oil—and the oil palm tree itself—found new niches outside Africa.

European Encounters with the Oil Palm

European visitors recorded many details about the oil palm and its produce in the sixteenth and seventeenth centuries. The author of the first botanical description of the oil palm is disputed, in part because early modern botanists were unapologetic plagiarists. Mathias de L'Obel offered what might be the earliest in the 1570s, based on stories and plant materials carried back by sailors. L'Obel called the oil palm the "little Indian nut" tree and claimed that palm oil was pressed from the kernel. Early descriptions by Charles de L'Ecluse (Carolus Clusius) and others made the same mistake.[3] An account of a Dutch voyage to West Africa in 1600 clearly identified the flesh of the oil palm fruit as the source of palm oil, but it also gave descriptions of palm kernel oil, perhaps adding to scholarly confusion.[4]

Once an authoritative explanation appeared in print, not even physical evidence or firsthand testimony could easily topple it. When a traveler brought an intact fruit bunch to Europe for Rembert Dodoens to draw in the early 1600s, the botanist treated the dried fruit flesh as an insignificant covering for the kernel, likening it to a coconut husk.[5] Confusion over the source of palm oil carried over centuries; a correspondent had to correct the prestigious Edinburgh *Pharmacopoeia*'s erroneous attribution of palm oil to the kernel in 1832.[6] Sir Hans Sloane, a famed naturalist (and owner of a fortune built on the backs of slaves in Jamaica), wrote: "Amongst the several vegetable Substances which afford Oil, which is so necessary for maintaining Life and promoting Manufacturers, I know none but the Fruit of this [oil palm tree] and the Olive-Tree whose Pulps are useful for these purposes." He was "so much surpriz'd with this Singularity" that he experimented with the fruit himself. Sloane suspected that kernels ground up by millstones in fact supplied the oil, and he was astonished to find that the pulp yielded so much oil. Sloane carried out his experiments on a "head" of fruit

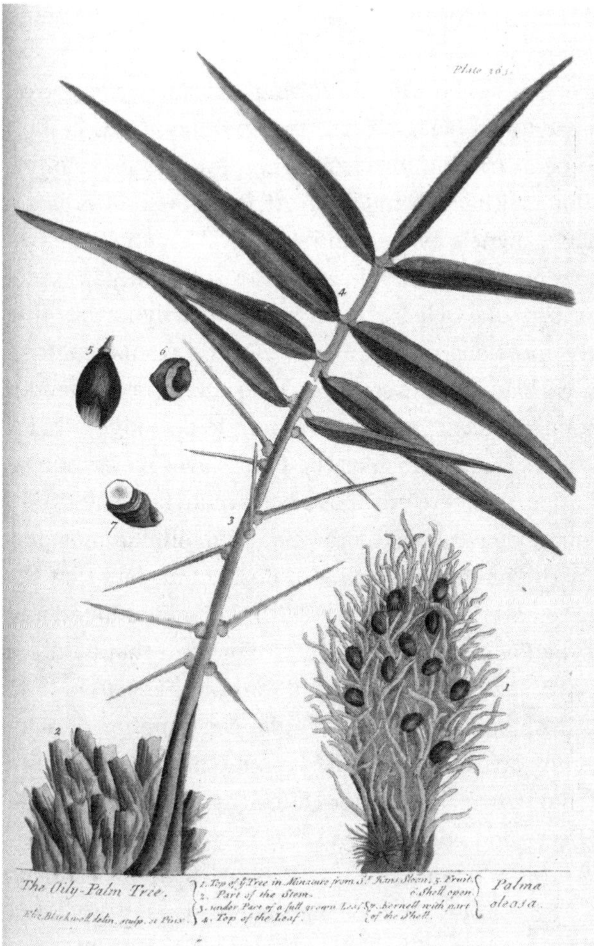

FIGURE 2.1
Elizabeth Blackwell's illustration of oil palm fronds and fruit, showing the fruit growing in clusters with a sectioned view of the shell and kernel. Her description of the "Oily-Palm Tree" or "*Palma oleosa*" reads: "1. This Tree grows to a good Bigness in its Native Country, the Leaves are a grass Green, and the Fruit a Chesnut Colour. 2. It grows on the Coast of Guinea. 3. The Oil, which is expressed from the Fruit of this Tree, is accounted good for all kinds of Pains and Weakness of the Nerves, Cramps in the Limbs, and strains and Bruises." *A Curious Herbal* vol. 2, plate 363. Biodiversity Heritage Library, contributed by Missouri Botanical Garden.

brought by "Mr. Staphorst an ingenious Surgeon to a Merchant Ship trading there [Guinea]."[7]

Sloane shared his samples and drawings with Elizabeth Blackwell, an accomplished botanical illustrator. Her drawing of an "Oily-Palm Tree" in *A Curious Herbal* (1737–39) captured important details about the plant and corrected old ideas about the source of palm oil (figure 2.1). She noted the oil's value in treating "all kinds of Pains and Weakness of the Nerves, Cramps in the Limbs, and strains and Bruises."[8] Blackwell's work is nearly forgotten today, though. The Dutch naturalist Nikolaus Joseph von Jacquin gets the most credit for his 1763 description and illustration, which came with a new Linnaean name for the oil palm: *Elaeis guineensis*. Jacquin sketched oil palms on Martinique during a 1750s tour of the Americas, though he noted the tree was uncommon in the region.[9]

While experts in Europe grappled with describing and classifying the oil palm, visitors to Africa had no problem making palm produce a part of their diets. Garrisons at European trading forts consumed prodigious amounts of palm oil and wine, along with yams and other African staples.[10] At Commenda [Komenda] Castle, owned by England's Royal African Company, commander Howsley Freeman noted the three staples constantly in use at fort were "powder, rum & palme oyle." The oil was primarily for burning in lamps and feeding the fort's enslaved workers.[11] Cut off from local food shipments by a conflict with an African state, another English commander complained: "Here is not a graine or corne or any fisch to be bought, that if it were not for palme berrys & such trash as the poor slaves pick amongst the bushes they would all of them ere now have been quite famished."[12]

Some European accounts disparaged the taste of palm oil, but most foreigners learned to enjoy it. Dutch writer Willem Bosman argued that the oil palm "claims the [preeminence] here; since with the help of Bread and Fish it subsists most of the People on the Coast." He warned that palm oil "is a little nauseous first to New-Comers here, but for him that is used to it, is no despicable Sauce." Bosman asserted that regular consumption of palm oil was "very strengthening and healthful; and I am inclined to prefer it in several Dishes before Oil of Olive."[13] Another Dutch traveler affirmed that palm oil was "verie delicate and good" when served with meat and fish.[14] English surgeon John Atkins complained about West African dishes made with meat "on the stink," excessively flavored with palm oil and pepper. But even Atkins admitted a liking for "black soup" seasoned with "Pepper, Ochre, and Palm-Oil." "At first I thought it disagreeable, but Custom reconciled it as the best [food] in the Country," Atkins wrote.[15]

André Donelha's 1625 account gave an early description of palm kernels, which he compared favorably to "a dry and well-ripened coconut." "It is a foodstuff which keeps for a very long time and is very useful to the blacks," he noted.[16] Captain Finch was less enthusiastic: to him the kernel was "a hard hornie substance without taste."[17] But neither kernels nor oil could match the importance of palm wine, a substance often overlooked today by experts preoccupied with palm oil. Frenchman François Froger praised palm wine as "very pleasant to drink when one is hot." He warned, "at the end of two or three Days, it is spoil'd, and easily inebriates."[18] If left too long, the fermenting sap becomes vinegar, making palm wine impossible to store or export. This was a challenge for European merchants, who were expected to offer alcohol to African business partners and even common laborers as a matter of course. One English merchant at Winneba begged his superiors to send rum: "The country people grumbles very much that I have not drams [of alcohol] to give them . . . [and] I am forced to give them palme wine." He complained, "the expense that I am at in palme wine almost distracts mee . . . for the young men will not touch a stone till such tyme as they are sure of some thing to drink."[19] The fact that imported rum from the Americas was more cost effective than palm wine speaks to the demand for wine and the difficulty of tapping it on a large scale.

High demand encouraged wine sellers to dilute their wares: "The people of the country generally mix this wine with water, to earn more gold at the market," complained a German visitor to the Gold Coast. Europeans got adulterated wine "unless they order it expressly in advance from a peasant whom they know, and pay twice as much as it costs in the market." The results were worth it, though: "Then they get such good and strong palm wine that one must truly say palm wine is better than many European wines."[20] Beyond its inebriating qualities, the author claimed that palm wine had medicinal value, helping to expel kidney stones.

If the cost and quality of palm wine was one problem for wine drinkers, overindulgence was another. An English officer at Dixcove complained about John Pimm, who refused to take his part on the nightly watch. Pimm "[told] me to my face I have no power to command him . . . I have not power to restraine him from his palme wine and keeping our eniemys company [at the Dutch fort in nearby Butri]." Another English commander complained that his men "all ran out of the fort att a time, and gott drunk with palme wine, that I had noe government amongst them."[21] An English bricklayer at Winneba dodged his work with complaints of illness, but "eats heartily and will swill palme wine till he splitts."[22]

Though often confused with wine from other palms (including the coconut and raffia), wine tapped from *Elaeis guineensis* became a symbol of tropical fecundity for Europeans. The Scottish poet James Thomson—despite having never seen an oil palm—praised palm wine as one of the many pleasures of the sweltering tropics in his popular ode to "Summer" (1727):

And from the palm to draw its freshening wine!
More bounteous far than all the frantic juice
Which Bacchus pours[23]

But palm wine was more than just a drink for Africans. The sugary sap is packed with vitamins and minerals, and in some areas, palm wine was an important source of potable liquid during the dry season. Palm wine also played an important role in religious beliefs, offered as a libation to spirits, deities, and ancestors—though rum, gin, and other drinks introduced by Europeans displaced palm wine at times.[24] Still, palm wine remained both a staple beverage and an important lubricant of commerce across Africa's Atlantic coast.

Feeding the Slave Trade

By the eighteenth century, more and more of those business transactions sealed with a sip of palm wine involved human cargo rather than gold, ivory, or spices. It was in the midst of the slave trade that the oil palm took the first steps of its global career, creating new markets for palm oil outside Africa and leading to a new foothold for the plant in the Americas. Palm oil sustained captives before and during the "Middle Passage" across the Atlantic. At times, it formed part of slaves' rations as they labored on American plantations. Palm oil played a cruel role in rituals used to mark and market slaves, transforming humans into salable property. Yet Africans brought knowledge of oil palms with them to America, and they used it to survive and resist slavery.

Holding people in bondage was a fact of life in much of Africa, as it was in every other part of the early modern world, including Europe.[25] In western Africa, children might be held as pawns for security on a debt, becoming slaves (or more commonly, a dependent household member) of the creditor's family if the debtor defaulted. Criminals might be sold as slaves in lieu of execution, removing individuals from the community without the spiritual dangers that bloodshed brought. Prisoners of war were often put to work as serfs on the farms of elite warriors and merchants, often assimilating into the host society over generations. The outright sale of prisoners as property—chattel slavery—dated back centuries in the Sahel, where merchants from northern Africa bought slaves to send across the Sahara.

African slave traders had no reason to see the victims of the trade as compatriots. As historian Patrick Manning has plainly put it, Africans "had no more common identity than did Europeans."[26] Like everyone else on the planet, Africans were divided by political allegiances, languages, religions, and other markers of identity. African elites along the Atlantic coast were some of the first customers of European slave traders: in the 1530s, European traders sold more captives to buyers on the Gold Coast than they carried across the Atlantic, a trade driven by booming demand for labor in gold mines.[27]

Unprecedented demand for slaves outside Africa soon eclipsed this regional trade, however. The conquest of the Americas created lucrative opportunities in mining and agriculture for those who could command large supplies of labor, labor that European migrants would not willingly provide. The colonizers enslaved Native Americans where they could, but warfare and diseases brought from the Old World decimated America's indigenous population. African captives had been at work in the Iberian Peninsula since the fifteenth century, and the Portuguese and Spanish extended this use of slave labor to the Americas after 1492.

A captive's journey into slavery in the Americas often began with palm oil, a fact captured in a vivid journal written by Johann Peter Oetinnger in the 1690s. Trained as a barber-surgeon, he was tasked with inspecting captives and deciding who was worth purchasing: "As soon as a sufficient number of the unfortunate victims were collected, they were examined by me: the healthy and strong ones were bought, whereas the *magrones*—those who had fingers or teeth missing or were disabled—were rejected. The slaves who had been bought then had to kneel down, twenty or thirty at a time; their right shoulder was smeared with palm oil and branded with an iron which bore the initials [of the trading company]."[28] Thomas Phillips's account of a slaving voyage in 1694 contains a similar story. Perhaps aware of how readers might react to the description of such torture, Phillips insisted that preparing the branding location with palm oil meant that the hot iron "caused but little pain."[29]

Once loaded on slave ships, captives had to be fed during the long Middle Passage across the Atlantic Ocean.[30] A 1729 text written for surgeons on slave ships noted the scanty rations given to slaves: "two Meals a Day, and [they] are scarcely allowed two Spoonfull[s] of Oil amongst a whole Mess, which serves ten." Palm oil and malagueta pepper were the two most common additions to yams and other starches. Stingy slavers might offer "a small Quantity of Palm-Oil, which is so little, that were they not to see it put in, they could not tell by the Tast[e] that they had any." Surgeons urged better food: "bring a Cruce of Palm-Oil round the Deck from Mess to Mess, and also Pepper, and let every one take as he pleaseth, for some will eat a great deal, others very little."[31] French sugar planter Jean Baptiste Labat instructed slavers to sail with generous rations of palm oil: "If the company was willing to spend extra for six barrels of lard, and two or three hundred pounds of palm oil to add to the salt that seasons the vegetables, it can ensure that its cargo of slaves arrives to the West Indies intact."[32] One writer claimed that captives "had died for hunger without this oyle," so vital was it to basic ideas about what constituted food in western Africa.[33]

Palm oil was sometimes made on ship, pounded out of fresh fruit bunches by African women carried on board for the purpose. Leftover kernels were sometimes fed to the captives, too. Slave ships departing Angola often bought additional sacks of palm kernels as part of their regular provisions. Rich in fat and protein, the nuts could be stored for long periods of time, and some accounts describe shipboard rations made from a mixture of manioc and flour made from ground palm kernels.[34]

On arrival in the Americas, prospective buyers inspected human cargoes as if they were horses: checking teeth, bones, and skin; looking for signs of disease. Sir Hans Sloane described the ways slavers prepared sick and starving captives for these demeaning inspections: "When a *Guinea* Ship comes near *Jamaica* with Blacks to sell, there is great care taken that the *Negros* should be shav'd, trim'd, and their Bodies and Hair anointed all over with Palm-Oil, which adds a great beauty to them."[35] The botanist Henry Barham confirmed that shaving and oiling captives was a regular practice, intended to make the enslaved "look smooth, sleek, and young."[36] In effect, palm oil was used "to sustain, promote, brand, heal, and finally commodify human chattel."[37] Palm oil thus also served to mark the end of the Middle Passage for captives.

Oil Palms in the Americas

European experts heartily recommended palm oil as a food for slave ships, but it did not become a staple food for most enslaved workers in the Americas.[38] A 1797 account of St. Domingue (modern Haiti), the most populous colony in the Caribbean, asserted that Africans never touched palm oil again after being smeared with the stuff at the auction block.[39] Early accounts show that Africans and their descendants did occasionally receive palm oil in other colonies, however. Henry Drax, a Barbados planter, instructed a manager to provide his enslaved workforce with "a barrell of palm oyle amongestt them" several times per year, providing it was "Easely to be procured."[40] Another source reported that a twenty-gallon barrel of oil could be bought in Africa for six dollars and sold in Barbados for more than twice as much.[41]

Drax and several other planters also tried growing their own oil palms, though we don't know precisely when the first seeds arrived in the Americas.

Richard Ligon, who spent the years 1647–1650 doing a poor job running a Barbados sugar plantation, gave an early description of what might have been oil palms. According to Ligon, a palm tree "so much admir'd for her two vertues of Oyle and Wine" was "newly begun to be planted" during his time on the island. Ligon described wine-tapping but said nothing about the fruit. His only description of palm oil states it came from "Barbary" and was highly valued by the enslaved.[42] Some scholars suggest these trees were coconut palms, but Ligon insisted he was "well acquainted" with coconuts from a stint in the Cape Verde islands; he had no reason not to name them in Barbados.

In the 1670s, Drax ordered his manager "to plant as you Can get the Nutts of 1500 or 2000 palm oyle trees."[43] He probably hoped to save on the costs of importing palm oil from Africa. About the same time, Colonel Colbeck had oil palms planted on his own 1,340-acre Jamaican estate. While Drax planted from seed, the palms on Colbeck's plantation were seedlings "brought over with some others from *Guinea* in Tubs water'd by the Way."[44] The botanist Henry Barham noted the presence of oil palms on Jamaica a few years later, differentiating this tree "from which oil and wine are got" from other palm species.[45] The common name for the oil palm in Jamaica was "abbay" or "abbey," derived from the Twi word *abe* used in the Gold Coast, where many of the captives sold to Jamaica originated.[46]

Planters often tried to compel slaves to grow food crops in their "idle" time.[47] Over the centuries, slaves won important—but tenuous—rights to manage their own "provisioning grounds," usually on the margins of plantations.[48] The sheer diversity of crops and herbs grown in these plots "astonished all who saw it."[49] Slaves fed their families with these gardens, and they often sold produce in urban centers to earn money. A 1774 report suggested oil palms found a home in provisioning grounds, noting the tree was "not so frequent in Jamaica as it deserves, being chiefly cultivated by the Negroes only."[50] Later writers noted that palm oil was "much esteemed for food and medicine" among the descendants of Africans in Jamaica.[51] Oil palms—whether carried over by planters or grown from kernels secreted away by captives—were a small part of a larger botanical assemblage of survival, joining yams, black-eyed peas, okra, and other African plants in creating "neo-African landscapes" in the New World.[52]

Yet in most corners of the Caribbean, oil palms and palm oil disappeared. A century after Drax planted his palms on Barbados, a visitor reported, "These Trees are very scarce in this Island, except at Drax's Hall . . . and even there they exceed not Twelve in Number." Slaves helped themselves to the "fine sweet Oil" from roasted palm fruit collected under the palms, but put no effort in cultivating their enslavers' trees.[53] While evidence is thin, wine-tapping may have held back the growth of oil palms in the America. Tapping standing palms weakens them and impedes fruiting; it could be done furtively at night to provide the enslaved with alcohol and an important ritual substance.

Slaughter tapping crossed the Atlantic as early as 1652: officials in Bermuda complained that "Idle negros in all parts of the Islands" were felling palms all over the islands to make wine. These weren't oil palms, but Africans were clearly putting their knowledge of *Elaeis guineensis* to work on American species.[54] Colonial officials thought wine-tapping was environmental vandalism, but for Africans, it was a way to exercise autonomy, engaging in familiar cultural practices (and of course enjoying a refreshing drink in the process). Sabotage against the planting class was an added bonus for the tapper.[55] Africans learned how to use many other American palm species for wine and food. Ligon described a man using climbing ropes to harvest fruit from a towering "Palmetto Royale" tree on Barbados, supplementing his own meager rations.[56] In Brazil, Africans turned to the pindoba palm for oil to dress their bland rations, adapting African oil-making practices to an American palm species.[57]

Whether tapped for wine or felled by land-hungry planters, oil palms didn't thrive in the Caribbean. By the 1850s, they were rare in Jamaica and had to be reintroduced to Dominica.[58] A Colombian researcher touring the Caribbean in the 1940s was surprised by how few oil palms he saw.[59] The oil palm's fate in Brazil was very different, however. Here, oil palms spread in garden plots and forest clearings, especially along rivers. The land constraints faced by the palm on Caribbean islands disappeared in the vastness of Brazil. The palm was so successful in colonizing Brazil that by the twentieth century, some experts theorized that the transatlantic exchange had gone the other way, from Brazil to Africa.[60]

While it's clear that the oil palm came to Brazil from Africa, archival records don't document its arrival. The Portuguese ordered colonists in Brazil to plant "palms" for the express purpose of making oil in 1612, and the earliest order noted useful fiber obtained from the palm. Watkins argues that this probably referred to coir fiber obtained from coconut husks, but Africans did make strong fibers from oil palm fronds. In any case, Portugal encouraged the transplantation of all manner of useful plants to Brazil, and the coconut and oil palms were both probably planted to supply the colonies with oil and other useful materials.[61]

Oil palms may have already been growing in Brazil before the government took an official interest in the plant, however. A Portuguese captain reported "an infinity of huge palms, of all castes [types]" in the vicinity of São Luis after Portuguese forces retook it from the French in 1619. The document specifically notes the oil value of "Guinea palms" and says they grew in considerable numbers.[62] Slave ships had been arriving in Brazil since at least 1574, giving Africans and Europeans ample time to transplant kernels. Still, the evidence for *Elaeis guineensis* in Brazil is ambiguous until 1699, when English privateer William Dampier delivered a clear description of "Dendees" (*dendê*). He was certain that these were "the same kind of Berries or Nuts as those they make the Palm-Oyl with on the Coast of Guinea, where they abound: And I was told that they make Oyl with them here also." Dampier added that in Brazil "They sometimes Roast and Eat them; but when I had one Roasted to prove it, I did not like it."[63] Like the Twi-derived term "abbay" in Jamaica, "dendê" points to an African origin, brought with captives from Angola speaking a Bantu language.[64]

As in the Caribbean, oil palms supplemented plantation rations in Brazil and became a part of provisioning grounds. But as Watkins shows, the key to the oil palm's success in Bahia was its relationship with humans and another plant: manioc. Like African yam, American manioc was grown with shifting cultivation. By cutting and burning the forest, slaves growing manioc for their subsistence created landscapes that could be colonized by oil palms, with seeds spread by people, birds, and animals. Individuals returned to mature groves to harvest, while they cleared new spaces for manioc further afield and started the cycle anew. Oil-making for food and soap offered enslaved and free Afro-Brazilians an important source of income as Bahian towns grew into bustling commercial centers.[65]

In Brazil and the Caribbean, oil palms also played an important role in religious practices among Africans and their descendants. *Candomblé* in Brazil, *Santería* in Cuba, *Vodou* in Haiti, *Obeah* and *Myal* in Jamaica: these and many other spiritual practices reflected the survival of African beliefs and their adaptation to the new social conditions of the Americas.[66] When they could, communities used ritual objects as they did in Africa: palm kernels, palm oil, and palm fronds all played important roles in ceremonies.[67]

But when oil palms weren't accessible, people readily adapted to American palm species. In Brazil, communities of escapees found freedom from plantation slavery in remote forest settlements, known as *quilombos*.[68] The largest quilombo (or rather group of quilombos) was Palmares, a complex

African kingdom built in the Brazilian forest. Johan Blaer, who commanded a Dutch force sent to attack Palmares in 1645, found evidence for a diet that reflected the preservation of African food traditions rather than practices learned from Native Americans. Palmares residents forged iron tools and cooking pots and pressed oil from local palms, deep-frying many foods.[69] Although some scholars suggest that Palmares was named after the oil palm, it instead reflected African adaptations to the pindoba palm, which also has oily fruit.[70]

In Suriname, maroon groups demonstrated a similar ability to adapt to new species. The Saramaka community, founded by escapees in the sixteenth and seventeenth centuries, used no less than five different species of American palms to make oil, among other uses.[71] A detailed account from a Dutch soldier in 1798 attests to the resourcefulness of these maroons, who cultivated and gathered an impressive array of foods in the rainforest environment.[72]

The oil palm's journey to the Americas—and palm oil's arrival in circuits of global commerce—was inextricably linked with slavery. Palm oil kept the trade in human captives moving across the Atlantic, and it nourished slaves in the New World. The oil palm took root in the Americas, but not as a commercial crop. European colonists largely ignored the tree, focusing their attention and money on cash crops like sugar, tobacco, cotton, rice, and indigo. The oil palm persisted on the margins of plantation society in the Caribbean, providing oil, wine, and kernels for generations of newly arrived Africans and their descendants. African Americans in Georgia made wine from palmettos as late as the 1930s, showing how robust African cultural practices were, even after the traumas of enslavement and the difficulties of adapting to new environments in the Americas.[73] Some communities took to using annatto in oil to preserve the distinct color of foods once cooked with red palm oil.[74] Although African uses of palms survived the journey across the Atlantic and thrived in the New World, the oil palm itself only truly took root in Brazil, seizing on a human-created ecology that replicated key aspects of the oil palm's African homeland.

"Those Most Skilful Botanists": Palm Oil as Medicine in the Atlantic World

Alongside its role as food, palm oil served as medicine for free and enslaved Africans on both sides of the Atlantic. It was no accident that the first botanical descriptions of the oil palm in Europe were produced in herbaries, texts designed to aid medical experts in the use of botanical substances.[75] Travelers in Africa often remarked on the medicinal qualities of palm oil; one said it was "considered by doctors a very beneficial balm."[76] A German account of the Gold Coast described the use of palm oil in treatments for guinea worm, a parasite that grows under the skin and must be slowly and painfully extracted. The palm oil, smeared over the wound and covered with an unidentified leaf, helped protect the wound from infection. The same combination of oil and leaf relieved pain and served as a general-purpose bandage: "This remedy they use for all their open injuries," the German noted.[77] Donelha's account noted that palm kernel oil was also "used as a medicine, and they put it in, while boiling, some strong-smelling herbs to make an ointment."[78]

Some European visitors were critical of African healers because of the close associations among medicine, religion, and magic in African cultures. Still, they took careful notes about specific substances used by African healers, accepting that a cure might be efficacious even if one didn't believe the reasoning behind it.[79] A priest at Christiansborg Castle wrote in 1703 that palm oil "keeps your stomach in a very good and healthy condition." Abandoning any idea of intellectual superiority, he asserted that Africans were "much better suited than we are [in the tropics], as regards their health care."[80] An English surveyor visiting the Gold Coast similarly praised "the Knowledge of those most skilful Botanists, the Negroes, who know well the Use of every Herb and Plant, and always apply them with such Success that the Cures wrought by them seem, at some times, to be little less than miraculous." His African informants told him that palm oil was "good for both the Back and Belly. And to prove that Assertion, they not only eat it with every Thing, but also daily anoint themselves with it, which mightily invigorates their Nerves, &c."[81]

Slave ships crossing the Atlantic provided another avenue for transferring African medicinal knowledge to Europe. Ship's surgeons, tasked with keeping crew and captives alive during the Middle Passage, freely mixed European and African medical practices during voyages across the Atlantic.[82] A French writer, Claude Biron, noted that palm fruit carried on slave ships to feed captives served double duty as medicine. Slaves who suffered from stomach ailments would be fed palm kernels, which the author described as the "best astringent" he knew. He explicitly states that he learned the practice from African healers: "imitating them I sometimes met with success."[83] The English surgeon John Atkins saw the habit of "constantly anointing [the body] with Palm Oyl or Tallow" as an important contributor to the good health in Africa. Atkins collected stories from informants about "African distempers" and noted that topical applications of palm oil were effective in treating all sorts of skin conditions that might appear during the wretched conditions of the Middle Passage.[84]

European naturalists and doctors embarked on scientific plant-collecting across Africa and the Americas, but they also "learnt about them from the slaves who used these in their daily lives."[85] Ligon described the medicinal use of "Negre-oyle" by slaves in Barbados. Imported from "Barbary," he described the stuff as "yellow it is as Bees wax, but soft as butter." When a person felt ill, "they call for some of that, and annoint their bodies, as their breasts, bellies, and sides, and in two daies they are perfectly well." He thought palm oil "does the greatest cures upon such, as have bruises or strains in their bodies."[86] Another writer noted that Africans sometimes secreted small amounts of palm oil with them on their forced voyages across the ocean; he urged planters to allow Africans to continue using it as a skin treatment.[87] A visitor to Jamaica remarked on the importance of palm oil in slaves' food, but added: "they oftener apply it by way of embrocation, for strains, or to discuss rheumatic aches, for which purposes it is very efficacious."[88]

Enslaved people kept some practices and the meaning behind them hidden. While religious beliefs varied widely across Africa, many societies shared a belief that illness or accidental injury were linked to the spiritual world. Even "trivial complaints, such as indigestion, diarrhea, or mild headaches" could be linked to spiritual forces at work in the natural world.[89] Historical records show that "African priests, herbalists, and magicians, initiated into their trade before passage to the Americas, pursued their vocation to the extent possible in their new urban or rural settings." While planters, missionaries, and government officials were wary of these healers, "African-based medicine, magic, and their associated plant pharmacopoeias have persisted and in many cases flourished in the Americas."[90]

African herbalists and healers were persecuted and feared by colonial authorities: plants could poison as well as heal. Moreover, the spiritual authority claimed by healers gave them significant influence over fellow Africans. Europeans clearly understood the linkage between medicine, magic, and political power in a religious tradition like Jamaican Obeah: to them, it was "mysterious and fear inducing," combining "the power to both heal and harm."[91] Obeah practitioners were accused of organizing slave rebellions.[92] The English ignored the important differences between herbalists, sorcerer-healers, and diviners, who worked within the same belief system but with different tools and methods.[93] Over time, European appreciation for the medical knowledge of Africans waned. By the nineteenth century, respect for African healing had turned to derision; white doctors in plantation zones were concerned that African cures would do more harm than good.[94]

Nonetheless, by that time, palm oil had earned a distinctive reputation in Europe as a healing substance. We don't know when the first shipments of palm oil arrived in Europe, but the early Portuguese expeditions to Africa probably brought some home. The first record of palm oil entering England dates to 1588, when Captain Welsh sailed to Benin and returned with palm oil and other goods. Welsh sailed to Benin again in 1590 and brought back thirty-two barrels of palm oil, evidently finding a good market for it in London.[95] Welsh's account affirms that the palm oil was intended for English customers, rather than being resold along the African coast. He also described a soap "which has the flavour of violets," undoubtedly made from palm oil.[96]

Palm oil became a familiar part of the healer's toolkit in England. A 1665 treatise aimed at alleviating the miseries of plague and other diseases among London's poor listed palm oil as an alternative to "Hogs-grease" for use in a soothing poultice. A 1692 book included palm oil as a standard ingredient for healing and soothing salves. No sources give any indication that palm oil was rare or especially costly. One guide for farriers even listed palm oil as a remedy for a horse with a sore back.[97] A 1678 pharmacy book gives the price of palm oil—*Oleum Palmae*—as one shilling per pound. That wasn't cheap, but it wasn't a prohibitively high price. Citrus oil sold at two shillings eight pence per ounce; nutmeg cost six shillings per pound.[98] By the early eighteenth century, the Royal African Company regularly advertised consignments of palm oil for sale in London, alongside other African produce like ivory, camwood, and beeswax.[99] London import records show about 4,300 gallons of palm oil arriving in the period from 1699 to 1744. While this figure is certainly less than the whole quantity shipped to England, it is consistent with the use of small amounts of palm oil as medicine.[100]

The English seemed especially fond of what one traveler called the "extraordinary Virtues" of palm oil.[101] "The common people sometimes apply it to chilblains; and, when used early, not without benefit," reported one source.[102] And although palm oil was not a luxury item, the trade was lucrative enough to attract dishonest merchants and counterfeiters. Buyers were advised to "Chuse your Palm Oil fresh and new, of a good Smell, and sweet Taste, so that it be as pleasant and grateful as the best fresh Butter we have, and of the highest Colour." Pharmacies sold stale palm oil that had lost its color so frequently that "some People fancy that the Palm-Oil is white." Fraudsters sold fake palm oil made of wax and olive oil, mixing in turmeric for color and orris root for its violet-like scent.[103] And while all sources agreed that palm oil should be fresh, not a single one suggested its use as food. Africans, one English writer noted, "use it instead of Butter; but we apply it only externally, as a Strengthener and Emollient, in all kinds of Weakness of the Limbs; in Pains, Cramps, Bruises, Strains, and Swellings."[104]

One curious anecdote appears in records from colonial Virginia. A certain Mr. Dennis responded to a letter from Mrs. Mills inviting him to work at her mill in July of 1663. On arrival, Dennis suffered an unfortunate attack at the hands of Mrs. Mills' friends: "Hee went upon this note & was by her and others Miserably whipt, Anoynted with stinkeing palm oyle all over his Body &c."[105]

We know nothing else about Dennis or Mills, or what they were doing with palm oil. What stands out about this episode is the word "stinking." Palm oil gets rancid and very smelly with time, but associations of palm oil with a stink were rare before the nineteenth century—a theme examined in chapter 3. One eighteenth-century description of a different kind of oil from an American palm noted it was like "the Palm-oil of Guinea, but has a disagreeable smell," suggesting that African palm oil wasn't typically encountered in a rancid state.[106] Pharmacy guides insisted that genuine, fresh palm oil had a "strong, not disagreeable smell."[107]

Yet by the end of the eighteenth century, Archibald Dalzel complained that palm oil sold in England was anything but fresh. The stuff he ate in Africa was "as different from that in an apothecary's shop as new butter from that which by age has become rancid and of different colours."[108] It's not clear that Dalzel's comment marks a shift in the quality of palm oil exported from Africa. It does coincide with the beginning a new phase in palm oil's commodity career, however.

• • • • • •

The oil palm crossed an ocean and colonized a new continent in the seventeenth century. The plant was conscripted by European slavers to sustain the forced migration of millions of Africans to the Americas, and it continued to feed and heal Africans living in captivity, as well as those actively resisting enslavement. Europeans added palm oil to their own repertory of healing substances, but the demand was so small that it probably had little effect on

Africa itself. Africans were already producing plenty of oil for themselves and to supply the ships of the slave trade. By the end of the eighteenth century, however, a new market for palm oil appeared in Europe: soap.

Africans had been making soap from palm oil and palm kernel oil for centuries. An anonymous Portuguese navigator sailing to Benin noted that soap made with ashes and palm oil had "great effect" in cleaning hands and linens, cleaning "twice as much as ordinary soap." Palm oil soap was "so highly valu'd by the Portuguese residing in those parts, that they will not suffer it to be exported to any parts of Portugal, lest it should undo the soap-boilers in that kingdom."[109] While some soap undoubtedly entered Atlantic commerce—including a brisk trade between western Africa and Brazil—African soap wasn't a major article of commerce. And by the 1790s, the English were experimenting with their own palm oil soaps.

Soap wasn't the most charismatic industry of the Industrial Revolution. Yet the dirty hands of workers needed washing, and so did mountains of wool and cotton feeding Britain's factories. The oil palm was about to enter the industrial age. Chapter 4 traces palm oil's changing roles in industrializing Europe. But first, we need to understand how new demand for palm oil reshaped Africa's relations with Europe, and in doing so, reshaped African societies and landscapes. As chapter 3 shows, the palm oil trade built on the legacies of the slave trade, leading to incremental change in some places and revolutionary developments in others.

3 From "Legitimate Commerce" to the "Scramble for Africa"

Abolishing the Slave Trade

Slavers carried two million African captives across the Atlantic in the last quarter of the eighteenth century, marking the grim peak of the transatlantic slave trade. The greatest number sailed under the British flag, yet that country reversed course and abolished the slave trade in 1807. It was the result of a long campaign condemning the moral outrages of the slave trade and slavery more broadly.[1] As Lord Grenville declared in an 1806 speech, abolition would reshape Britain's ties with Africa: no longer arriving as "robbers and pirates, carrying off the helpless inhabitants," British traders would carry on "a just and equitable traffic." It would, proponents insisted, lead Africans toward "the pursuits of civilized life."[2]

Europeans had bought much more than slaves in Africa: gold, ivory, palm oil, and many other things flowed out of the continent. Even at the height of the slave trade, one British ship managed to load 1,280 tusks, one ton of redwood, and 25,000 gallons of palm oil in a major slaving port.[3] Merchants who dealt in these "legitimate" products insisted that trade would thrive after abolition.[4] Along with settlers in Sierra Leone and Liberia—two colonies founded to test the "civilizing" power of commerce and Christianity—most merchants put their hopes in cash crops like cotton, sugar, indigo, and tobacco.[5]

Yet these proved disappointing. Pro-slavery writer James MacQueen mocked Sierra Leone's dependence on its African neighbors for even basic foodstuffs like rice and palm oil, warning that the colony was "a wild dream" doomed to failure.[6] By the late 1830s, though, settlers in Sierra Leone and neighboring Liberia were at the forefront of a new business, one that had already begun to replace the slave trade farther east in the Niger Delta: palm oil (map 3.1). When Liberian colonists declared independence in 1847, their new national seal proudly featured what one settler called "the Royal Oil Palm, the hope and strength of the country."[7] The Maryland Colonization Society boasted that the "wonderful productiveness of the palm tree . . . and the boundless extent of territory in which it grows spontaneously, and the myriads of inhabitants which swarm these fruitful forests, ready to labour for the smallest consideration" meant that palm oil could be made and sold cheaply to a world increasingly hungry for fat. They predicted "that this oil will yet form one of the heaviest articles of traffic in the commercial world."[8]

The statement captures common Western misperceptions that saw Africa as a fertile, untapped continent; Africans as untapped labor ready to work for low wages; and oil palms as a natural and largely wasted resource. Beyond indulging in racist caricatures, Europeans and Americans misread oil palms as evidence of Africa's natural bounty—trees planted by God in a garden of Eden—rather than as signs of intensive agricultural activity in the recent past. Across Africa's Atlantic littoral, centuries of slaving, warfare, disease, migration, and a changing climate had radically modified demography and land-use patterns, creating new villages and palm groves and leaving others abandoned. Settlers and traders weren't stepping into a primordial forest, but a landscape with a long history of human occupation.[9]

And even though the palm oil trade grew rapidly, it didn't bring a swift end to the slave trade or slavery. African elites continued to sell slaves alongside palm oil, reflecting continuity with the past rather than a sudden break.[10] Historians often dismissed the increased production of palm oil as a change of little significance in African societies. In these tellings, the real entrepreneurs were political elites who controlled trade routes, adding palm oil to their repertory of commodities.[11] In fact, the growth of the export trade encompassed a diverse collection of societies and historical experiences. Oil loaded on a European ship might have been made by women gathering "wild" fruit, or by gangs of slaves on a plantation.

European demand for palm oil did matter: chapter 4 explores the many ways that palm oil served industrializing Europe. But palm oil became one of many more or less interchangeable commodities; the key fact wasn't European demand but rather the ability of Africans to create a surplus for export. African societies adapted to the palm oil trade by expanding existing systems of production, and by creating new ways of making oil and organizing labor and land.[12] Some places were only marginally affected by change, but others saw profound developments. A few African states thrived in the era of "legitimate trade," using palm oil exports to amass wealth and arms.

Yet trade fueled conflicts with local rivals as well as with European merchants. Through the mid-nineteenth century, heavily armed African states often had the upper hand. When one British governor provoked war with Asante in 1823, Asante routed the British and their allies and decapitated

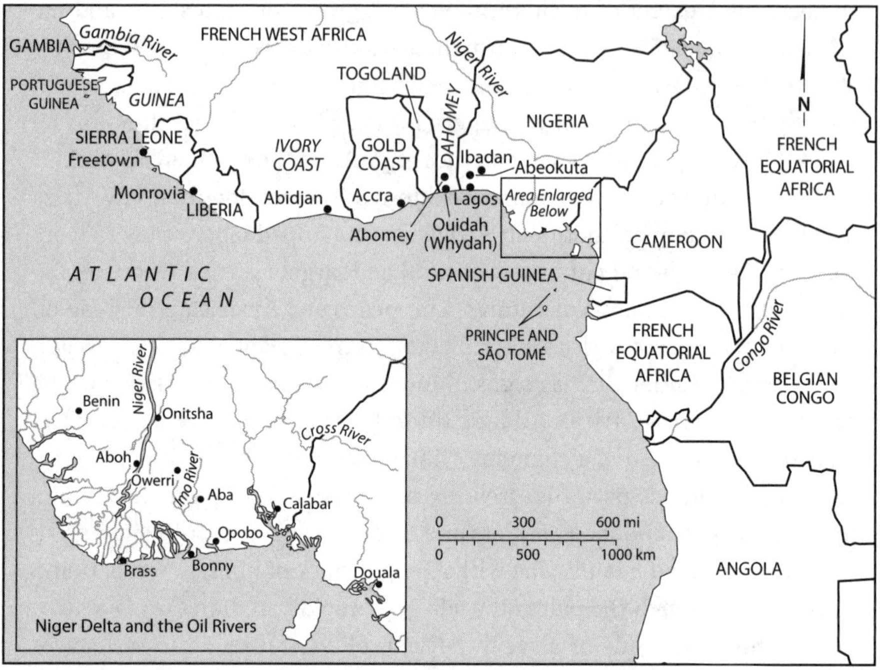

MAP 3.1 Western Africa after the "Scramble for Africa" (1885). Inset shows the Niger Delta region (including the "Oil Rivers" to the east), the most significant palm-oil exporting region in the nineteenth and early twentieth centuries.

the governor. Malaria and other diseases also took a horrifying toll on outsiders, making the permanent occupation of African territory a costly proposition.[13] But the balance began to shift after 1850: conflicts between European and African powers were increasingly settled by heavily armed European steamships. As the last section of this chapter shows, conflicts over trade sparked a "Scramble for Africa" that swept through the palm belt and ushered in seven decades of colonial rule (map 3.1).

Slaves and Palm Oil Together

The slave trade persisted long after 1807, as Britain struggled to enforce its ban and to convince other nations to adopt it. American plantations still demanded labor, and many capitalists—like those who arranged the notorious voyage of the slave ship *Clotilda* to Alabama in 1860—flouted legal penalties that included execution to procure slave labor. The *Clotilda*'s owners and many others were confident that they faced little danger. Britain's West Africa Squadron captured more than 1,600 slave ships during its sixty-year career, but thousands more successfully defied the blockade. One merchant complained that in the 1830s, he typically saw "20 and more vessels waiting for their cargoes of slaves" off Whydah. The ships could sell off their goods and buy slaves on land, carrying on trade "in perfect safety, unless they had slaves actually on board."[14] "Legitimate" traders complained that slaving proceeded "with the utmost vigour" across much of western Africa into the 1840s, causing a great "scarcity of palm oil" as their African partners abandoned oil for slaving, and as violence kept oil producers away from markets.[15]

Palm oil provided no great incentive to give up slaving by 1807. Prices in Europe were high, but the quantities traded were rather small, not nearly enough to replace slaving income for African elites or European shipowners. When quantities grew in the 1820s, prices fell accordingly (graph 3.1). MacGregor Laird wrongfully asserted in 1837 that the "collection of palm-oil is lazily and indolently followed" in Africa, but he recognized that it was more laborious than slave raiding. "The capture of a man partakes of the exhilarating nature of a hunt," he argued, while palm oil was "devoid of excitement, and becomes the sober tedium of business."[16]

The slave trade persisted in part because it hid behind palm oil and other legitimate commerce.[17] Britain negotiated a series of treaties allowing its navy to board vessels suspected of slaving activities, but it was difficult to prove that a vessel was a slaver, short of finding captives in the hold. In the early days, captains simply lied: a British commander wrote in 1821, "I have known the master [of a ship] to swear he came for palm oil, though his hold be stowed with water casks and farina [cassava], sufficiently attested a different object."[18] Iron manacles were clear evidence of slaving, but other equipment was ambiguous. Slavers needed big pots to prepare food for captives, but palm oil traders used them to refine oil and liquify it for pumping below deck. Casks of fresh water kept captives alive, but they also served as ballast that could be refilled with palm oil.[19] The voyage of the American-flagged *Mary Cushing* in 1838 offers one example. Sailing from Havana, the captain showed British officers documents affirming that the ship and its casks were intended for palm oil. When the *Mary Cushing* landed in Brazil, it delivered three casks of palm oil. The real cargo was 388 captives, twenty-six of whom died on the voyage. The ship was caught red-handed in 1839, freeing more than 400 captives.[20]

The notorious slaver Theodore Canot chided the British for hypocrisy. They condemned the slave trade, yet happily sold the goods that sustained it: "England, to-day, with all her philanthropy, sends . . . to convenient

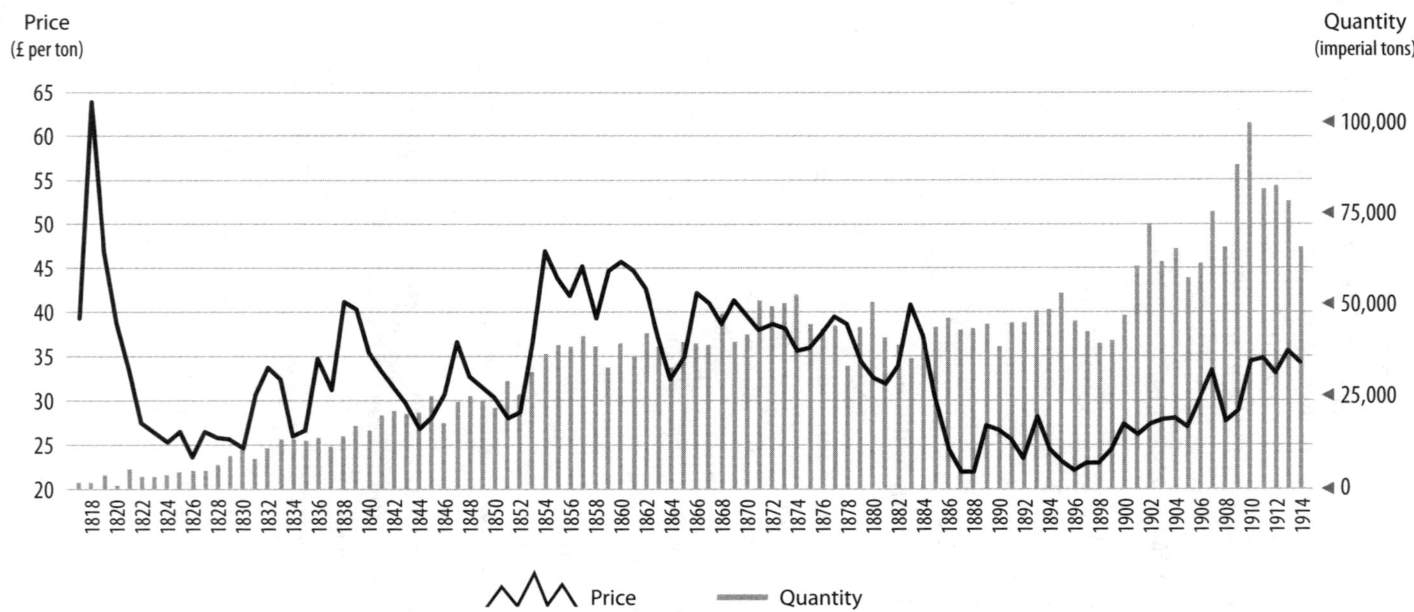

GRAPH 3.1 Chart illustrating UK palm oil imports (tons), and average UK prices (£ per ton) between 1818 and 1914. Data from Lynn, *Commerce and Economic Change*; and Latham, "Palm Oil Exports from Calabar."

magazines of *lawful commerce* on the coast, her Birmingham muskets, Manchester cottons, and Liverpool lead, all of which are righteously swapped at Sierra Leone, [Accra], and on the Gold coast, for Spanish or Brazilian bills on London. Yet what British merchant does not know the traffic on which those bills are founded, and for whose support his wares are purchased?"[21] In one 1827 deal, Canot's Cuban partners sent a ship laden with gold and 200,000 cigars to the Rio Pongo. Canot bought British-made fabric with the gold in Sierra Leone and sold the cigars along the coast. The fabric and other goods bought 220 captives, 217 of whom survived to be sold into slavery in Cuba. In another episode, Canot swapped palm oil for fabric and other trade goods he needed to buy more slaves.[22] British traders admitted that they "sell to any one, we never inquire."[23] At times, slavers might buy an entire palm oil ship and fill it with captives, sending the crew home on another vessel.[24]

Contrary to the claims of abolitionists, the trade in slaves was perfectly compatible—even complementary to—legitimate commerce. King Opubo of Bonny ignored British requests to stop selling slaves to the Portuguese, but he entered the palm oil trade to keep British ships visiting his port with desirable trade goods.[25] British merchants complained that when the French slaver Gaspard sailed into Old Calabar in 1833, "all legitimate commerce ceased, and a general scramble of robbery and plunder commenced to supply him with slaves."[26] When the slave ships left, palm oil sales resumed.

At Whydah, slave dealers were "the first to involve themselves in the production of export oil on a large scale." Captives were put to work producing the very commodities that paid for the capture and sale of themselves and their compatriots.[27] A missionary described a visit to Don José Dos Santos, "who, although a great slave-dealer, is also a palm-oil purchaser to a great extent." Dos Santos had his own palm plantation but bought much of his oil from local producers: "His yard was filled with traders, some with only a gallon, others having slaves loaded with large calabashes of oil; while dozens of his own slaves were counting out cowries to pay for the produce."[28] Another slave trader with connections to Brazil, Domingo Martinez, said "the slave trade and palm-oil trade helped each other." He "did not know which was the most profitable," and planned on continuing both lines of business.[29]

The Mechanics of the Trade

Slaves had been their own form of transportation, taking themselves, their food, and extra products like ivory to the coast. But calabashes of palm oil couldn't carry themselves. Animal haulage was impossible: horses, camels, and other pack animals died quickly in the palm belt. Human carriers could manage about forty or fifty pounds as a head-load, and an imperial gallon of palm oil weighed about nine pounds. The standard load was thus five gallons.[30]

Henry Broadhead, a commander with Britain's anti-slavery squadron, thought that transporting palm oil required even more slave labor than making the oil. A typical ship might buy 60–80,000 gallons of palm oil per voyage (240–320 tons). "When all this is brought down on the heads of natives, each man carrying two or three gallons, you may form an idea of the number of hands employed, independently of the collecting and the cultivation," he said.[31] In the 1830s, a slave was worth about as much on the coast as a ton of palm oil. A slaver might employ one armed guard for every five slaves, but otherwise the "merchandise" transported itself. Moving five tons of palm oil required about 300 carriers, to say nothing of guards.[32]

The Niger Delta ports (for our purposes, a region encompassing the "Oil Rivers" to the east of the Niger Delta proper) had a distinct advantage over other parts of the oil palm belt: extensive water transport.[33] Whydah, Accra, and other export centers relied on oil carried overland with head porterage. In contrast, the Niger and rivers to the east reached more than a hundred kilometers into the interior, though many smaller streams and tributaries were seasonally too shallow to navigate or were choked with mangroves.[34] The rivers feeding into the Lagos lagoon provided similar transportation services in western Nigeria, once carriers brought the oil to riverbank depots.[35]

The slave trade had already transformed Niger Delta states from communities of fishers and salt-makers into collections of powerful trading "houses" or "wards." Led by powerful men—and sometimes, women—these houses employed slaves as canoe paddlers, soldiers, and farmers. In all the trading states, the ability to deploy a fleet of canoes proved no less important in the palm oil trade than it had been during the slave trade.[36] Trading houses sometimes promoted and freed enterprising slaves, and several prominent leaders had slave origins. This mobility was largely reserved for men, however. Most enslaved women faced a life of domestic drudgery or hard agricultural labor.[37]

Palm oil might pass through a dozen hands on its way to the coast, as one trader sold her produce off to another. Large trading houses bought palm oil on riverbanks, transferring it from calabashes to big puncheons provided by European merchants. The traders jealously guarded access to oil markets; one merchant complained that coastal powers "do not suffer the inland natives to bring down the oil to the coast." Bonny sent its canoes more than a hundred miles upriver to buy oil, as far as Aboh on the Niger River. "The price they pay for it is extremely small," griped one Briton.[38] Palm oil sold in Aboh for £4–5 per ton in the 1830s. Europeans paid £12–20 per ton on the coast, and in Liverpool it sold for £30–40.[39]

Huge canoes manned by forty or more paddlers could carry up to twelve puncheons, nearly nine tons of oil.[40] A British visitor saw up to one hundred such canoes sailing for the oil port of Brass at one time. King Eyo of Creek Town could reportedly muster 400 canoes at the peak of his power.[41] The wealth needed to buy canoes and enslaved paddlers kept smaller competitors at bay; the trading houses were also perfectly willing to use the muskets and cannons carried by their canoes against commercial rivals, including Europeans trying to sail upriver.[42] The trading states were rightfully suspicious when Europeans began probing Africa's waterways with steamships in the 1830s. King Eyamba V of Duke Town told a group of Britons hoping to venture up the Calabar River: "I hear your countrymen done spoil [the] West Indies. I think he want come spoil we country all [the] same."[43]

Like their African counterparts, European slavers were best positioned to enter the palm oil business, having access to ships and credit relationships with suppliers. Of the seventeen captains who sailed out of Liverpool to buy palm oil in 1809, every single one of them had been a slaver.[44] Ships usually spent several months on the African coast, arriving in the rainy season when palm oil and foodstuffs were available. Many hired African sailors to navigate, handle the ship and its small boats, and load cargo. Kru seamen had a long-established reputation as sailors, and many also worked as skilled coopers, assembling puncheons that had been knocked down to save space on the outward voyage.[45] Malaria and other tropical diseases encouraged Europeans to remain on ships as much as possible, leaving the land-side business in the hands of Africans. One writer warned, "No seaman should be allowed to sleep *one night* on shore, if it can possibly be avoided."[46] Captain Gentle Brown noted the risks of spending time loading oil at Bonny: "The last voyage I lost none of my people, the former voyage only one, but the voyage before that 11 out of 19 died of fever."[47]

Newcomers often carried on a "coasting" trade. After anchoring near a town and firing a cannon to summon traders, these ships bought small consignments of oil as locals paddled out to trade.[48] More established firms used coastal "factories" (trading stations, rather than manufacturing sites) to collect oil, often relying on African or Afro-European agents to manage trade between visits from ships.[49] William Hutton testified that his factory at Whydah accumulated enough oil to load eight or ten ships a year by 1846.[50] By the 1850s, British firms began towing old ships to the Niger Delta. These "hulks" served as floating factories, allowing firms to collect palm oil and sell goods year-round.

Figuring out what goods to sell in exchange for oil was a tricky business.[51] Every port of call demanded different products. A sample of goods needed to buy ten "crews" of palm oil (roughly one-third of a ton) in Bonny in 1822 included salt, gunpowder, soap, hats, fabric, and a mug. Despite the resemblance to barter, the palm oil trade was—like the slave trade—a monetary affair. Africans and Europeans priced oil and trade goods in standard currencies. Those ten crews of Bonny oil were priced in "coppers," one of many types of metal bars, rods, and rings used as money. Oil-makers in the rural Ibibio hinterland insisted that Bonny traders pay them in coppers rather than in goods, a testament to the extent of the money economy.[52]

A small mollusk from the Indian Ocean, *Monetaria moneta*, turned out to be a vital part of the palm oil story in Africa. Colloquially called the cowrie, the shell of *M. moneta* and similar species had long been used as currency in Asia and Africa. Imports of cowries surged in the nineteenth century, and Captain Brodie Cruickshank credited the shell with boosting mid-century exports of palm oil. Strings of cowries were divisible down to a single shell, allowing people to buy and sell small pots of oil. Small-scale producers could save cowries to buy cloth, alcohol, firearms, and other goods previously reserved for slaving elites.[53] Cruickshank reported that when cowries were scarce, "no oil is brought to the market unless in such small quantities as may be required for immediate consumption." He understood that the cowrie was "tantamount to ready cash, which [the African] can apply in any manner he thinks fit."[54]

Egba traders insisted on being paid in cowries for their palm oil by the 1850s. Accepting trade goods in barter would have meant extra expense handling bulky produce and losses in converting the goods to currency, and they needed cowries to buy slaves from the north.[55] Cowries became the preferred medium of exchange across a vast stretch of western Africa. But by the late nineteenth century, the value of the cowrie had plummeted, diminishing its utility as currency. Europeans caused inflation by importing huge amounts of the less desirable East African cowrie, rather than the prized Maldivian variety.[56]

Trade ultimately relied on social capital as much as it did on money and equipment like ships and canoes. Merchants needed to trust their business partners; they in fact referred to the credit system used in the palm oil business as "trust." Some traders continued a slave-trade era practice of offering pawns as security for goods offered on trust, but without a ready market for slaves to dispose of pawns, most relied on personal ties and the promise of future trade to maintain trust.[57] As one British merchant wrote about his first voyage, traders at Bonny refused to board his ship because they "did not know me."[58]

Initially only kings or trading house heads could do business on trust, but as the volume of trade increased, Europeans offered credit to more junior traders.[59] A British critic remarked in 1851, "With the utmost confidence a fellow nearly naked will ask you for three, or four, or even five thousand pounds worth of goods on credit, and individuals are often trusted to that amount." In his view, "Not one in ten however that asks for trust is worthy of credit to the amount of so many farthings."[60]

Europeans could turn to kings, chiefs, and "secret societies" like Calabar's Ekpe society for satisfaction when trading partners defaulted or cheated them.[61] Ekpe members carried out rituals in masks, meting out fines, prison terms, and even death sentences to lawbreakers. Over time, membership in this and similar societies in other towns opened to slaves and even foreigners who could pay the initiation fee. Bishop Samuel Crowther wrote in horror: "Strange things are done by some European merchants in Africa, which Christian friends in England could hardly believe were possible." One merchant joining Ekpe took part in "all the idolatrous rites required on the occasion," costing him £300.[62]

When trust broke down, European merchants resorted to "chopping oil" from unlucky passers-by, a practice in which Europeans took "forcible possession" of oil-laden canoes, "often only after a desperate struggle." Once the oil was secured, the merchant "carefully tests and measures the oil so distrained upon, and gives the owner an order for the same quantity upon one of his absent debtors." In the eyes of the European, this wasn't theft: a paper receipt given for the "chopped" oil could be taken to the king responsible for the debt or the debtor, who would repay the victim, minus a commission.[63]

Merchants argued that chopping oil was simply a way of forcing a debtor to honor a contract.[64] Critics in Europe saw it as barbaric behavior, brought on by tropical conditions or prolonged contact with "savage" peoples.

Violence, along with hard drinking, sexual promiscuity, and the low-class origins of many British traders earned them the sobriquet "palm oil ruffian." The stereotypical ruffian was a "tyrant" who managed his crew only "by threats of having their brains blown out by the ever ready pistol." The ruffian's honesty toward African business partners was "simply regulated by the power of the natives to avenge ill-treatment."[65] One ruffian "candidly owned that he thought it hypocrisy for any man engaged in the palm oil trade to profess religion. He could not carry it on and keep God's commandments."[66]

When "chopping" got out of hand—and it often did—Africans responded by boycotting trade. Europeans knew "that Africans obeyed promptly any order 'to stop trade' issued by their Government."[67] Consul Livingston warned in 1870, "Native traders are united, and almost always win." Livingston overstated the case for African solidarity, but he nonetheless recognized that in the short term, European merchants needed to buy oil more than African traders needed to sell it. Livingston complained that eagerness to fill a ship's hold led to unsavory practices: "The whites cannot trust one another, and lose; and some begin to doubt whether these whites are the superior race."[68]

But British merchants had an ace-in-the-hole: the British navy. Until the Oil Rivers Protectorate was declared in 1884, merchants in the Delta technically enjoyed no special protections from Britain. In practice, British firepower settled intractable disputes. Several Niger Delta towns suffered naval bombardments after relations between African kings and British representatives broke down.[69]

Slavery and Gender in Palm Oil Production

One of the consequences of "legitimate commerce" was the expansion of slavery in Africa itself.[70] This wasn't unexpected: many early abolitionists opposed the slave *trade*, not the institution of slavery, and some explicitly called for African elites to become plantation masters.[71] Nonetheless, by mid-century, most abolitionists held that legitimate commerce would lead to the gradual extinction of slavery in Africa itself. They could not have been more wrong. "Where there is most commerce there is the most slavery," noted one British officer.[72]

Defining slavery in the historical record isn't straightforward. Like societies around the world, African societies practiced many forms of "unfreedom" and dependency. Reports from Europeans "anxious to eliminate slavery, and not always familiar with the intricacies of local usage," tended to use the word "slave" as a catchall. People who would be better described as pawns, serfs, or servants all fell under the same "slave" label.[73] Danish minister Andreas Riis reported that by mid-century, "the labouring people are mostly slaves" on the Gold Coast, describing people growing food and making palm oil under a wide variety of social conditions.[74] Among the Krobo of the eastern Gold Coast, household slavery was common but—at least according to European sources—fairly benign. "The slave is part of the family, his master eats and drinks with him; he works with him on one plantation," reported one visitor in 1867.[75] Krobo and Fante elites in the Gold Coast received a steady flow of captives produced by wars to the north. By the 1840s, they were so satisfied with the price of palm oil that they "no longer bothered with the overseas slave trade," preferring to use captives in palm oil production.[76]

Krobo oil production was carried out on a grand scale, with teams crushing fruit in stone-lined pits up to twelve feet across.[77] Large-scale production also emerged to the east in the Bight of Benin. The collapse of the Oyo Empire in 1817 unleashed decades of war among Yoruba states, and Yoruba elites used prisoners of war to make themselves "into proprietors of great estates."[78] Dahomey also amassed huge numbers of slaves for agricultural work, keeping the roads and borders under close surveillance for runaways.[79] The king settled captives around the capital, Abomey, where troops could suppress rebellions with deadly efficiency.[80]

A British missionary, William Clarke, reported the widespread use of slaves in palm oil production in Yorubaland around the 1850s. One enterprise had "more the appearance of a manufactory than other departments of labour," and at many sites, "perhaps fifty persons or more were engaged in labour."[81] We have few written accounts of palm cultivation and harvesting before the late nineteenth century, but some other reports like this confirm that enslaved workers made large amounts of palm oil for export. Groups of enslaved workers pounded or treaded palm fruit while "keeping time to the measured cadence of a song by a kind of overseer, who stood outside and regulated their movements" (figure 3.1).[82]

Though households produced much of the oil exported from the Niger Delta region, slave labor was important there, too. In Aboh, Ontisha, and other towns along the Niger, elite men and women amassed slave workforces numbering in the hundreds to make palm oil.[83] A British merchant noted that slaves were employed in "cultivating the palm tree, and making palm oil," though he failed to elaborate on what "cultivation" really meant.[84]

FIGURE 3.1 Pencil sketch, by Édouard Auguste Nousveaux, of men crushing palm fruit in a pit, overseen by a figure in Western attire. To the right, more workers cook fruit or refine oil in pots over fires. The text reads: "The making of palm oil" and "Forest of palms two leagues from Whydah." Metropolitan Museum of Art, 67.539.326.

Ibibio palm oil producers in the Niger Delta typically assimilated slaves into households as laborers: "any stranger bought as a slave [was] adopted into the family purchasing him."[85] The same was true for many captives in Igboland, though some Igbo communities checked social mobility by discriminating against slave origins across multiple generations.

By the end of the nineteenth century, a few Niger Delta elites owned thousands of slaves. Most were put to work transporting goods or producing food and palm oil. Yellow Duke, a Calabar trader, owned perhaps 3,000 slaves by the 1880s, and he was himself a former slave.[86] British consul Richard Burton dourly noted in 1865 that palm oil seemed to be encouraging slave-trading across the region: "I do not think [demand for slaves] has fallen off at all. . . . The man is bought for a few shillings, and sent up to collect palm oil." Burton argued that most slaves suffered terribly: "Provisions are very expensive; it is worth while for a native gentleman to buy those men at a cheap rate, and work or starve them to death in a few months so as to buy others."[87]

Regardless of whether a society produced palm oil with free or enslaved labor, that work was gendered. Men climbed trees; women (usually) cooked fruit and extracted oil. The association between women's work and palm oil put off some male elites, like the Egba war leader Sodeke, who responded to a British query about palm oil by asking, "how can a man sell oil like a woman?"[88] But elite men found that there was plenty of money to be made by capturing women's labor and redirecting it to export production.[89] Among the Ngwa Igbo, historian Susan Martin noted, "the key to a man's success in commercial palm production lay in control of women's labour."[90]

Work was gendered differently across western Africa. For example, Ibibio and Igbo men considered pounding palm fruit in a mortar to be manly work, but the rest of the process was for women. Yoruba men thought that pounding fruit was women's business, no matter how hard it was.[91] Samuel Johnson, born in Sierra Leone to Yoruba parents, reported that typical Yoruba gender norms prevailed around the city of Ibadan: cutting palm fruit was men's work, while making oil was women's work.[92] But in other Yoruba towns, observers clearly stated that "the manufacture of palm oil is the work of men."[93] A photograph taken in the Lagos hinterland, probably from the 1880s, shows only men at work in a palm oil workshop.[94] What was happening?

Gender roles proved flexible in the face of economic opportunities. The men employed in making palm oil were often slaves, and their social status took precedence over gender. A free-born Yoruba wife owned her own produce, while a man's slaves had no inherent right to the fruits of their labor.[95] The simultaneous growth of palm oil exports and the decline of slave exports meant that Yoruba elites had growing numbers of male slaves and new ways of extracting value from them. Over time, elite Yoruba men came to see agricultural work as slave's work, reversing cultural values that had once prized farming as a noble task.[96] Similar processes unfolded across the oil palm belt. In central Congo, Mongo elites forced male slaves to perform women's work in palm oil making, "a demeaning task that differentiated them from other men in the community."[97] Among the Fante and other Akan peoples of the Gold Coast, male slaves became the major source of labor for the palm oil export industry, while women continued to make oil for domestic use.[98] An 1844 sketch of palm oil production on a Whydah plantation (see figure 3.1) clearly depicts men carrying out the laborious task of mashing fruit, as well as the feminine tasks of tending pots and cooking fires.

Enslaved men's labor was conspicuous to European writers. The labor of enslaved women was less visible, hard (for Europeans, at least) to distinguish from household labor in general. But even when enslaved women worked within the context of household production, they could use palm oil as a path to freedom. A British consul noted that in the Lagos hinterland, enslaved women made and sold oil on their own account: "by care and frugality [they] soon amass sufficient cowries to pay a heavy sum for the redemption of themselves and their children."[99] Enslaved men and women around Lagos could buy the right to harvest palm trees outside the slaveowners' working hours, giving them ownership over the oil they made and sold.[100]

Most of the powerful coastal traders in the pre-abolition period were men, with a few notable exceptions. This pattern gradually changed as the palm oil trade expanded and offered new opportunities to upstart traders. The Yoruba trader Madame Tinubu, who got her start in the slave trade, had a workforce of at least 360 people doing nothing but carrying palm oil and ivory.[101] In Bonny, the growing prominence of women as palm oil traders forced men to adjust social norms. Prohibitions on women's travel upriver were lifted to allow them to buy oil at interior markets. Two women, Fat Mammy and Orumbie, became the titled heads of trading houses.[102] While women could not attend all the secret rituals of the Ekpe society and similar organizations, they could sometimes purchase the titles and privileges available to men if they had enough money. In other cases, women organized their own societies and titles, creating a parallel power structure. And women continued to dominate the small-scale trading of oil, which supplied domestic consumers as well as larger export buyers. Historian Francine Shields notes that the importance of women in the trade was best illustrated by their absence: during a dispute with the British in 1858, the government of Abeokuta banned female traders from traveling to Lagos. The result was a near-total halt to the palm oil trade from the Egba territory around Abeokuta.[103]

The most significant shift in gendered work in the oil palm industry came with a new kind of palm oil. Until the 1840s, most palm oil for export was made same way African women made oil for domestic use. Little changed as far as technology and process, though the brass pots ("neptunes") and iron cauldrons bought from European traders were more durable than traditional ceramic vessels.[104] As producers recognized that Europeans didn't care about the taste of oil, they began to simplify the process.[105] The result was rancid "hard" oil. Unless palm fruit is thoroughly cooked, enzymes in the fruit break apart fat molecules, releasing free fatty acids (FFAs). FFAs add a sharp taste to oil and react with oxygen and water to make unpleasant-smelling chemicals—the main cause of "rancid" odors and tastes in all fats.

Hard oil was probably developed independently several times across the coast. In one technique, producers made oil as usual but extended the initial fermentation of the fruit, softening it to save labor in pounding. If consumed promptly, the oil was still edible. Producers could save more labor by using the most rudimentary refining methods, such as rinsing the palm pulp only once with cold water to separate the oil. Some hard-oil methods used no heat at all, letting the fruit ferment until the oil oozed out. In the western Niger Delta, Urhobo men dug pits eight feet long and two or three feet wide (and as many feet deep) and lined them with planks to pound or stomp fermented fruit. Ngwa men might use a hollowed-out tree stump as a giant mortar, and many producers used old canoes (later, purpose-built "oil canoes").[106] A late-nineteenth-century traveler in Liberia noted that people prepared edible "soft" oil ("the finest and the most favourite in the kitchen") year-round, but turned to hard oil during the peak fruiting season, when the soft oil method "cannot meet the supply."[107]

Relying on fermentation alone was "a heathenish way to get the oil," according to one critic, but the hard-oil method had important implications for men's productivity.[108] Making a ton of soft oil required up to 420 workdays, divided between men and women performing gendered tasks. A ton of hard oil made without cooking took about 132 workdays, with no women's labor required.[109] One historian estimates that households in Togo spent twice as much time making hard oil as they did soft oil, but they produced ten times more oil using the hard process.[110]

Young men without wives or slaves could invest their own sweat equity in pruning, climbing, and harvesting palms, sometimes working distant groves set aside by communities for the purpose, to make export-grade oil.[111] And in some regions, hard oil was simply a necessity for entering the export trade: getting enough firewood or fresh water to make soft oil at scale was impractical.[112] Hard methods also eliminated the scale constraints imposed by the size of cooking pots and the availability of women's labor.[113] Owners of plantations near Arochukwu created pits as much as ten feet across to allow large gangs of men to pound fruit, a technique seen across major exporting regions.[114]

Foreigners complained about the smell of hard oil ("a stink absolutely indescribable"), but they still bought it.[115] The physical properties of hard oil proved useful. Soft oil tended to seep out of puncheons, while hard oil was solid or semi-solid in tropical heat. Puncheons filled with hard oil floated better than those with soft oil, making them easier to handle on rivers. Even small traders moving oil in calabashes or clay pots saw benefits in handling hard oil: "if the carrier falls or stumbles, sweet soft oil which is liquid is wasted, whereas hard oil remains and can be gathered up again."[116] Hard oil was well suited to puncheon rolling, a trade that emerged in the Gold Coast and Dahomey to take advantage of improving roads.[117] In Dahomey, a special term denoted slaves assigned to roll puncheons.[118]

Unfortunately, the historical record isn't clear on precisely when hard oil appeared. Ship captains and merchants who wrote about palm oil through the early 1840s made no mention of a foul odor. They often differentiated between fresh palm oil and the stale oil sold in Europe, but the latter might have spent a year sitting in a ship's hold.[119] By the late 1840s, European sources agreed that all palm oil "as it occurs in commerce [in Europe], is in that state in which the ordinary fats are called rancid."[120] Some Europeans distinguished between trade oil and the "finer variety" eaten by locals, but they chalked up the difference to adulteration. It is not until the 1850s that we see clear descriptions of "hard" oil. William Baikie reported that dark, discolored oil sold at a discount to bright red oil in the markets around Aboh in 1854, though he made no comment on the edibility or smell of the darker oil.[121] William Cole's contemporary account identified "trader's oil" as a distinct product, made by men stomping palm fruit in a canoe. Cole contrasted it with tasty soft oil, noting: "[A] nicer relish is seldom required to render the bachelor's yam palatable."[122]

Many regions continued to export "soft" oil instead of, or in addition to, hard oil. Lagos was famed for the quality of its oil, made by briefly fermenting fruit and then boiling it in a plaster-lined pot. The thick pot allowed producers to regulate temperatures and conserve fuel. The feet used to crush fruit in a canoe were "carefully washed," and the resulting "thick gamboge-coloured scum" was repeatedly skimmed, sieved, and heated until the pure oil turned "almost blood-red." Much of the work went on overnight, taking advantage of cooler morning temperatures to help separate the oil and water. The result was a labor-intensive product, prized in Lagos as well as in Europe for its fine color, odor, and taste.[123]

The premium paid for soft oil overseas can't explain the decision to stick with soft oil, however. It typically sold in Liverpool for 20 or 25 percent more than hard oil, not enough to make up for the larger investment in labor.[124] The key in Lagos, as in inland parts of the Niger Delta region, was the local market. Domestic consumers usually paid more than export buyers, leaving Europeans with the unsold leftovers (a theme explored in more depth in chapter 6).[125] The marginally higher returns from soft oil were often worthwhile for women, who could not easily redirect their labor to gathering fruit or other cash-earning tasks.[126]

The Palm Kernel Trade

As palm oil exports reached new heights in the 1850s, they were joined by another commodity: the palm kernel. Most accounts credit Europeans with "discovering" the value of the palm kernel, asserting that they were "formerly thrown away, or burnt, though the negroes were well acquainted with the fact that they contained oil."[127] "Until very recently [1870] the nuts were thrown aside as worthless," wrote one authority, "and they form in some places useful mounds of rising land. The mud floors of houses and yards, and roads leading to wharves, are paved with them, and they set almost as hard as cement."[128] In fact, palm kernels had variety of important uses across Africa, and they had long been used as food and in soap, medicines, and lamp oil.

Unfortunately, Britain's record-keepers lumped kernels in with "nuts and seeds for expressing oil" until the twentieth century, making it difficult to track early imports.[129] Most accounts credit the British firm of W. B. Hutton & Sons with the first imports in 1848 or 1849.[130] One Cape Coast merchant circulated information about kernels along with samples of kernel oil and a candle made from it in 1842, but found no interested buyers.[131] French firms bought kernels from Dahomey from at least 1847, however. One account claims that kernels arrived in Marseille as early as 1832 and another that British firms imported small quantities practically from the beginning of "legitimate trade."[132] An Englishman did receive a patent for extracting palm kernel oil in 1832, using methods "communicated to him by a foreigner residing abroad" that were similar to African processes.[133]

No matter who shipped the first consignments, much credit is due to two African pioneers: Charles Heddle and Samuel Herring. A trader of Scottish-Senegalese descent, Heddle had gotten his start in the peanut business. By the early 1840s, he moved to Sierra Leone, where he claimed to have shipped the first cargo of palm kernels, worth £4, in 1846.[134] By 1860, kernel exports were worth nearly as much as Sierra Leone's palm oil exports, and by 1880, the colony earned four times as much from kernels as it did from palm oil.[135] When Heddle died, he was "by far the richest man in West Africa."[136]

A man of his times, Heddle embraced the rhetoric of "legitimate commerce" and Victorian notions of a "civilizing mission" in Africa. But he was also a committed opponent of European racism: he argued that huge volumes of oil and kernels produced entirely by hand offered "a most emphatic denial to the often repeated assertion that the negro will not labour except on compulsion."[137] He depicted palm kernels as a quintessentially democratic commodity, accessible to anyone: "Every member of every family above the age of three years will not only have employment but remunerative employment, during every hour of every day in the year." Unlike palm oil, which required male harvesters and plenty of hard labor in its preparation, anyone could crack kernels. The promise of year-round work would create "the habit of labour," which Heddle was sure "will bring with it other habits, and new wants."[138]

Like their neighbors in Sierra Leone, Liberians were early participants in the kernel trade. This was in part due to their own tastes: African American settlers found that kernel oil was a closer substitute for the lard and butter they were accustomed to than palm oil. Kernel oil sold for twice the price of palm oil in Liberian markets.[139] Samuel Herring, born into slavery in Virginia before emigrating to Liberia, developed a butter-like product out of kernel oil in the 1840s. His "African lard" attracted excitement in abolitionist circles, but it never became a successful commodity. Herring also invented a machine to crack kernels.[140] He claimed credit—probably too much—for "the discovery or rather practical application" of palm kernel oil, and thus with its "subsequent demand by the French traders."[141] Toward the end of his life, Herring published a pamphlet enshrining his claim as "the sole originator, and Introducer of the great Palm Kernel Trade in 1850–4," noting that he had started the business in 1848. He wasn't shy about self-promotion, calling himself "the instrument employed by Divine Providence." Unfortunately for Herring, he was also ahead of his time: consumers in the North wouldn't start eating margarine and other novel fats made from palm kernel oil for another two decades. Without a foreign market for his kernel oil, Herring turned it into soap for sale in Liberia, running a successful business up to his death in 1895.[142]

Making kernel oil, even with Herring's machines, made little sense on a commercial scale in Africa. A report from the late 1850s noted that a bushel of kernels sold for fifty cents in Liberia, and two bushels made one gallon of oil. But a gallon of oil was only worth seventy-five cents: the raw materials were worth more than the finished product.[143] In the long run, it made more sense to export intact kernels, as kernel "cake" left after extracting oil had no value in African markets.

Most of the work in preparing kernels was already being done to make palm oil, and cracking them required no capital beyond a pair of rocks. "[W]hat a loss of time and labour!" exclaimed one British trader at the sight of women cracking piles of kernels by hand.[144] But this missed the point: not all time and labor were equivalent. Women didn't climb trees to cut fruit. Kernel cracking was also intermittent. A woman could start or stop in a moment, and children could help.[145] Over a year, intermittent work added up to a lot of kernels, about 167 workdays for one ton.[146] Women often kept ownership of kernels, even when their husbands claimed ownership of finished palm oil.[147]

Transport problems held back the kernel trade in some regions: moving a ton of kernels required at least thirty porters, and a canoe load of kernels was worth much less than a comparable volume of palm oil.[148] But environmental factors were also important. On the western end of West Africa, thin-fleshed fruit was common because of the long dry season. Producers got heavier kernels but less palm oil from the average fruit bunch, compared with fruit from the rainier Niger Delta. The western palm population also appears genetically distinct from those found east of the "Dahomey Gap," further affecting kernel size. These natural forces, coupled with huge domestic demand for oil (see chapter 6), meant that kernels rather than oil came to dominate the export trade in a region stretching from Sierra Leone to Lagos by the end of the nineteenth century.[149]

Environmental Transformations

Europeans buying oil rarely saw where oil palms grew or how the oil was made. In the 1840s, a British official on the Gold Coast declared: "A gallon of palm-oil has probably never been collected from a tree that was planted for the purpose of obtaining this article." Like many Europeans, he insisted that the oil palm "grows wild, and the nuts are collected here and there, where a chief has no other employment for his slaves; but nothing like a settled system of industry is connected with this trade."[150] Palm oil merchant William Hutton expressed his own "great doubt" about whether "the palm tree is planted much for the production of palm oil."[151] But the typical merchant "merely saw the puncheons of palm oil" arriving on canoes; "his object is to get out as soon as possible."[152]

The information that did trickle down to palm oil buyers on the coast was fragmentary. Captain Midgley attested in 1842 that "palm trees have been planted" in the land behind Bonny, for example.[153] But words like "planted" and "plantation" were ambiguous. A "plantation" could describe a small farm, or a large-scale enterprise worked by slaves. At times, it may have described "wild" groves managed by humans. As historian Martin Lynn noted, "it would be wrong to see a strict dichotomy between 'farmed' and wild palms." Lynn argued that oil palm cultivation is best seen as a "continuum." Africans did plant palms, but they also used palms left behind by earlier cycles of cultivation.[154]

Scattered descriptions show that African landscapes—including oil palm landscapes—were changing during the nineteenth century.[155] The rise and fall of states; the ravages of war, disease, and drought; and patterns of shifting cultivation all left their mark on Atlantic Africa. One clear example was a late seventeenth-century description of area near Mount Cameroon. The islands offshore were rich with oil palms and yielded plentiful oil and wine, but "there is not one [palm] to be seen on the opposite continent." By the time Richard Burton visited in the mid-nineteenth century, the mainland was covered with oil palms.[156] The change may have been due to a wetter climate, but it likely owed much to new settlements creating emergent palm groves.[157]

The clearest evidence for systematic oil palm agriculture comes from the region stretching from the Gold Coast to the Yoruba states. European travelers described farmers in several societies replanting palm seedlings or sowing kernels.[158] The Rev. Riis testified that he had seen "natives planting palm trees" for the export trade in Krobo territory in the 1830s and 1840s. He thought the work was mostly done by slaves.[159] Two decades later, Rev. E. Schrenk stated that "There is a large plain near the Volta which once had no trees on it about 100 years ago, and the whole of that plain is now a palm forest." When asked whether the palms had in fact been planted, rather than spread by natural means, Schrenk confirmed that they were planted.[160] Palms in Dahomey were also planted, transplanted, or thinned from natural seedlings, resulting in shorter, easy-to-harvest palms. A "well-tended" palm in Dahomey could be harvested three times a year, and a man might harvest from fifty trees in one day.[161] This pace would have been impossible in "wild" groves, with trunks towering twenty meters or higher.

Krobo men used a special form of land tenure, the *huza*, to acquire and manage new land for palm plantations. Departing from kin-based systems of tenure common across western Africa, the *huza* was a tract owned by investors. They carefully measured the land with ropes, bought it from its owners, and divided it as private property. Farms were cleared and settled by workers, who might be relatives, slaves, or migrants from another ethnic community. Danish observers were impressed with the system, noting that it allowed the Krobo to "extend the cultivation of oil palm year after year, thus building up oil palm forests."[162] These "forests" were in fact extensive plantations, in contrast with the "isolated, often island-like plots" of palms and other crops cultivated by neighboring groups like the Akuapem.[163]

Krobo men used their profits to buy more land, engaging in what one anthropologist called the "bloodless conquest" of the region.[164] Demand for labor to work these plantations helped prop up slave prices on the Gold Coast in the aftermath of abolition.[165] Plantations worked by slaves and migrants also appeared west of the Krobo among the Fante. "A generation ago, there was no property in palm-trees in Africa," one observer reported. By the mid-nineteenth century, there were "now *plantations* . . . and lately the title to some has been disputed in courts on the Gold Coast."[166] While "wild" palms had traditionally been common property, Fante landowners now asserted individual ownership over trees. They barred tenants from planting new palms, which could mark ownership of land. "Custom does not permit any person to be improved out of his land," one noted jurist wrote. Tenants typically paid the landowner with half of the palm oil harvest, though tree owners were only entitled to a quarter of any palm wine tapped. A special form of tenure allowed Fante landlords to collect rent in a fixed quantity of palm oil, incentivizing tenants to get the most out of the palms.[167]

A similar process unfolded across the Yoruba states. Land was controlled by lineage heads, who assigned it to families, slaves, and newcomers. Tenants held usufruct rights to the land but not trees. An expression recorded by the Yoruba historian Samuel Johnson warned: "The grantee is to look down not up." An Oyo prince reportedly caught one of his tenants illicitly cutting palm fruit and offered to spare the man if he "could climb up again and replace the nuts; this being impossible he clubbed him to death!" Farmers were prohibited from planting oil palms or other permanent crops, which marked ownership. Lineage heads couldn't harvest all of their trees themselves, though. They allocated harvesting rights to wives, children, slaves, and retainers. In exchange, the beneficiaries paid the owner with cash or up to half of their finished palm oil.[168]

Yoruba farmers "cultivated" the palm by regularly pruning it; they "never cut [it] down in clearing a farm, except when from age the tree has ceased to bear." "Male" trees—infertile specimens—could be cut for their wine.[169] Access to palms became increasing contentious in the nineteenth century as Yoruba commoners competed with elites to make and sell palm oil. At

Badagry, religious leaders tried to restrict access to communally owned groves by declaring a "fetish" over the trees, preventing commoners from harvesting them.[170] Africans liberated from slavery and returning home from Brazil or Sierra Leone also created problems for elites. When Yoruba rulers granted them land to settle, the settlers acted as if the grant gave "individual rights of absolute ownership in property, although local land tenure recognised no such thing."[171]

Palm cultivation was most clearly established in Dahomey, where visitors regularly commented on sights like a "splendid palm-oil plantation."[172] Dahomey was a favorite subject for abolitionists and pro-slavery writers alike: the former argued that legitimate trade would civilize the warlike kingdom, while the latter insisted that the state's propensity for violence and human sacrifice reflected innate barbarism that justified slavery.[173] Dahomey's leaders proved both camps wrong, balancing tradition and innovation as they shifted resources toward palm oil.

Thomas Hutton received permission to open the first palm oil factory in 1838 from King Gezo, who was still a notorious slave exporter.[174] The French firm of Régis soon followed, providing slavers with European manufactures in exchange for palm oil. Dahomey had exported palm oil since the 1790s, and European and Brazilian merchants convinced Gezo that it was a worthwhile business, even if its main function was to attract merchants with desirable British goods.[175] Royal officers, Afro-Brazilian settlers, and European traders all established plantations using slave labor.[176]

Gezo began collecting taxes in palm oil, and he ordered a new oil palm planted at every place a child's umbilical cord was buried, ensuring that the peasantry had plenty of young palms.[177] He also banned the felling of palms for wine.[178] Gezo's own plantations were worked by war captives from Yorubaland and northern regions. Europeans were impressed by the "fine farms" of oil palm and maize these slaves cultivated.[179] Surveyors planted locust beans alongside oil palms "to demonstrate the king's interest in those trees."[180] As a British visitor noted, palms were even individually numbered "with a view to revenue." Punishments for stealing fruit or tapping wine from the king's palms were severe, including being beaten with sticks and dragged through sand.[181] Gezo's successor Glele remained committed to the palm oil trade, using it to secure arms and trade goods needed to maintain the militaristic slave society at the core of the Dahomean state.

A French account from the 1860s described "immense plantations of palms" owned by elites but added that "every free black possesses a few feet of them." Even slaves could share in the prosperity brought by oil palms:

they "pilfer a little everywhere," making oil to sell and gathering kernels to crack.[182] Large slave plantations were politically dangerous, however, and the fear of revolt may have checked the expansion of Dahomey's palm plantations. Environmental factors also limited the industry: outside of heavily settled areas, young palms succumbed to grassland fires. Communities living on the Abomey plateau also confronted a lower water table than groups living farther south, and as a result, they saw lower fruit yields. Water to cook and refine palm oil was scarce, forcing some families to carry palm fruit to riverside "factories" where oil was made. Aja farmers often managed their palms as "wineries" instead of making oil, tapping the trees to supply Dahomey's capital and other towns with palm wine.[183]

There is less evidence about changes in cultivation in response to the export trade from the Niger Delta, the most significant exporting region. The trading states fiercely guarded access to the interior. As the Brass trading chief King Boy put it, the attackers who killed Richard Lander during his 1834 attempt to sail up the Niger felt "the river was theirs and white men should not go up."[184] Lander's compatriot, MacGregor Laird, wrongly asserted that palms stretched endlessly into the interior, with huge quantities wasted—but he never saw more than the riverbanks.[185] Later travelers recognized the connection between human settlements and oil palms in the Niger Delta hinterland. As one British official noted, the number of palms around a town was a good proxy for its population.[186] People planted and protected the trees they needed, while truly wild palms fell to drought, elephants, fire, or the shade of forest trees.

Men and women interviewed in the region in the 1960s spoke of a distant past in which no one bothered to own or even climb palms; women gathered ripe fruit fallen on the ground. The export boom changed that, leading to new land-use rules.[187] Patchworks of tenure and usufruct systems developed, including individual, lineage, and communal ownership rights, often overlapping. In one town north of Benin, residents stated "that all oil palms had owners formerly, but that they are now common property, except in so far as claims dating back to the period of individual ownership can be established."[188] In contrast, some Igbo elders insisted that oil palms were originally "a wild communal resource," only becoming the subject of property claims once the palm oil trade began to grow.[189] While some communities responded to the growing export trade by repealing communal rights to oil palms, others strengthened them to ensure that all men had a fair shot at harvesting fruit.[190]

Individual ownership usually belonged to the tree planter and their heirs, and children might receive an oil palm as a birthright, on the burial site of the placenta or umbilical cord.[191] Among the Igbo, such a "tree of status" was the inalienable property of a free-born child, forming "the foundation for the socially ambitious."[192] Wealthy traders used slave labor to plant more trees as private property.[193] As early as 1835, British traders warned that "no young trees are planted" in the Bonny hinterland while slave raids were underway.[194] Later in the nineteenth century, Aro traders put slaves to work farming oil palms on large plantations.[195] Elite Igbo men came to think it "almost a taboo for a free-born man to climb the palm tree or process palm oil."[196]

Evidence from across Igboland shows that poorer households as well as slave-owning elites produced oil for the export trade; it's not possible to quantify which group made the greater contribution.[197] What is clear is that neither commoners nor elites enjoyed easy access to an infinite number of palms. Like their counterparts across the palm belt, farmers had to balance the land and labor requirements of food and oil production, and many palm oil specialists relied on purchased food for their survival. There was less room for the palm oil industry to grow than Europeans imagined, an idea examined in greater depth in chapter 6.[198]

Toward the "Scramble for Africa"

The 1850s were a high point for the palm oil trade: European prices reached a peak of nearly £50 a ton, thanks to a tallow shortage caused by the Crimean War. Over the rest of the century, prices steadily fell, reaching a low of £19 in 1887. The goods Africans bought with their earnings also got cheaper, but that didn't make up for the fall in palm oil prices.[199] Traders sought out higher volumes of oil, and later, kernels to compensate. But for the people who actually made palm produce, this meant more work for less money. Africa's palm oil exports to Britain surpassed 50,000 tons for the first time in 1871; thirty years later, they fluctuated at about the same level, illustrating the lack of enthusiasm for expanding palm oil output as long as prices remained low. Some producers shifted toward hard oil, but low-quality oil fetched lower prices, continuing the downward spiral.[200]

But Europeans wanted more and more oil and kernels. MacGregor Laird claimed in 1839 that an area the size of Ireland east of the Niger was covered in palms; "the palm-nut now rots on the ground unheeded and neglected."[201] Burton similarly accused Dahomey of neglecting the "mine of wealth" at its doorstep.[202] Traders on the Gold Coast asserted "there should be eight times as large a quantity of kernels as palm oil," demanding that

Britain wield its power to build railroads to rescue "hundreds and thousands of tons of [oil and kernels] which lie rotting in the interior."[203] In the European (and especially British) imagination, palm trees were "wasted by the tens of thousands" because of a lack of labor, transportation, and good governance.[204] The solution was "free trade," which really meant free reign for British capital and the destruction of the coastal elites' monopolies on trade. Bolstered by anti-slavery rhetoric, the pursuit of free trade became an excuse for conquest in Africa. The result was a "scramble" for Africa that ended with the continent divided among European powers.

Lagos was the first major port to fall under direct British control. The city was a leading exporter of slaves in the early nineteenth century, benefiting from wars across Yorubaland. Yoruba-speakers liberated from slave ships and settled in Sierra Leone began to arrive in Lagos and towns around the interior waterways in the 1840s, bringing Christianity and the ideals of "legitimate trade" with them. Missionaries, especially Britain's Church Mission Society (CMS), built churches and schools across the area.[205] The newcomers were skillfully manipulated by Akitoye, who had been ousted as *Oba* (king) of Lagos in 1845. Akitoye portrayed himself as a champion of abolition and the palm oil trade. Through missionaries and traders, he urged Britain to help him regain control of Lagos.

Missionaries warned that Kosoko, the man who overthrew Akitoye, was a slave trader and a threat to both Christianity and commerce. "Will men endanger their lives by cultivating the soil? Will they seek the palm-nut in the woods when they know a price is set upon their heads?" asked one missionary.[206] Kosoko refused to sign an anti-slavery treaty in 1851, giving the British consul a pretext to attack. While Kosoko repulsed the first assault, a second attempt put Akitoye back in power.[207] Akitoye signed an anti-slavery treaty and agreed to new terms that banned the "trust" system used by palm oil traders in the Niger Delta—a system that gave African traders a great deal of leverage over European partners. Nonetheless, the Oba and his successor, Dosunmu, faced constant headaches as palm oil merchants battled over trade and debts. Europeans appealed to Britain's consul, rather than the Oba, as the real authority. Frustrated British officials decided to annex Lagos in 1861, though one fretted that this "original sin" would lead to more conquests. He was right.[208]

"It is said that Commerce will put an end to Slave Trade," wrote British Prime Minister Lord Palmerston in 1860. Yet "it is equally true that Slave Trade puts an end to Commerce," he asserted. While he agreed that free trade "ought not to be enforced by Cannon Balls," Palmerston argued for assertive military action in Africa.[209] Free trade was now a cause worth killing for. When the king of Porto Novo blockaded palm oil exports to Lagos in 1861, British warships shelled the town, and later attacked with rockets. The consul "hoped that the interior people would take note."[210] People in and around Yorubaland had been paying attention, of course—particularly the Egba. This Yoruba-speaking group founded Abeokuta in 1830 and welcomed some returnees from Sierra Leone in 1841. They were accompanied by CMS missionary Henry Townsend, who helped build political and commercial ties with Britain.

A decisive Egba victory over a Dahomean invasion in 1851 overshadowed earlier bouts of anti-Christian persecution in Abeokuta, and the CMS dramatized the conflict as a battle between freedom and slavery, Christianity and barbarism. The fact that most Egba weren't Christian, owned slaves, and continued to buy and sell them didn't matter: Abeokuta became the "sunrise within the tropics" in the British imagination.[211] CMS agents worked fervently to promote cotton growing in Abeokuta and other parts of Yorubaland, enjoying a brief moment of success when the American Civil War (1861–65) sent cotton prices skyrocketing.[212] But in the end, palm oil and kernels, not cotton, became the key "legitimate" exports.

Despite frequent wars, huge amounts of oil and kernels flowed into Lagos from Abeokuta and other Yoruba states. One visitor reported passing "thousands of persons bearing palm-oil and other commodities" on a road north of Abeokuta in 1860.[213] Palm produce sales in fact kept Yoruba states well supplied with weapons. Madame Tinubu, a former slave trader turned to palm oil, supplied Abeokuta with arms during the war with Dahomey. Ibadan's war chiefs complained that their foes to the south, the Ijesa, had acquired a Gatling gun. In 1884, they begged their contacts in Lagos to send Snider rifles: "We promise to pay in palm oil and palm kernels which are collecting for the purpose," they declared.[214] Trade itself was a powerful weapon, too. Yoruba states used tolls on road and river traffic as a source of income.[215] Blockades were used against rival powers as well as European merchants, and they became more frequent as palm produce prices stagnated during the 1870s and 1880s.[216]

Despite its role financing warfare, Christians in Europe and Africa expressed their faith in the "civilizing" power of the oil palm. "This tree has proved to be a more powerful opponent to the slave trade than even the unconquerable navy of Great Britain," claimed one author. David Livingstone passed out palm kernels during his famous travels across central Africa.[217] The Yoruba bishop Samuel Crowther was confident that palm oil exports would not only end the slave trade, but also "will prove a silent but a sure death blow in course of time to the existing domestic slavery." Crowther knew of "upwards of a thousand slaves [who] have purchased their own liberations, through extra labour and earnings," mostly in palm oil and kernel production.[218]

Slavery was outlawed in the British Empire in 1833, which created problems for Britain's new subjects in Lagos and the Gold Coast (annexed in 1867). Domestic slavery was widespread and vital to the economy. Some individuals took advantage of official abolition decrees and claimed freedom in cities and mission stations, but officials didn't encourage mass emancipation. Former slaves faced severe social barriers: without patrons, they had no access to land or an income.[219] Still, palm oil traders in Lagos complained that they lost money as slaves walked away from farms to freedom after 1861.[220] A British merchant complained that "A good many of the trees are loaded with stuff that never will be garnered," because so many slaves had fled.[221]

The linked spread of Christianity, British colonialism, and anti-slavery ideas created conflicts as British power crept into the Niger Delta region. In western Igboland, "slave owners watched with disgust as their slaves trooped to the Christian churches and industrial centers." Interior communities saw British demands for the abolition of human sacrifice (typically of slaves) as a prelude to the destruction of slavery as an institution.[222] Ibibio chiefs and family heads used palm trees as a weapon against emancipation and Christianity. Elite men "hostile to the Gospel" banned converts from harvesting palm fruit, or put a "Ju-Ju mark" on trees to designate them as property of men's secret societies.[223] Farther west, Isoko converts faced similar persecution: Christians were told, "You have your choice, give up Christianity and gather palm-nuts, or remain Christians and leave the palm-nuts alone." In one village, only seven men kept the faith; they fled and established a new Christian village.[224]

Like Yorubaland, the Niger Delta region increasingly fell under British influence. Initially, British naval commanders made diplomacy on the fly, responding to conflicts between British merchants, slavers, and African leaders without any coherent strategy. William Dappa Pepple, an ally of British palm oil traders in Bonny who had taken the title of *Amanyanabo* ("owner of the land") after winning a struggle against pro-slaving rivals, complained: "One white man come and make book [treaty] and another white man come tomorrow and break it."[225] After a crisis broke out in Bonny following Britain's seizure of four Spanish slavers in 1836, British officials decided that the Amanyanabo was in fact a king and could negotiate on behalf of Bonny. Bonny and Britain signed a treaty outlawing slave exports in exchange for a subsidy of $2,000 per year, paid for five years.[226]

The leaders of Old Calabar signed treaties with Britain outlawing the slave trade in 1841 in exchange for a subsidy, like Bonny.[227] But French slavers visited the next year, and a French warship reportedly threatened to bombard the towns when they refused to sell slaves.[228] The slavers settled for cargoes of palm oil, but to keep the French from pressing upriver, King Eyo made the momentous decision to invite the British in. Britain named a consul for the Bight of Biafra in 1849, and though his formal powers were limited, the consul played an important role in regional politics, intervening on the side of British traders.[229]

As William Dappa Pepple discovered, British policy favored merchants over Britain's ostensible allies in Africa. British troops helped capture and deport a Juju priest responsible for inciting violence against European merchants (allegedly at the behest of pro-slaving houses in Bonny).[230] Yet when rivals drove the Amanyanabo from power in 1854, the British consul only offered safe passage to exile.[231] "King Pepple's" supporters in London printed a pamphlet condemning "the wrongs sustained by him at the hands of the British government," blaming his downfall on a conspiracy between British merchants and other Bonny houses. "The real crime of which King Pepple had been guilty, was the entering into competition with British subjects in the trade of Palm Oil."[232] William Dappa eventually returned to Bonny as Amanyanabo, but his own house had been massacred and plundered, leaving him relatively powerless.[233]

Relations soured across the Niger Delta in the 1860s as the "palm oil ruffians" entered a "savage trade war" in response to declining prices. Competing to fill their ships with oil, the merchants granted too much credit to traders, leading to defaults and retaliatory "chopping" of oil.[234] Initially, the trading houses could hold their own: one source claimed that Bonny's houses could each muster 2,500 men with muskets in 1864, backed by an "abundance" of ammunition and cannons.[235] The rivers around Brass regularly saw attacks on merchant ships and factories in this period, often followed by British gunboat reprisals.[236] Consul Livingstone complained: "Loud is the call for the immediate presence of the consul in a man-of-war to punish the lawlessness of the black goose, but not the lawlessness of the white gander."[237]

New steamship services, which began in 1852, added to the competition. Steamers sold cargo space to all comers, driving old sailing ships and the firms that owned them to the scrapyard. The change created opportunities for African traders to sell cargoes directly to Europe: one boasted that he could sell oil for half the going price in Liverpool and still earn fifty percent more than if he had sold his oil on the African coast.[238] By the 1880s and 1890s, surviving European firms entered a series of mergers. As one historian put it, "Over-competition was the disease, and monopoly the only cure."[239] Trading monopolies along with an oligopoly in steamship services held down prices in Africa and soon checked the ability of Africans to sell palm produce directly to Europe.[240]

Monopoly firms tried to secure their positions through treaties with local chiefs and kings. Conflicts over such "treaty-hunting" led to the 1884–85 Berlin Conference, which partitioned the continent among European powers. The conference was initially about the Congo River basin, where competing treaty-hunters had sparked an international crisis. The 1885 Berlin treaty settled the matter by giving Congo to Belgium's king Leopold II, in addition to declaring Europe's intention to eradicate slavery. The treaty laid out the rules for claiming the rest of the continent, insisting on "effective occupation." To seize a territory, European powers had to prove—treaty or not—that they actually governed a place. In the oil palm belt, European factories on the coast became launching pads for invasions of the African interior.[241]

The Royal Niger Company (RNC), chartered in 1886, claimed the western half of the Niger Delta palm belt. The RNC's predecessor firms had been working since the 1870s to bypass coastal traders, chugging up rivers in armored steamers and signing treaties with kings and emirs in the interior. The man behind the RNC, George Goldie, was convinced that African middlemen had too much power to set the terms of trade, encouraging reckless competition among Europeans.[242] The RNC paid lower prices for palm produce at its trading posts up the Niger River, and—once chartered—collected taxes on trade and enforced its own law with a private army.

To the west, France and Britain attacked inland states across the Bight of Benin to secure valuable trade routes, and to keep each other—and Germany—from monopolizing trade. France annexed Porto Novo in 1883, challenging Britain's colony at Lagos. French agents roamed across Yorubaland and regions to the west, countering RNC treaty-hunters on the upper Niger. Abeokuta signed a treaty with France in 1887, laying the groundwork for a railroad to Porto Novo that would divert Yorubaland's palm produce away from Lagos. Egged on by merchants and shipping firms, the British took a more assertive stance in Yorubaland.[243]

Ijebu, situated northeast of Lagos and in control of key routes to Ibadan and beyond, had long refused to cooperate with Britain. No doubt aware of how Christian converts pulled Abeokuta into Britain's orbit, Ijebu leaders denied access to consuls, missionaries, and even Ijebu returnees from Sierra Leone.[244] On flimsy pretexts, British forces attacked Ijebu in 1892 and quickly captured the capital. The Egba and other Yoruba groups got the message and opened trade roads, ultimately placing themselves under British protection.[245]

Foiled in western Yorubaland, the French had more success in Dahomey. The Dahomean king, Béhanzin, was furious about France's annexation of Porto Novo: it had long been a vassal of Dahomey and was a vital link to Lagos. Recognizing the economic intent of France's imperialism, Béhanzin's forces destroyed oil palm plantations around Porto Novo in 1890. Fearing the loss of the region's key exports, the French negotiated a peace settlement. Béhanzin used the time to buy rifles, cannons, and machine guns. When fresh hostilities broke out in 1892, however, the modern weapons weren't enough to protect Dahomey. Invading French troops set Yoruba slaves free as they advanced, turning them loose on farmland and food stockpiles. Dahomean troops deserted to defend their homes. By early 1894, the war was over: France was in control of the country, and Béhanzin was a prisoner.[246]

African leaders who chose to negotiate rather than fight discovered that treaties weren't worth the paper they were written on. In Cameroon, German firms thought they had secured the region's trade in an 1884 treaty signed by the Duala kings Bell and Akwa. The treaty recognized Duala rights over the interior, including the right to levy taxes on merchants. When Europeans finally sailed upriver, they "had much to say about the low costs of palm products and ivory as compared with the prices paid to the Duala middlemen on the coast." Rather than buy palm oil for eight times more than what the Duala paid in the interior, European merchants convinced the Germans to abrogate the treaty. Kings Bell and Akwa were powerless to stop them.[247]

The tragic fate met by King Jaja of Opobo was one of the more dramatic examples of how African states resisted and succumbed to imperialism. Jaja's life was steeped in palm oil: kidnapped from Igboland as a young boy, he was sold downriver to Bonny's Anna Pepple House. Instead of laboring as a canoe paddler or a palm oil collector, Jaja was chosen to work as a trader. His peers soon chose him to lead the house, recognizing his charisma and business savvy. The British consul mused: "In a short time he will either be shot or he will beat down all his rivals."[248]

Jaja's meteoric rise led to a civil war with the rival Manilla Pepple house in 1869, and Jaja retreated to the east after a dramatic cannon duel in the Bonny marketplace. In a stroke of strategic genius, he founded a new town, Opobo, astride key canoe routes to palm oil districts.[249] By 1873, Jaja had a near-monopoly of trade and signed treaties with Bonny and Great Britain, recognizing Opobo with Jaja as its king.[250] Over the next decade, Jaja annexed more land in the interior and levied stiff duties on trade.[251]

British firms tried to boycott Jaja in 1885, but he found other outlets for his oil and retaliated with his own boycott of leading merchants. The British consul demanded that Jaja reopen trade and stop levying duties, ultimately calling in a gunboat. Jaja made the fateful decision to board the ship in 1887 to meet the consul. Acting without authorization, the consul gave Jaja an impossible choice: sail to Accra to face trial for obstructing trade or face an unwinnable fight with an armored warship. Jaja chose peace. He died in exile in the West Indies. British officials admitted this was an outrageous kidnapping, but they didn't repent or relinquish control of Opobo, once Jaja's successors signed a new treaty.[252]

Most other Niger Delta towns fell into line, signing new treaties with Britain. Towns that fought back, like Brass, faced overwhelming military force. Brass's traders had been cut off from upriver markets by the RNC, depriving the city of food in addition to palm oil.[253] The town's leaders thought it better to fight than starve, and attacked an RNC base in 1895. RNC and British forces crushed the town. The consul thought it all horribly unjust: "Jaja was deported . . . because he was a big monopolist. . . . now we have wiped the floor with the Brass men because they have endeavoured to go for the biggest monopolist of the crowd—the [RNC]."[254]

By the late 1890s, British forces were pushing inland across Nigeria's oil palm belt, and they would spend the next twenty years conquering states and suppressing rebellions. Benin fell in 1897, after anti-British chiefs massacred a colonial delegation and provoked a brutal reprisal. A 1901 expedition destroyed Arochukwu, breaking the power of Aro traders and their "Long Juju," an oracle with great influence over southeastern Nigeria's palm oil region.[255] Punitive expeditions left devastation in their wake throughout Igboland and regions to the north and east. The decentralized nature of many Niger Delta states played an important role in preventing coordinated resistance to British imperialism. But as the satirical poet Hilaire Belloc famously suggested, much came down to technology:

> Whatever happens we have got
> The Maxim gun, and they have not.[256]

. . . . . .

An early historian of the palm oil trade couldn't have been more wrong when he wrote that the shift to legitimate commerce in palm oil "was without consequence" for African societies.[257] Land tenure, social status, gender roles, technology, politics, and landscapes all changed in response to the export industry. Most of these changes didn't shatter the foundations of African social or political systems, but they were part of a transformation that accelerated in the twentieth century.

Importantly, the wealth created from palm trees attracted imperial predators. Every part of Africa's oil palm belt except for Liberia was claimed by European powers by 1900. Cheerleaders for colonization made glib claims that the continent's wealth had only begun to be tapped. Oil palms served as an example of "neglected" resources that only European capital, technology, and initiative could bring to the world market. One particularly absurd article suggested that palm oil was so abundant it could be "mined" from the soil under unharvested trees.[258]

Writers who actually spent time among oil palms and the people who worked them were more cautious. An 1859 pamphlet noted that in the Bight of Biafra, "the palm oil trade absorbs all the available labour," dispelling the notion that Africa was teeming with idle hands.[259] A British consul warned in 1872 that the palm oil trade, as it was then constituted, was nearing its peak output, given the number of workers and palms available.[260] A soldier who fought with the RNC mused in 1892 that "The palm oil trade is perhaps fully developed." He and others looked north of the palm belt, where shea nuts, cotton, and peanuts beckoned.[261] Low prices for palm oil in the 1890s discouraged palm oil exports at the very moment when colonial power was reaching deep into the oil palm belt.[262]

That was about to change. As chapter 4 shows, European industries developed new ways of using palm oil and kernels during the nineteenth century. From early uses in soap and candles, palm oil found its way into tinplating, industrial lubricants, and new chemicals of the industrial age. European industries used palm oil, but they were never reliant on it: palm oil was a substitute for other commodities and subject to replacement itself. Africans had little power to shape this market. Yet by the end of the century, new products like margarine created big openings for palm oil and kernels in a growing industrial food system. Fierce competition between soap and margarine firms sent prices soaring to new heights, inspiring African producers laboring under colonial regimes to embark on a massive expansion in exports.

4 Oil Palms and the Industrial Revolution

Before abolition, most palm oil rode along with the slave trade in the most literal sense. A surviving freight bill for the *Hawke* offers one example: the vessel loaded 359 captives at Gallinhas and Bassa and discharged 328 survivors at Dominica in 1777. On the return trip to Liverpool, the vessel brought back six puncheons of palm oil.[1] Small shipments like this amounted to a few thousand gallons annually, and British port officials levied stiff duties on these imports. Given its medicinal uses, palm oil was classed and taxed as a drug.

The Industrial Revolution radically changed this situation for palm oil and the people who made it and used it. No longer an exotic medicine, palm oil embarked on new roles in soap, candles, and a seemingly endless array of industrial novelties. Britain led the way in palm oil consumption, though France and Germany took a considerable share of Africa's exports by the end of the nineteenth century. Germany became an especially important consumer of palm kernels, crushed for oil and turned into cattle feed. Industrializing Europe could use as much palm produce as Africa could supply, on one condition: that it was cheap enough to compete with rival sources of fat.

In its industrial life, palm oil was a substitute commodity. New technologies let manufacturers bleach, refine, and transform palm oil, interchanging it with other fats. In many cases, palm oil disappeared from view, subsumed in finished products. Yet at other times, palm oil was extremely visible. Some advertisers invoked ideas about Africa, slavery, and "legitimate commerce" to sell their wares. European consumers encountered palm oil in new ways over the course of the nineteenth century, including a particularly unappetizing use on railways that cemented the oil's reputation as inedible grease. By the end of the century, Africa's palm produce represented a significant share of Europe's fat supply, and it was—despite northern revulsion at stinking palm oil—starting to enter the industrial food system.

Soap and the Industrial Revolution

Soap forms when alkali chemicals (obtained from plant ashes in the preindustrial era) react with fats. An alkali breaks apart the triglyceride molecule that all fats are made of, bonding to one of the three fatty acids in the triglyceride. The result is soap, a molecule that cleans because it attracts both oil and water. Early accounts showed that Africans had long used palm and palm kernel oils to make soaps with ashes from palm fronds and other plants, and most observers praised their quality and cleaning power.[2] One European complained that Africans on São Tomé washed so vigorously that the main river on the island was "nothing but Sudds, enough to poison our Men."[3] Portuguese traders valued palm oil soap so highly that their king "will not suffer it to be exported to any parts of Portugal, lest it should undo the soapboilers in that Kingdom."[4] The ship captain who complained about women washing clothes with "nasty, stinking palm-oil soap" was a rare exception.[5]

Around the 1770s, British shepherds started using palm oil as a dressing to protect the skin and wool of sheep before shearing. Palm oil wasn't yet cheap, but it was cheaper than butter, the traditional dressing. Each sheep needed up to a pound of the stuff. Though palm oil left wool "somewhat yellow when clipped," the color readily washed away with soap, leaving the fleece "a purer white than common."[6] After shearing, wool-combers used palm oil to lubricate the fiber as it was carded. In time, spinners, weavers, printers, and dyers started using palm oil and soaps made from it throughout textile manufacturing.[7]

The dirty hands and clothes of human beings presented an even larger market than textile manufacturing, however. European soap-makers found that palm oil's bright color gave mundane tallow soaps a pleasant yellow hue. The oil's signature violet scent helped perfume the soap, saving the expense of costly essential oils. It was a welcome ingredient for Britain's manufacturers of "yellow soap," a product that was the most popular in the country by the beginning of the Industrial Revolution. Early recipes didn't call for much palm oil, just enough to create "a beautiful colour" and to cover up the unpleasant smells of tallow and rosin, a conifer sap used as a filler. (Rosin was the original coloring agent in yellow soap.)[8] Made with a "hot process" that was amenable to large-scale manufacturing, these hard bars of soap lasted longer in daily washing than soft, cold-process soaps made at home or in small workshops.[9]

Palm oil's reputation as a skin balm carried over to soap, and by the early nineteenth century, retailers advertised "palm oil soap" as a luxury product. A London perfumer sold balls and bars of soap made from "that sweet fragrant article Palm Oil" at the princely sum of a shilling each.[10] Reputable dealers stressed that they sold *real* palm oil soaps, distinguishing their product from fakes colored with turmeric and scented with orris root.[11] Companies hawked medicinal "palm oil soaps" well into the latter part of the century, but the product gradually disappeared as branded soaps with familiar names like "Ivory" and "Lifebuoy" captured markets after the 1850s.

In the 1820s, French chemist Michel Eugène Chevreul published his discovery that soap was in fact a new chemical rather than the emulsification of fat and alkali that most scientists had assumed. He identified "fatty acids" as distinct components of oils and fats that bonded to alkali materials to form soap. He isolated two types of fatty acid: "elaine" (which flows as a liquid) and "stearine" (a pearly white solid).[12] Chemists building on this work soon showed that the plant and animal fats all relied on the same chemical structures. The palmitic acid Edmond Frémy identified in palm oil in 1840 also appeared in beef tallow and—as Chevreul discovered in a grisly experiment with an exhumed corpse—human fat.[13]

Initially, industry didn't have much use for the chemists. Artisans already knew that different kinds of oils behaved differently in soap. They could separate liquid and solid fractions by squeezing firm fats under pressure, or use changing temperatures as a "chemical sieve" to sort harder and softer fats.[14] A soap boiler didn't need to know what a fatty acid was to realize that palm oil and tallow had similar properties in soap. But as the chemists proliferated, their ideas started to reshape the soap industry, which was in the process of transforming from a craft to a factory enterprise. New factories often featured glass roofs, under which shallow tanks of palm oil and other fats sat bleaching in the sun. Companies hired chemists and experimented with new chemicals—sulfuric acid was in vogue for a while as a bleaching agent in palm oil—and they combined new scientific insights with artisanal experimentation, developing more efficient ways of manipulating fats.[15]

As the price of palm oil fell in the 1820s, manufacturers used more and more palm oil to the point of replacing tallow entirely. These bars "had more the appearance of a piece of clay than that of soap," owing to adulterants added to stretch the palm oil. Yet they were cheap and washed well enough to win British customers.[16] By 1837, almost every soap maker visited by tax investigators in and around Liverpool used palm oil in their soap.[17] Palm oil soap was about three times more profitable than the tallow-based yellow soaps preferred by London's soap boilers.[18] By the 1850s, the Liverpool region displaced London as Britain's premier soap manufacturing center, thanks to the port's direct trade with Africa and local supplies of alkalis.[19]

Palm oil competed with tallow because it was cheap, and it was cheap because Africans started making lots more of it. But the price in Britain also fell because soap firms joined forces with merchants to repeal taxes. The merchants argued, "when the duty was laid on, it [palm oil] was considered as a drug." The stuff had since become "a necessary ingredient in making Yellow Soap." At 12s. 2d. per hundredweight in 1807, the palm oil duty was five times higher than the tallow duty. Campaigners pointed out that Britons needed to buy a lot more things from Africa—like palm oil—if "legitimate commerce" was going to eradicate the slave trade as promised. The soap boilers noted out that they already paid excise taxes on soap manufactured in Britain and shouldn't be taxed on raw materials.[20] The campaign worked: the duty was slashed to 2s. 6d. by 1819, a better rate than tallow.[21] The duty fell by half again in 1834.[22] Together, soap sellers and merchants claimed to be "doing silently and slowly, but surely, the work of emancipation in Africa."[23]

It was certainly slow work: Britain's Parliament held inquiries in the 1840s to find out why the slave trade seemed alive and well three decades after abolition. Merchants assured Parliament that buying even more palm oil was the solution. As one speaker put it, "the best mode of destroying the traffic in man is to encourage the cultivation of a vegetable producing a valuable oil." An added benefit for Britain was the fact that palm oil was "entirely paid for in our manufactures." The same could not be said for Russian tallow, palm oil's main competitor.[24]

Parliament abolished the import duty on palm oil in 1845, but campaigners continued to fight the soap excise. By this point, the reformers linked free trade and abolition with broader ideas about "civilization"—in Britain as well as in Africa. As one chemist famously declared, the "quantity of *Soap* consumed by a nation would be no inaccurate measure whereby to estimate its wealth and civilization."[25] By this measure, Britain was only slowly civilizing itself. According to official records, soap consumption inched up from 6.55 to 7.12 pounds per capita between 1821 and 1846, despite the price of soap falling by half.[26] Reformers printed pamphlets in the 1840s teaching the urban poor how to wash themselves and their clothes, linking soap consumption with material and moral progress.[27] "Give the people cheap Soap," promised one pamphleteer, and "they will purchase it eagerly, and use it their homes' comfort, and to the general improvement of the race."[28]

Elites fretting about disease and the moral character of the urban poor made soap into a leading "Victorian Fetish."[29] Advertisers often used racist caricatures of Africans to sell soap, depicting black children turning white after a good scrubbing. And soap figured prominently in accounts of Britain's "civilizing" work in Africa. One merchant testified in 1842 that palm oil soap made in the United Kingdom sold in "very large quantities" in Bonny, a mark of trade's civilizing impact.[30] Missionaries noted with satisfaction that Africans were "getting into the habit of going about more cleanly and decently clothed," thanks to the soap and cloth their palm oil purchased.[31] Having (allegedly) crushed the slave trade, the palm oil trade created "the feeling of property and the desire for accumulation" among Africans. "[T]hey are everywhere the germs of nascent civilization."[32] One Liverpool writer admiring puncheons stacked on the docks mused about "what the Kroomen's cooperage and the greasy sap [sic] of the African Tree have done for civilization and for Christianity."[33] Statements like this had the whole situation backward, crediting trade with the creation of consumerism in Africa. In fact, African demand for consumer goods was long established, had underpinned the transatlantic slave trade, and was the sole reason Africans produced palm produce for export.[34]

Lever Brothers—a relative latecomer to the British soap business—appealed to Victorian notions of cleanliness, progress, and empire building in its advertising, discarding the abolitionist sentiments of the 1840s. Lever's first big success, "Sunlight Soap" (1884), was a physical representation of changes in chemical knowledge and global commodity chains over the course of the nineteenth century: composed of 41.9 percent palm kernel or coconut oil, 24.8 percent tallow, 23.8 percent cottonseed oil, and 9.5 percent rosin, the firm bars of soap worked in hard or salt water.[35] Lever Brothers used colorful advertisements and paper wrappers on its soaps, part of a broader shift away from bulk commodities stored behind a grocer's counter toward packaged, branded goods.[36] But with a few exceptions, soap labels didn't even name palm or kernel oil as an ingredient. Trusted names sold soap, not feelings about the ingredients.

French soap makers lagged behind the British in using palm oil. The first puncheons reached Marseille in 1823, but had to be re-exported because nobody wanted them. A businessman urged the Marseille soap makers, famous for their olive oil soaps, to try palm oil in 1837. A few firms dutifully experimented, but French consumers found yellow soap rather unappealing. Régis, the biggest French trading firm on the West African coast, worked long and hard to convince soap makers to find uses for palm oil. Tax breaks helped, and a new bleaching process developed in 1852 finally let the Marseille firms use palm oil in white soaps (see graph 4.1).[37]

Across the Atlantic, the United States provided another growing market for palm oil. American soap makers campaigned against a 15 percent tariff on palm oil, arguing that it made their soap too expensive to compete with

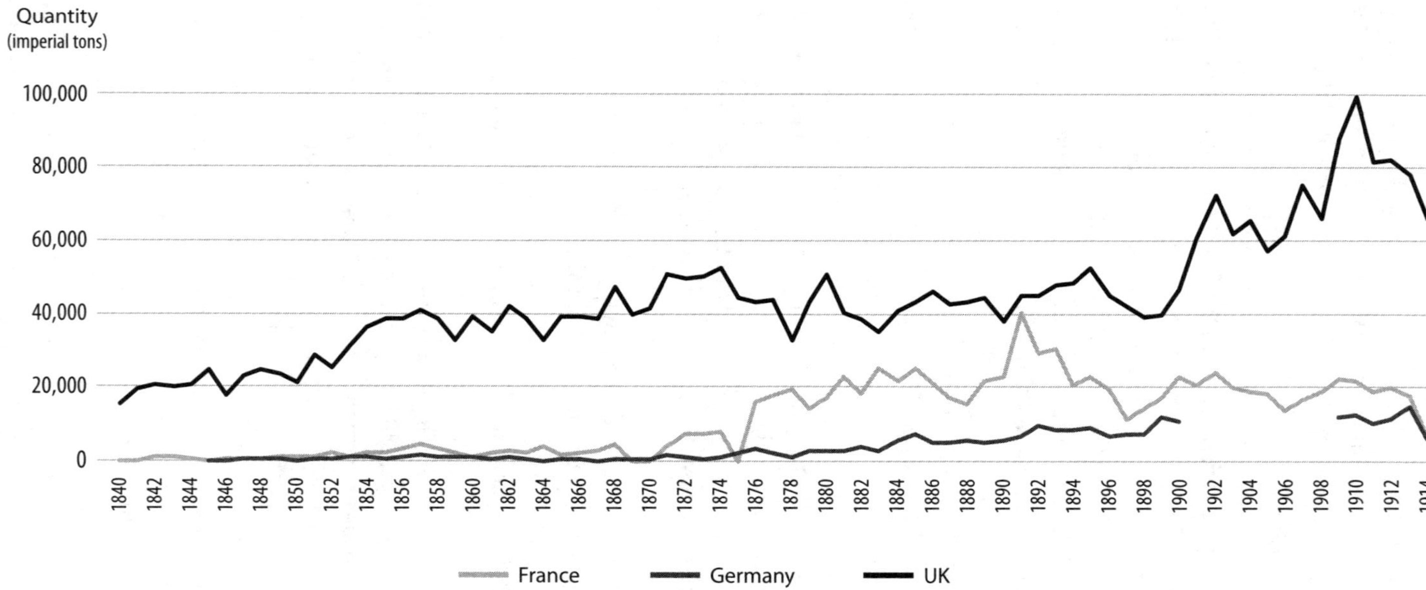

GRAPH 4.1 British, French, and German palm oil imports (tons) from Africa, 1840–1914. Figures do not show reexports, which amounted to nearly half of British imports by 1914. German figures to 1900 are for Hamburg only, and for all of Germany from 1909 to 1914 (prior to 1909, figures include coconut with palm oil). Data from Kaiserlichen Statistischen Amt, *Statisches Jahrbuch für das Deutsche Reich* (Berlin: Puttkammer & Mühlbrecht, 1914); Harding, "Hamburg's West African Trade"; Frankema, Williamson and Woltjer, "An Economic Rationale"; Lynn, *Commerce and Economic Change*; and Latham, "Palm Oil Exports from Calabar."

British imports. Whalers retorted that the tax was needed to protect *their* industry.[38] In the end the soap makers won, and lawmakers struck palm oil off the 1833 list of dutiable goods.[39] "Palm oil soaps" were common in American shops by the 1840s, but as the soap industry grew into an industrial giant, firms couldn't get enough of it to compete with domestic supplies of animal fats. Big firms like Procter & Gamble used palm oil mainly for its color and scent, or to improve the firmness and lathering qualities of soaps made from lard and tallow.[40]

Candles and Chemistry

Procter & Gamble, like many other soap manufacturers, was as famous for its candles as it was for its soap. The two trades used the same raw materials, and in the preindustrial era, soap boilers might pour their better tallow into candle molds before using the rest for soap, or vice versa. Chevreul's work and the chemical discoveries that followed had little immediate impact on soap, but they revolutionized the candle industry.[41] Tallow candles, made by congealing hot fat around a wick, had been expensive and imperfect. They melted in hot weather, dripped messily when lit, and produced smoke and foul odors. Beeswax candles burned bright and clean, but not even the rich could afford them on a daily basis. Spermaceti candles, made from a waxy material harvested from sperm whales, appeared in the late eighteenth century, but they still cost twice as much as tallow candles.[42]

Palm oil wasn't an obvious solution to the high cost of candles. In its raw state, candle makers found it "not sufficiently concrete and indurated to be manufactured into candles."[43] In other words, it melted into a puddle soon after the wick was lit, or even sooner in hot weather. Chevreul had the answer, though. When separated from liquid "elaine" and glycerine, his tallow-derived "stearine" formed a white, solid substance that burned brightly for hours without drips, smoke, guttering, or snuffing.[44] But Chevreul's imperfect process wasted much of the raw material, making his candles impractical to mass produce. An unfortunate manufacturing trick also limited the market for the new "stearic" candles: a little arsenic added to improve their appearance led to mass hysteria about "corpse candles" that could poison a room.[45]

Two inventors patented better methods for extracting stearic acid (and unknown to them, palmitic acid) from palm oil in 1836, but their acid process left an indelible dark color in finished candles. The British found them rather unpleasant to look at. An advertisement for one such "workman's candle" admitted it was "ugly," but insisted that they "burn well, and without guttering."[46] In the late 1840s, Price's Patent Candle Co. (PPCC) perfected a way of distilling fatty acids from palm oil without discoloration. One chemist rhapsodized that palm oil was "purified by adversity. The vitriol [sulfuric acid] and the heat have purged away the dross, and limpid streams of the pure fat acids run from the taps of the stills."[47] Palm oil soon became PPCC's staple material.

The firm had gotten its big break in 1840 with an arsenic-free "composite" candle made by mixing pure stearic acid with cheaper stearin pressed from coconut oil. They were white (the sign of good taste in candles) and affordable, though they were a bit "greasy and snuffy."[48] There was no Mr. Price; the company had been founded a decade earlier by London merchants who thought it demeaning to enter the stinky business of candle making under their own names.[49] Pure palmitic acid extracted from palm oil solved the firm's problems with "composite" candles and launched an industrial empire.

By the late 1850s, PPCC could make white, nearly odorless candles from the most rancid palm oil. The company sold its purest candles as "Belmont sperm," better—and cheaper—than spermaceti candles. PPCC had no hesitation about buying coconut oil, palm kernels, or animal fats for the right price, though. A company chemist noted that the "products of a hundred climes" on land and sea were now in direct competition. But palm oil remained PPCC's favorite from the end of the import duty in 1845 up to the 1880s. It cost less than tallow in most years and yielded more solid fatty acid per ton.[50] In 1853, the company built a factory outside Liverpool to get closer to the palm oil supply.[51]

Palm oil became a key raw material for firms like PPCC in the nineteenth century, but it did not circulate as an undifferentiated commodity (in spite of the efforts to chemists to treat it like one).[52] Ports of origin served as brands, distinguishing qualities of oil. Lagos was renowned for its pure, yellowy soft oil. Bonny, Calabar, and "Congo" oils tended to be harder in texture and "dirty dark red," while Saltpond and Dixcove oils were "grey and brown" due to adulteration.[53] The free fatty acid (FFA) content of palm varied widely from place to place. Both hard and soft oils needed to be heated in the British climate to get the fat flowing out of puncheons, and dirty oils sold at enough of a discount to compensate for the cost of refining. PPCC estimated in 1852 that it took 4,800 tons of palm oil to get 4,000 tons of clean material, a loss it was willing to accept, given the price of tallow and other alternatives.[54]

FIGURE 4.1 A trade card for "Price's Distilled Palm Candles." An aproned chandler with candle-molds at his side uses a candle to burn the rope holding an African man in bondage. To the right, a Southern planter pulls on the rope in front of a weeping figure. A device for distilling fatty acid sits in the background, and men carrying calabashes of palm oil walk to the left. This image was reproduced by kind permission of London Borough of Lambeth, Archives Department, Ref. 04248.

PPCC's shift to palm oil led to one of the best-known marketing campaigns in mid-century Britain. PPCC eagerly adopted the rhetoric of "legitimate commerce" and antislavery to promote its new palmitic acid candles. "Every candle of 'em that's burnt helps to put out a slave," quipped one writer.[55] British merchants pointed out that American spermaceti and "adamantine" candles (the latter infringing on PPCC's patents) were staples in the illegal slave trade; perhaps 150 million candles were swapped for African captives between 1807 and 1865.[56] One not-at-all subtle advertisement featured an aproned candle maker surrounded by molds handing an African a liberty cap. At the same time, he burns away the rope holding the African in bondage with a palmitic candle (figure 4.1). The message was clear: buying PPCC candles would fight slavery in Africa.

Critics sneered at the idea. How could buying a candle "promote the extinction of the slave trade," as PPCC claimed?[57] But the message sank in, and by the 1850s, many Britons—or least British writers—took it for granted that using palm oil was the best thing consumers could do to speed the demise of slavery. As another writer stressed: "Men have bought Price's candles because they felt they *ought* to buy them."[58] This antislavery stance was purely marketing, however. PPCC's director boasted, "every additional ton of palm oil that is imported into this country is a blessing to Africa."[59] But the company had no loyalty to African producers, buying whatever the cheapest fat happened to be. Meanwhile the firm provided an impressive range of services to its English workforce, earning accolades from social reformers.

PPCC's success owed much to the company's work in commercializing the by-products of candle making. While the palmitic and stearic acids in palm oil served in candles, leftover oleic acid went to the booming textile factories of northern England for soap and fiber combing. PPCC developed new techniques for preserving the glycerin left behind when it extracted fatty acids from triglycerides, selling glycerin to pharmacists and hawking its own moisturizing, glycerin-enriched soaps. Glycerin became an essential skin treatment and the base for tinctures, extracts, and cosmetics. Some scientists thought it had a great future as a preservative, keeping slabs of meat fresh for months at a time.[60] By the twentieth century, the pure glycerin contained in palm oil was worth more than the fatty acids.[61] This was in part because a chemical derived from glycerin, nitroglycerin, created a truly explosive market for palm oil. Alfred Nobel's invention of dynamite allowed miners and engineers to safely handle the volatile liquid, blasting away mountains of rock across the globe.

Greasing the Wheels of Industry

Palm oil washed hands and clothes and illuminated the rooms of industrializing Europe, but it worked at a more fundamental level by keeping the wheels of the Industrial Revolution in motion. With its thick consistency in European climates and high levels of antioxidants, palm oil emerged as a favorite lubricant. While textbook accounts often link palm oil with cotton machinery to highlight continuities in Africa's ties to Atlantic capitalism before and after abolition, Lancashire's spinning machines in fact relied on much lighter oils. Several inventors in the 1830s tried and failed to sell palm oil lubricants to the cotton industry.[62] Palm oil's real niche was in lubricating heavy machines, including the steam engines driving locomotives and factories.

Britain's new railways provided palm oil with its most iconic use outside soap and candles. When the Liverpool and Manchester Railway opened in 1830, carriages whisked passengers and goods along at speeds unimaginable in a world of canal boats and oxcarts. The investor Henry Booth praised it for effecting a "sudden and marvelous change . . . in our ideas of time and space." Besides being an investor, Booth was an inventor. In 1835, he received a patent for a new axle grease featuring palm oil, taking advantage of the growing quantities of the stuff for sale in Liverpool.[63] Booth combined 6 pounds of palm oil with 3 pounds of tallow, adding water and soda to make a soapy grease. The grease went into a box above the train axle; as the axle turned it rubbed the grease and melted it. Pure tallow was too stiff and pure palm oil too soft, melting away in as little as 40 miles. A blend had just the right consistency, lasting 1200 miles or more. Saponification with the soda kept the fats mixed and neutralized free fatty acids that might corrode the axle.[64] With a few modifications, the recipe would keep trains rolling smoothly for more than a century.

Locomotives and other steam engines also used palm oil grease in moving parts of the engines, such as pistons. One estimate from the late 1850s suggested that the nearly 20,000 locomotives and 5,000 steamships at work in the United Kingdom and the United States each needed "25 lbs. of grease a day, or more." This estimate is probably too high, but it does give a sense of the scale of the market for palm oil as a lubricant.[65] Even mundane oxcarts and farm wagons needed grease, and although much of this came from domestic tallow, palm oil found some uses here, too. Britain's axle-greasing needs ran to well more than 17,000 tons annually in the 1860s, though only 25 to 35 percent of this weight was actually fat, the rest being water trapped in the grease emulsion.[66]

Petroleum cut into this market from the 1860s onward. Mineral oils proved better at lubricating fine machinery and stood up to steam exposure in engines. But palm oil grease didn't disappear overnight.[67] As competition with petroleum and other fats drove down prices, manufacturers reaped the benefits, choosing whatever was cheapest. Although their American counterparts quickly switched to oil lubrication, British railways—with no domestic sources of petroleum—continued to use palm oil grease for cargo wagons until the 1940s.[68]

Axle grease was never a giant market for palm oil, probably no more than 10 percent of Britain's annual palm oil imports in the 1860s. Yet it cemented one particular idea in the British, and more broadly, European imagination: that palm oil was disgusting stuff. Few people got the chance to sniff palm oil in dockyards or soap and candle factories. But on the railways, palm oil grease was a highly visible and smellable part of the experience. Eager passengers on the first trains had "their attention riveted on the man with the yellow ointment" who walked down each side of train at every stop. "If the axle is getting hungry, he digs a wooden knife into his grease box, takes up a tempting lump of cream, puts it into the axle box, shuts down the cover, and trots on to the next pair of wheels."[69] Other writers marveled at seeing rats opening up grease boxes "to feast on the palm oil and tallow inside."[70]

The sensory experience was unsettling: palm oil had been a luxury ingredient in soaps and medicinal creams, something to rub on chapped skin and to leave a hint of violets on one's hands and face. Now people encountered it as a horrific-smelling lubricant on a noisy, smoky train. An author with Charles Dickens's magazine wrote with some surprise that the palm oil he encountered in railway depots didn't smell anything like violets, like textbooks said it should: instead it was "that sickening-looking stuff which the railway porters dab into the wheel-boxes."[71] One writer quipped that the grease-making sheds at railway centers were "interesting, but should not be witnessed before dinner."[72]

Despite the unappetizing association between palm oil and railway grease in European industry, palm oil went on to play an important role in the emergence of an industrial food system. Railways and steamships made it possible to transport foods over vast distances, connecting producers and consumers around the globe in an economic and ecological web.[73] The products of oil palms would eventually wind up in many new industrial foods, but first they helped make a key container for other foods: the tin can.

Tinplate and Tin Cans

Few people in the industrialized world ate much palm oil until the 1920s, but practically everyone ate things that came in a can.[74] Tin coatings made canned food possible, creating an envelope that protected iron cans and the food inside from air, water, bacteria, fungus, acid, and rust. Palm oil's role in tinplate was critical but completely invisible to consumers. When a British colonial official mentioned tinplate as the major use for palm oil behind soap in an 1891 address before the Royal Geographical Society, he met with "skeptical amusement." The learned crowd couldn't imagine what palm oil had to do with tin cans. The official couldn't explain it either: he had read the fact but had no idea what purpose palm oil served.[75]

Tinplating dates back to the 1670s in Britain, when an entrepreneur brought the secrets of the industry (then concentrated in Saxony) to Wales.[76] The "tinman" and a crew of assistants dipped thin iron sheets in acid, then moved them to a pan of hot oil. The oil cleaned away the acid, leaving a film that prevented oxidation. Next, the tinman moved plates through a series of molten tin baths of increasing purity. A layer of oil floated on top of the tin to protect it from oxygen. The "washman" took over midway in the process, wiping droplets of molten tin off the sheets with a brush. He then fed the sheets into a pot of the purest tin, and then into another bath of oil. On the far end of the process, a group of "girls" scrubbed the sheets with mild abrasives to remove residual oil. Finally, another set of women hand polished the sheets with sheepskin, the final step before the tinplate was ready for sale.[77] Similar processes were used for terneplate, coated with a mixture of tin and lead, as well as for zinc plate. Other workshops rolled, folded, and stamped the shiny plates into finished cans, biscuit tins, roofing sheets, and more.

Welsh tinmen began using palm oil in place of tallow around the 1830s.[78] Aside from cost, the main advantage for palm oil was its pleasant smell: "Tallow is equally well adapted for all practical purposes," one authority noted, "but the unpleasant smell . . . makes it impossible for workmen to employ it with any comfort."[79] As African exporters shifted to hard oil in the 1850s, tinmen insisted on using the freshest stuff, settling on Lagos soft oil as the standard. Tinmen didn't know why, but palm oil's balanced chemical composition of saturated and unsaturated fatty acids was perfect for their work: it can withstand high heat without burning, but it gradually breaks down into free fatty acids that "wet" metal sheets and absorb metallic oxides.[80]

As tinplate production boomed in the second half of the nineteenth century, inventors developed new processes that reduced demand for palm oil. Most important was a rolling machine that squeezed molten tin and palm oil off iron plates at the end of the process. The rollers left plates with a smooth surface, and it brought—as one worker noted with disapproval—"a great saving in palm oil—[and] one girl and one list-boy less to pay."[81] Automated rollers soon replaced workers who skillfully moved plates from bath to bath with tongs. At the same time, a chemical flux made with zinc chloride replaced the initial bath of palm oil in the production line.

Workers fought zinc chloride and the new machines for years. They insisted that "All the best plates . . . are still made by the palm-oil process," blaming fluxes for rusty plates and health problems among workers.[82] In 1907, Howell Lewis testified that he had worked with palm oil for fifteen years without complaint, but gave up the trade after seven more years working with zinc chloride: "the fumes arising from it affect the men so much that the result is asthma." Men used to work in tinplating to age sixty or even sixty-five, but now "you cannot find a man tinning to-day at the age of 45."[83] Workers alleged that canned foods would be poisoned by flux, too. Zinc chloride flux was perfectly safe in well-made tinplate, and the truth was that palm oil wasn't as benign as workers thought. As it gradually decomposed in the hot baths, it surrounded workers with acrolein, a toxic irritant found in cigarette smoke that forms when glycerin burns.[84]

The shift to zinc chloride flux didn't mark end of palm oil's role in tinplating, however. As rolling machines replaced workers in the 1870s and 1880s, palm oil continued to serve as a lubricant and antioxidant. While zinc chloride floated above molten tin as plates entered their fiery bath, they passed through a layer of palm oil on the way out. The oil protected the tin bath and the newly tinned plates from the atmosphere.[85] As U.S. manufacturers ramped up production under tariff protection in the 1890s, some Welshmen called for a return to palm oil and more labor-intensive manufacturing, predicting that "superior quality" would win customers. Americans, many of whom were Welsh immigrants using Welsh machines, similarly claimed that the "palm oil process" left more tin on the sheet and made a durable—but more expensive—product.[86] By the 1910s, however, only a few specialty shops used the old method. In 1914, Welsh factories only needed 3,500 tons of palm oil to make more than 900,000 tons of tinplate and related products. Across the Atlantic, US Steel used more than 2,700 tons in its tinplate works to turn out a similar amount of plate, and consumed more palm oil for cold-rolling other metals.[87]

Electroplating further diminished the market for palm oil in tinplate later in the century, but cold-rolling proved to be a durable niche. Lightly heated, palm oil melts over metal as it passes through powerful rollers. The oil forms a film that helps prevent damage to the plate and rollers. Other oils and waxes tend run off the metal, produce bad odors, or simply cost more than palm oil. Despite efforts to find substitutes, palm oil to this day still "has no real equivalent substitute in cold-rolling."[88]

Palms and Petroleum

The tinplate story is one more example of how palm oil emerged as an industrial substitute in the nineteenth century. All of the tasks it performed had been done by other substances, and this meant that cheaper or better-performing replacements would displace palm oil, too. In the hands of chemists, palm oil was merely a collection of fatty acids and glycerin. These chemicals could be interchanged with a global smorgasbord of plant and animal fats by the 1870s. The color and smell that had once been prized by manufacturers in soap became a liability, forcing manufacturers to employ costly bleaching and deodorizing processes. Though highly productive, Africa's palm producers had to compete with farmers spread across the planet, as well as with ranchers and whalers. This fact became painfully clear in the 1870s and 1880s, as prices for palm oil dropped to new lows—£19 a ton by 1887.[89]

Historians often point to the late-nineteenth-century arrival of petroleum, gas, and electricity to explain this crushing drop in the price of palm oil. Though chemists had already tapped coal and shale deposits for waxes, oils, and gases, the sheer amount of petroleum bubbling out of new wells in Pennsylvania in 1859 was something new.[90] Chemists could now turn cheap, plentiful fossil fuels into wonderous substances, with catastrophic results for farmers. Synthetic rubber made from petroleum ate away at markets for natural latex. Cheap aniline dyes made from coal tar wrecked the market for natural dyestuffs. Mineral oils shoved aside plant and animal fats in lubricants, lighting, and a warehouse's worth of minor uses like inks and polishes.[91]

Even so, the impact of petroleum on palm oil has been overstated. Railway grease was ultimately a small market for palm oil, and palm oil persisted long after petroleum lubricants arrived. Petroleum only nibbled at demand for natural fats in the soap industry, the biggest industrial consumer of palm oil throughout the nineteenth century. Early petroleum-based soaps were only suitable for tough laundry cleaning, too harsh for skin.[92]

Palm oil did lose market share in the candle trade, but not because new energy sources did away with the need for candles. As an 1884 advertisement for PPCC asserted: "Contrary to the expectation of many, Candles have not been displaced by Gas, Petroleum, or the Electric Light, but are now being consumed more largely than ever."[93] Gas lights proliferated in outdoor spaces beginning in the 1810s, but they only slowly made their way into houses. Oil lamps burning rapeseed or whale oil, camphene, or kerosene were costly and cumbersome to fill. Camphene lamps were notorious for exploding; kerosene was less volatile but still caused burns and fires. Given these drawbacks, it's not a surprise that the humble candle remained atop writing desks and bedside tables well into the twentieth century. According to one estimate, British candle consumption in 1850 was nearly ten times larger than in 1750, and it held steady at around 100 billion lumen-hours per year until electric lights reached most homes in the 1920s.[94]

Petroleum did erode demand for palm oil *within* the candle industry. PPCC, those champions of palm oil and legitimate commerce, were among the first to distill paraffin wax from fossil fuels. The high cost of paraffin meant that the firm mixed it with fatty acids in most candles.[95] Even when the cost of paraffin fell in the twentieth century, manufacturers continued to add fatty acids to make harder, less smoky candles. Pure stearic and palmitic candles remained popular among miners and explorers: in a pinch, you could eat the candle. PPCC sold 80,205 tons of candles and other products by 1886, about the double the quantity sold in the 1860s. This included candles, soap, textile oils, and specialty machine lubricants made by blending petroleum products with natural fats. The firm reportedly used 7,000 tons of palm oil annually as late as 1880.[96]

We don't have any sources describing African reactions to the rise of petroleum in the nineteenth century. Some foreigners expressed worries about the fate of the oil palm industry, noting competition from "earth oils of America."[97] But most were optimistic. One writer pointed to the stability of palm oil exports from Africa, despite plunging prices in the 1870s and 1880s and growing competition from rival products. "[If] there is a commodity from which the circumstances and conditions of the growth of the palm-tree, the harvest of its crop, and the manufacture of its oil, should be able to accommodate itself to any variation of the market, it is palm oil," he declared.[98] Meanwhile, a big "variation of the market" was still to come for the oil palm: its jump from an ingredient in industrial containers for food to being an industrial food itself.

Moving up the Food Chain: Oil Palms and the Industrial Diet

Food provided a practically limitless market for palm produce, and it remains the leading factor in palm oil consumption today. But getting palm produce into the food chain outside Africa wasn't straightforward. One early pioneer was Samuel Herring, the Liberian settler introduced in chapter 3. Herring and other emigrants developed a taste for palm kernel oil, finding that its bland taste made it a better substitute for butter and lard than did palm oil. Although the details are vague, Herring built a machine to extract kernel oil, producing a food he called "African Lard." Resembling a white butter, it had an "admirable beauty in its appearance, with a taste, when used for cooking purposes, unexcelled by that of the best lard."[99] Abolitionist and missionary publications excitedly reprinted accounts of Herring's invention, noting that the food had "such richness, clearness, and adaptedness to table purposes that one would not suppose it to be a product of Africa."[100] Herring sharply contrasted it with palm oil: "there is as much disparity both in their appearance and taste, as there is between lard and butter."[101]

Herring's African Lard probably never appeared in a foreign grocer's shop. But that didn't mean that palm produce remained outside the industrial food chain. Palm kernels started feeding Europeans around 1850 by way of an intermediary: the cow. Crushing palm kernels for oil leaves behind a protein-rich "cake." Farmers use oil cakes from all sorts of seeds and nuts as animal feed and fertilizer. The firm of G. L. Gaiser started shipping kernels to Germany in the 1850s, where they operated a crushing mill. The company convinced local farmers to start feeding kernel cake to their cows.[102] Kernel cake helped raise the fat content of milk and put "a finish on fattening stock." One authority claimed that cows fed with kernel cake "could do with a third less pasturage." Kernel cake "sells itself," asserted one trader.[103]

Cattle feed was part of a larger assemblage of plants, animals, people, and machines. Kernels cracked with rocks by West African women traveled thousands of miles to Hamburg, where steam-powered machines crushed them and forced out the oil with hydraulic pressure. Much of the cake fed Dutch and Danish cows, whose butter wound up on the tables of wealthier households in Britain and beyond.[104] Initially, the oil went to soap and candle factories. By the 1870s, however, manufacturers began to churn kernel oil and other fats with leftover skim milk to make a new food: margarine. French chemist Hippolyte Mège Mouriès patented margarine in 1869, responding to Napoleon III's call for a cheap butter substitute. Using "oleo oil" extracted from tallow, Mège Mouriès claimed his "oleomargarine" was "identical in chemical composition with butter" and therefore perfectly acceptable as a substitute.[105]

This claim was part of a broader "chemical redefinition of food" that positioned fats as mere collections of fatty acids. Palmitic acid from an oil palm was no different than palmitic acid found in tallow or cream churned into butter. Scientific language helped manufacturers introduce unfamiliar and sometimes unappetizing ingredients as commodity foods for the masses.[106] Although critics complained that "all the greasy rubbish of the world" was being dumped on consumers in the form of margarine, chemists rallied around margarine and other substitute foods: "not the slightest objection can be raised against the substitution of cheaper animal or vegetable fats for the expensive cow-butter, and it is rather desirable that this industry should extend," one chemist asserted.[107]

Efforts to make products "palatable" would pay off in the 1880s, when Loders, a British firm, patented a deodorizing process for kernel oil. Loders sold the neutral oil to margarine makers and industrial bakeries. Refined kernel oil resisted oxidation much better than butter did, and it let foods like crackers and cookies (encased in tin boxes, of course) stay fresh for months or even years.[108] Marseille-based Rocca, Tassy and de Roux sold their own palm kernel oil to customers across Europe from the 1870s on, though their famous *Végétaline* all-purpose cooking fat (1896) was supposedly made only with coconuts rather than chemically-similar kernel oil. Consumers wouldn't have been able to tell the difference anyway. One expert marveling at the dawning "synthetic age" of food opined: "What the eye does not see, and what the mind does not know, occasions the heart no grief."[109] While invisibility was a key selling point for palm and kernel oil, some firms boldly sold their wares as "nut butters," claiming that highly refined plant fats were cleaner and healthier than animal products.[110]

Palm kernel oil easily disappeared into the food manufacturer's arsenal of interchangeable fats. Palm oil suffered from a serious image problem, however. One writer urging food uses in 1883 acknowledged that "the familiar uses of palm-oil in candle-making and for railway grease will cause my suggestion to shock the nerves of many delicate people." Chemistry textbooks insisted that palm oil was "even when fresh, always more or less in that condition in which other oils are called rancid." As late as the 1950s, Britons disparaged palm oil as "axle grease," unfit for human consumption. And they were right, in that most of the oil arriving in Europe had high free fatty acid levels and unpleasant smells and tastes.[111]

That didn't stop unscrupulous butter sellers from using palm oil to color their wares as early as the 1860s.[112] As European cattle ate less grass and more grain and oilseed cake, they made whiter, less desirable butter. A little palm oil gave cheap butter or margarine a lovely yellow color, and by the early 1900s, manufacturers were slipping bleached palm oil into margarine as a primary ingredient. Much of the "Dutch butter" sold in Britain was a blend of margarine and butter, the former increasingly made with palm or palm kernel oils.[113] One expert asserted that Britons were eating "quite a considerable amount of pure palm oil refined and sold . . . not always under the name of palm oil." By 1914, Britain reexported about half of the palm oil it received from Africa and got much of it back from German and Dutch companies in the form of refined oil, margarine, and other foods. Dutch and German margarine firms were reticent about their use of palm oil: they "like to sell their product upon the excellence of the product and not upon a statement of the contents," remarked one expert.[114]

Palm oil had important qualities for margarine beyond color. Its natural blend of fatty acids offers similar physical properties to butter: solid but spreadable around 25° C, melting in the mouth.[115] Palm oil also became an ideal feedstock for a process patented in 1902: hydrogenation. By adding hydrogen atoms to fatty acids, manufacturers could tweak the physical characteristics of fats. Liquid, unsaturated fatty acids became solid, saturated ones. While a fully hydrogenated fat is about as appetizing as a block of wax, they can be mixed with unsaturated fats, or more commonly, left partially hydrogenated. This lets manufacturers fine-tune the melting point, creating products with textures like—or unlike—anything nature produces. In a fat like palm oil, partial hydrogenation targets the fatty acids most prone to turning rancid, increasing the shelf life of packaged foods. The process also strips undesirable smells and tastes from oil (though it can create new ones, too). One chemist succeeded in hydrogenating inedible hard palm oil into something usable as a cocoa butter substitute, though the method proved too costly to scale up.[116] Soap and margarine firms fought a vicious legal battle across two continents over patents and improvements to hydrogenation, seeing the potential for remaking the world's fat market.[117]

Hydrogenation allowed companies to purify palm oil and adjust its consistency, fine-tuning fat blends for winter or summer temperatures. Austrian shoppers snapped up cheap, white "cooking fat" made from bleached and hydrogenated palm oil when it arrived on the market around 1913, using it in place of lard.[118] One chemist praised the new ability to "transmute . . . vegetable oils into all sorts of substances useful to man" and predicted that the plant-fat industries would "be understood as of greater importance to the human race than the great steel and iron industries."[119] Exaggerations aside, palm and palm kernel oils made an important contribution to the European diet by 1914. Prices shot up as margarine competed with soap and other industries for the fats, and manufacturers clamored for more edible palm oil.

Palm oil's entry to the food system was more visibly contested in the United States than in Europe. Thanks to lobbying by dairy farmers, the U.S. government imposed a tax of two cents per pound on margarine in 1886, raising it to ten cents in 1902 if the margarine was "artificially colored." Clever manufacturers tried to skirt the tax by coloring with palm oil, but they ran headlong into palm oil's negative reputation as rancid railway grease.

In a series of lawsuits, manufacturers argued that palm oil was a bona fide food ingredient in margarine. But a government chemist found that palm oil samples sent by the firms were "rancid, of bad taste and smell, and . . . wholly unfit for use in oleomargarine." Even refined palm oil would result in a food "deleterious to public health and offensive in taste and odor."[120] The courts ruled that palm oil was an artificial coloring. In 1914, revenue agents seized an Ohio factory owing more than $2 million in taxes on account of its "illegal adulteration and coloring" with palm oil. The U.S. Government ruled that palm oil was in fact a food in 1930, but in doing so, it extended the excise tax to naturally colored margarine.[121] While U.S. consumers were an important market for the new palm oil industry in Southeast Asia (see chapter 7), palm and palm kernel oils provided only a minor share of the calories in American margarine and cooking fats before 1950.

Oil Palms and Europe's "Ghost Acres"

In 1961, food scientist Georg Borgstrom described a peculiar historical phenomenon: the reliance of European and American societies on resources produced in distant ecosystems. These "ghost acres," as he called them, allowed industrialized societies to consume in ways their own lands could never support. A forebearer of today's "ecological footprint" calculations, the notion of "ghost acres" allowed historians to quantify the ecological contribution of trade and empire to industrializing countries. One recent estimate suggests that Britain consumed products drawn from about fifteen million "ghost acres" in 1832, growing to as much as sixty-three million by 1870, and up to 186 million acres by 1907.[122] The latter figure is more than three times the size of the island of Britain.

Victorian writers were well aware of this ecological dependency. One declared in 1858—with a good deal of exaggeration—that if palm oil had never been imported, "every particle of fat now converted into soap and candles for the use of the poor and the trading classes, would have been required for the purposes of the railway."[123] Yet palm oil and kernels made only a modest contribution to Europe's ecological balance sheet during the nineteenth century. They were substitute products, filling in for more expensive commodities. And they were not unlimited, as Europeans calling for more "legitimate trade" with Africa often imagined. The Crimean War (1853–56) was one illustration of this: the war knocked Russian tallow out of the global market. As a PPCC director remarked, the shortfall was "putting the screw on the oil collectors all over the world."[124] British imports of palm oil increased from 31,000 to more than 40,000 tons during the course of the war, but the supply plateaued. It took another decade to break 50,000 tons, and three more decades to crack 60,000. Low prices for palm oil were key to British demand for it, but those low prices discouraged Africans from investing more land and labor in palm produce. Drought may have also played some role in holding down output, but the data don't show an obvious pattern.[125]

In 1909, Britain imported more than 100,000 tons of palm oil from Africa, though almost half immediately left for other ports in Europe and America. If we take the "ghost acres" metaphor literally, this amounted to roughly 280,000 acres tacked on to Britain.[126] But palm oil replaced tallow, and Britain's domestic tallow output peaked at 100,000 tons by 1900. Another 90,000 tons from Russia, Australasia, and the Americas supplemented this domestic supply. Africa thus provided another Britain's worth of tallow in the form of palm oil, to say nothing of kernels. Ranchers would have needed to find at least 3.2 million acres to host enough cattle to replace that palm oil.[127]

But the oil palm's full impact on industrializing Europe wasn't easily counted in tons or acres. Labor rather than land was the key constraint on production in Africa. Because oil palms grew in forest-fallow systems, palm exports were eminently sustainable, unlike the extractive practices that characterized many of Europe's "ghost acres." In Britain and Europe, palm produce followed circuitous routes. Kernels fed tallow- and butter-producing cattle, easing the burden on Europe's ecosystems. Leftover skim milk, practically a waste product in the early industrial era, was transformed into margarine with palm and palm kernel oils. Cheap tin cans made with palm oil allowed food to move safely across continents, gliding on railways greased with more palm oil. The total impact was greater than the sum of its parts.

Disappearing Act

In 1899, about 70 percent of margarine was made with animal fats. The figure was 6 percent by 1928, and palm and palm kernel oils made up half of the fats going into margarine in key producers like the Netherlands.[128] By that point, on average, Britons got more than a third of their daily calories from fat, much of it in the form of margarine.[129] Yet the more palm produce Europeans consumed, the less they seemed to notice it. A sailor working on a ship hauling palm oil from Africa to Liverpool provided one telling example in a 1938 letter: he asked an expert to explain what palm oil was, and what it could possibly be used for. He and his crewmates had only the vaguest ideas about their oily cargo, though they certainly ate margarine and dined from tin cans on board.[130]

Advertisements for margarines and shortenings with names like Vegétaline, Palma, and Palmine caught the eyes of shoppers with "exotic scenes of palms, blue lagoons, and happy natives," the latter usually depicted as crude caricatures.[131] But the palms depicted were often the familiar coconut, not the oil palm with its spiky bunches of fruit. It was the same story for most soaps. The few exceptions proved the rule. Advertisements for Lever's "Lifebuoy" soap noted that its orange-red hue was "the color of its pure palm oil."[132] Lever and other firms sometimes rebranded palm oil as "oil of palm," giving it a more refined sound. The American "Palmolive" soap company called its eponymous ingredient an exotic "oriental oil."[133] When Palmolive mentioned Africa at all, its ad copy and imagery pointed to Egypt, not Nigeria. In one fanciful ad, the company claimed: "3,000 years ago, [Egyptian] slaves blended the tropical tree oils of Olive and Palm" to clean and moisturize skin. Africa's century-long connection with northern industry was reduced to an "ancient beauty secret."[134]

Despite the importance of palm oil in European industries, neither palm oil nor palm kernels were *essential* commodities for the industrializing north. The same chemical technologies that made palm oil versatile opened the door to competition from all sorts of plant, animal, and marine fats. Palm oil made up more than 54 percent of the value of West Africa's exports to Britain in the mid-nineteenth century, but it was part of a starkly unequal relationship.[135] All of Britain's trade with West Africa—palm oil and kernels along with gold, ivory, pepper, wood, and everything else—represented less than 1 percent of Britain's global trade in the nineteenth century.[136] This growing inequality had serious consequences for Africa by the end of the nineteenth century, as European countries used their military might to seize control of Africa's palm groves and reshape the oil palm industry.

5 Machines in the Palm Groves

Concession Fever

By the turn of the twentieth century, the industrial world was hungry for fat. Dirty hands needed washing; powerful machines needed greasing. More importantly, the growing acceptance of margarine as a food meant that more and more fat from the tropics was headed for northern stomachs. Before his own turn to the margarine industry, soapmaker Sir William Lever griped that he had “no chance against the Butterine maker” in battles for raw materials. Consumers were willing to pay more for food than for soap.[1] Firms making lower-value products like greases and polishes had to “ransack the ship’s hold for leakages from the palm-oil cask.”[2]

This should have been good news for oil palm farmers: prices for palm oil and kernels finally rose, pulling the industry out of a long slump. But the recovery came just as Europe’s imperial powers consolidated their grasp on Africa. Except for Liberia, all of the oil palm belt was under European control by 1910. Industrialists like Lever applauded these invasions: “The native himself cannot develop [palm oil],” he declared, condemning the “wasteful, extravagant, and costly” methods African farmers used to make oil and crack kernels.[3] He and others looked to new machines—for which more than twenty British patents were issued by 1914—that flailed, shredded, squeezed, spun, and pressed palm fruit with more power and speed than human hands could ever match. Moreover, they promised to make oil purer than anything made in an earthen pit, carried in calabashes. By 1910, scientists had proven that rancid palm oil was caused by enzymatic fermentation; promptly cooking fruit at high temperatures to deactive these enzymes was the key to edible oil.[4]

But machines for palm oil weren’t new. In 1852, a Liberian colonist set up a screw-press imported from Liverpool, inspiring “emotions of astonishment, and admiration” among local people who ridiculed “their own imperfect method of making oil.”[5] In 1864, Europeans predicted the demise of handmade oil in Africa once the right machines arrived, freeing up labor to harvest more trees.[6] But early machines were too expensive and unwieldy for African oil-makers. A rich entrepreneur could make all the palm oil they needed by buying slaves or acquiring wives, who fulfilled other functions outside the palm oil season.[7] Most new machines of the twentieth century weren't developed for Africans, though. These steam-driven monsters were designed to serve European capital.

To entice investors, colonial administrations offered business concessions that granted land and special powers. In theory, colonizers granted concessions in partnership with local chiefs and landowners; but in practice, the interests of local people were ignored. An outbreak of "concession fever" in the Gold Cost in 1900, for example, led chiefs to sign away rights to more land than the entire colony contained.[8] Some concessionaires wielded extraordinary power; companies in the Congo Basin and the interior of Nigeria raised private armies and negotiated treaties with African kings and chiefs.[9]

Lever Brothers, Jurgens, and other soap and margarine firms hunted for concessions in Africa after 1900, lured by reports of "an inexhaustible supply" of palm oil and kernels.[10] While these companies had a particular interest in palm oil as edible fat, oil palms had special attractions for investors when compared with other tropical commodities like rubber, coconuts, or cocoa. These tree crops took years to mature, making machinery less important than land and labor. In contrast, oil palm groves looked like ready-made plantations. "[The] capital engaged will be immediately remunerative," promised one colonial official.[11] With prices for edible fats surging in the years leading up to 1914 (see graph 3.1), managing and mechanizing Africa's palm groves looked like an eminently sensible idea to European capitalists.[12] But as these capitalists discovered, Africa's palm groves were not what they seemed. There were plenty of trees, to be sure, but making money out of them proved more difficult than Europeans imagined. Harvesting labor, rather than the mere presence of palm trees, turned out to be the key factor. Only European firms that could rely on state-backed coercion could get enough of this labor at prices deemed "reasonable" to company directors.

Oil Palm Concessions in France's African Empire

Cocoa, coffee, rubber, and cotton all attracted capital to French West Africa, but French reports stressed the sheer abundance of oil palms to prospective investors. In 1902, one writer claimed that an average hectare of "forest" in southern Dahomey contained 128 trees, and only three of them were *not* oil palms. He thought Dahomeans used perhaps 70,000 hectares of palms, leaving 230,000 free for Europeans to exploit.[13] Another writer worked through estimates for Ivory Coast. After taking into account the varying age and health of trees, losses in "native" oil-making, and domestic consumption, he concluded that 50,000 tons of oil and more than 125,000 tons of kernels were "lost in the bush," waiting for European machines to recover.[14]

The pace of investment was slow in French West Africa, however. The long-established merchant firms had modest means, and the most successful were British and German and therefore politically suspect. Colonial wars of "pacification" continued into the twentieth century in several regions, discouraging investors. And although France considered "vacant" land to be state property, early officials were in no rush to alienate large tracts in French West Africa. In Ivory Coast, Governor Clozel argued that vacant land didn't really exist, recognizing the importance of forest fallow for African farmers.[15]

Clozel's successor, Angoulvant, had no sympathy for the idea of "forest fallow." He echoed the claim that palm produce was "lost" in the forests and called on European capitalists to modernize the industry, sweeping aside local concerns.[16] The groves weren't as empty or unused as Europeans thought, of course. Fishermen from the Ivoirian village of M'bato protested one 1,000-hectare concession granted to a French company, noting that they collected kernels in the grove during the dry season. The seasonal use accounted for the lack of houses and routine maintenance. One official griped, "These people have more palms than they can use." In another Ivoirian village, Locodjo, fishermen admitted, "We aren't made to climb the palms, we are fishermen." But they still protested a concession that threatened their access to oil palms. The government told them it had no interest in taking away palms that villagers kept weeded and pruned, but the people declared: "Leave us our palms!"[17]

Despite protests, foreign companies won concessions to exploit oil palms in French West Africa. The machinery that was supposed to revolutionize the industry failed to perform well, however. A plant installed by the Fournier company near Cotounou in 1908 used hydraulic pistons to smash fruit and squeeze out oil, replacing hundreds of men and women pounding out fruit in mortars. Yet the machine broke down almost immediately. The company continued buying fruit "to avoid shaking the confidence of the natives," but wound up making inedible hard oil with their fermenting stockpile when the machine was fixed months later.[18]

Even when machines worked, working palm groves proved harder than expected, as the American trader and inventor Earle W. J. Trevor discovered. His Pericarp Syndicate opened a factory at Bingerville in 1914 using patented machines of Trevor's own design, and the syndicate soon opened a second plant at Abidjan.[19] His concession gave him rights "to all the fruit not required by the natives for domestic purposes," but no ownership of the trees themselves. "It would be unfair to make a contract by which a native could not take all he wanted for his own domestic use or manufacture," he noted. Trevor explained that he managed conflicts informally, having his men harvest palms that were farther away from villages. Villagers were free to make and sell all the oil they wanted, though Trevor refused to buy it.[20]

The company hoped that locals would trade oil-making for fruit harvesting, but when they refused, Trevor recruited men from the north who had no ties with the local communities. "The natives like this work and do not like the mining," Trevor insisted.[21] When pressed about how the men were recruited—and whether they were indentured or enslaved—Trevor replied: "It is a French colony." What he meant was that formal indenture wasn't required: recruiters used incentives and coercion to bring migrant laborers from the north, men who couldn't slip away into a village and take up farming because of ethnic, linguistic, and religious differences with the locals. Some were sold by elites into this new kind of servitude. With imported labor, the Pericarp Syndicate produced about 100 tons of palm oil monthly. While Trevor boasted of shipping oil with only 2 percent free fatty acid (FFA), he noted that some of his shipments had as much as 80 percent FFA.[22] The root cause was the Pericarp Syndicate's inability to get fresh fruit fast enough to keep its machines running at an efficient pace.

This experience proved typical for oil palm concessions in France's West African colonies. Local people refused to sell fruit to the mills at the prices offered, and industrialists found that harvesting "wild" groves with their own employees was more labor-intensive than they anticipated. There was no harvesting machine and no prospect of ever making one. By the late 1920s, only about twenty-eight of the forty or so large concessions in French West Africa were still operational, collectively working less than 30,000 hectares. Only a handful of these continued to work palm concessions, with most pursuing coffee, cocoa, or rubber.[23]

In French Equatorial Africa, with a smaller population dispersed over a vast area, colonial officials offered companies gigantic concessions and incredible power over the people living on the land.[24] While rubber and mining attracted the most capital, reports of endless palm groves interested Lever

Brothers in 1910. Instead of negotiating his own concession, Lever bought what was once the largest concession in the world, the Compagnie Proprietaire du Kouilou Niari (CPKN). Holding rights to six million acres, CPKN seemed like a bargain at £70,000. An expedition sent in 1911 delivered bad news, however: reports had wildly exaggerated the number of oil palms, which "grew in scattered clumps, in no state to supply an oil mill."[25] Lever built a mill anyway, but couldn't find workers to cut fruit. He disassembled the machines in 1913 and shipped them to Belgian Congo.[26]

Agronomist Yves Henry complained that Africans refused to "take long term responsibility" for working oil palms because the government was "reluctant to apply coercive measures." The only way to make machinery work, Henry concluded, was "the payment of a price which removes all interest in treatment by the native." But this was out of the hands of anyone in Africa and depended on consumers of palm oil and kernels in Europe and on competing fat industries around the globe. Writing in 1918, Henry argued that the only path forward for a mechanized palm oil industry was within a plantation, where managers could control the palms and the people who harvested them.[27] Yet while French colonies remained open to capitalists into the 1920s, colonial administrators increasingly pinned their hopes for oil palm exports on small-scale farmers.[28]

"Some Light Compulsion" in Germany's African Empire

Germany's time as a colonial power in West Africa was short: not even three decades, from the Berlin Conference of 1885 to Germany's loss of Togo and Cameroon in 1914. Although the two colonies were very different in size and environment, colonial officials used coercive policies in both to provide European companies with land concessions and low-cost labor.[29] Prosper Müllendorff, editor of the *Kölnische Zeitung*, responded to early criticism about these practices in a 1902 speech. "I do not pretend that there has not been some light compulsion used on the workmen." But he insisted that coercion was necessary to keep concessions running, and he claimed that Africans were better off for having earned cash wages.[30]

German scientists working in Cameroon and Togo were early advocates of a plantation system in the oil palm industry, arguing that labor costs were compounded by relying on "natural" palm groves with tall trunks and variable yields.[31] In a 1902 paper, scientist Paul Preuss argued that a hectare of palms run as a plantation would pay more than a hectare of cocoa, the crop that was luring many Africans away from oil palm. But this calculation assumed that machines were processing the fruit. A skeptical French expert calculated that simply cracking the kernels from that hectare of palms (2,235 kilograms) without machines would take four people an entire year. Extracting the oil (1,095 kilograms) would take two people three-quarters of a year.[32] Without machinery, plantation cultivation made no sense.

Germany's Colonial Economic Committee offered a prize for the first practical palm oil mill, jumpstarting interest in mechanization. Germany's Haake company won the prize in 1902 with a set of machines that cooked and shredded fruit, separated pulp from the kernel, squeezed out the oil, and then dried and cracked the kernels. Two German plantations in Cameroon and a French concession in Dahomey installed the machinery to great fanfare. But they were far too big and costly for African elites to consider; the intended customers were European capitalists with large tracts of land. To feed the new machines, administrators ordered chiefs in Cameroon to plant twenty-five oil palms for every house in a village, and fifty for each new house built.[33]

Farmers weren't enthusiastic about raising seedlings and harvesting fruit for the mills, however. The prices offered per bunch were low compared with prices for finished oil. Men avoided produce-buying stations out of fear of "being caught by a recruiting agent and forced to work on some distant plantation or on the government railway."[34] By 1913, at least 18,000 Africans were working—often involuntarily and under harsh conditions—on Cameroon's fifty-eight European-owned plantations.[35] Among those plantation owners were the Dutch soap and margarine firms of Jurgens and Van den Bergh. Their directors were cautious about investing in Cameroon, despite the "light compulsion" provided by government. The two firms partnered on a 3,000-hectare concession in 1908 using the Haake mill, planning to gradually replace fruit bought from villagers with a plantation. The project stalled and was abandoned when the First World War broke out, however.[36]

In Togo, Germany conquered a population already busily at work in the palm oil and kernel trade. Like neighboring Dahomey, Togo already had some oil palm plantations founded by local traders and Afro-Brazilians, many of them former slaves who returned to their homeland as entrepreneurs.[37] Colonial officials focused on rationalizing African agriculture, introducing new machines, plant varieties, and farming techniques.[38] But several companies did get big land concessions, including one firm that started an oil palm plantation at Agou. The firm installed a stamp mill adapted from oilseed crushing (figure 5.1). The process was the same as African palm oil methods, but with steam power replacing human muscle, and steel-shoed hammers taking the place of wooden pestles.[39]

FIGURE 5.1 The Humboldt company's stamp mill for pounding palm fruit, similar to the machine installed at Agou. A belt drives the upper camshaft, which raises and drops hammers on fruit in the angled trough. Soskin, *Die Ölpalme*, facing 48.

At least 960 people labored to feed the machines at Agou. When British forces seized the plantation in 1914, only 184 were willing to return to work, giving some indication of the popularity of the experiment.[40] The British soon received a joint letter from King Kofi and King Leleklele recounting the plantation's troubled history. A certain Dr. Grüner arrived in 1898 "asking me to give them some land to buy but I and my old [elders] don't agree," the kings wrote. Grüner "said that if we failed to give him, he will let his soldiers fight with us, and kill us with gun." Grüner took "all the land from our hand," leaving the people with "no place for us to plough." Hoping to make their case in colonial courts, the kings hired a lawyer for the princely sum of £25, but he never came. The British officer who received the letter noted that King Kofi brought "the actual bag of money forced on him some nine years ago which he has left untouched."[41]

Though they removed the German staff, the British didn't give the land back. Instead, they let villagers farm the plantation in exchange for weeding and cleaning the palms, hoping to study the experiment in palm mechanization. British managers complained about villagers gathering palm fruit for their own use and admitted to threatening villagers with eviction to keep them pruning and weeding.[42] After a year, the results proved disappointing. A British officer reported, "the oil produced by the natives in Togoland is very good quality, and the oil produced by our machinery [at Agou] was very little better than that produced by the natives." The machinery got 12 percent more oil out of palm fruit than did the local methods, but this hardly justified the expense of the machines and the whole plantation system.[43]

The main lesson learned from Germany's oil palm experiments was that a steady pool of labor was required for any mechanized mill. No matter how good the machines or stands of oil palms were, they were useless without people to cut fruit. Neither Togo nor Cameroon provided a model for the victorious British and French, as it became clear that the German projects relied on a politically unacceptable amount of violence for their very existence.

Concessions in British West Africa

Unlike their French and German colleagues, British officials were cautious—overly so, argued some contemporaries—about granting concessions. Sir William Lever had multiple bids for concessions in Nigeria rejected; he blamed haughty attitudes toward industrial capitalism among colonial officials, along with what he thought was misplaced respect for African land tenure systems.[44] But Lever, Jurgens, and other firms had acquired stakes in oil and kernel trading companies before 1914, and they were successful in winning concessions in two other British colonies, Sierra Leone and Gold Coast.[45]

The race for concessions began in earnest in 1907, when Lever—armed with patents for a new machine for processing palm fruit—approached the government of Sierra Leone. The company's plan was straightforward and brash: Lever Brothers would build a factory and light rail network in the midst of palm groves that were "practically wasted."[46] The government would grant a ninety-nine-year monopoly on fruit purchasing within a circle twenty miles across. The company stressed, "We desire that no compulsion should be used in influencing the people, the aborigines being absolutely free to carry their produce to our mills or elsewhere as they pleased." But Lever only planned to offer £1 per ton for palm fruit. The governor did the math: about 2.5 pence per day for a fruit cutter. A government head porter—hardly an enviable job—earned 7 pence.[47] The governor nonetheless agreed to the proposal with the lease reduced to twenty-one years, and a surveyor offered encouraging news that "palms grow in greater profusion than in any place I have yet seen" around the proposed site at Yonibana.[48] But now Lever stalled, refusing to respond to Sierra Leone's counteroffer.

Unlike the French and Germans, British officials in West Africa made no claim on "vacant" land if Africans asserted rights to it. As one governor put it, "The ownership of the palms is not in dispute: they are the property of the native inhabitants."[49] Europeans were free to negotiate with African landowners, and some did. One group acting on behalf of British soap firms leased more than 36,000 hectares in central and western Gold Coast, setting up a network of small mechanized palm oil mills as "Palm Oil Estates Managers Ltd." (POEM).[50]

Officials in Gold Coast and Nigeria told Lever that he was welcome to negotiate leases and set up machinery, but they refused to grant monopoly rights over anything more than a small circle around a mill. Lever set up three kernel-crushing mills in 1910 to test the waters, but they were unprofitable messes. Lever's managers didn't trust Africans to manage the machines, and the cost of hiring Europeans heaped up operating expenses.[51] There was no local market for kernel cake, and the puncheons used to carry kernel oil leaked, causing "tremendous wastage." When Lever Brothers sent specialized tankers to carry the oil, they incurred the wrath of the monopolist Elder Dempster steamship line, which threatened to cancel freight rebates it used to reward its big customers.[52] Rival traders also bid up kernel prices, hoping to force Lever to quit—which he did.

The experience convinced Lever of the importance of a monopoly to keep competitors from starving his machines. He reopened discussions with Sierra Leone and other colonies in 1911, noting that his machinery was "now perfected."[53] Some top officials in London urged the colonies to accept Lever's proposals, predicting that his machinery "will eventually so reduce the price of oil so that its extraction by native methods will no longer be remunerative."[54] But African elites ferociously opposed monopoly concessions. One writer warned his countrymen that leasing palm groves was a "policy of killing the goose that lays the golden eggs," depriving Africans of food and wine from their own trees.[55] "Do not idly think the country is being developed for you," wrote a Sierra Leone columnist.[56]

Lever never got his monopolies in Nigeria, but British authorities made sure he and other big investors could lease tracts in Sierra Leone and Gold Coast without the kind of competition from smaller firms that had doomed the kernel-crushing venture. To keep out amateurs, British authorities announced that concessionaires would have to spend at least £15,000 on machinery and buildings in the first year. This was a fateful decision: the POEM mill-builders in the Gold Coast had spent only £48,714 on their fourteen separate concessions. The German mill installed at Agou cost only £3,500.

The £15,000 policy forced Trevor's Pericarp Syndicate to set up in Ivory Coast rather than in Gold Coast.[57] Under these conditions, Lever Brothers finally signed an agreement for one concession at Yonibana and started paperwork for a second. Other European firms joined the rush, but the outbreak of the First World War in 1914 prevented them from breaking ground.[58]

The Sierra Leone and Gold Coast concessions that did begin operations experienced the same problems with labor as their counterparts in French and German colonies. Europeans significantly overestimated how many fruit bunches a man would cut in day, as well as underestimating how much wages would cost. One Lever manager suggested a man could cut 100 or even 150 bunches a day. Sierra Leone's governor thought fifty was a more realistic number, and a Lever employee working at Yonibana noted that fifteen or twenty bunches was all that the mill could expect from a man. He reported: "we have been successful in obtaining several tons of fruit only, and this mainly through the good offices of Chief Fulla Mansa," who signed a contract to supply fruit through his slaves and retainers. The Sierra Leone government threatened to withdraw annual subsidies worth £160 if chiefs failed to deliver more fruit cutters. Officials dangled a carrot alongside this stick, noting that the mill would pay £6,000 in wages for a year's worth of fruit.[59] The governor urged the mill to offer higher prices, even temporarily, but Lever Brothers refused to budge. Palm oil had to be cheap and profitable from the start.[60]

After asserting that the chiefs could not conceive of what a ton of fruit looked like, the company measured out a heap of fruit as a demonstration. The chiefs collected their own heap of comparable size, and had it turned into oil and kernels by hand. The results were clear: Lever Brothers offered £1.5 per ton, but the finished oil and kernels were worth £4. The price of fruit had to come closer to £4 before the chiefs would consider selling to the mill. POEM had the same experience in Gold Coast: they offered men 1s. 6d. for cutting twelve bunches per day, but men demanded nearly the full value of the finished oil, 2s. "You cannot induce him to believe that by selling that fruit without the trouble of extracting its oil . . . he is not losing 6d.," complained one European. He added: "however ridiculous it may appear, I think it is this belief which is behind the whole difficulty of production."[61]

The charge, in effect, was that African men were acting irrationally. They couldn't see the value of labor-saving machinery: "To the native, time and labour do not count and are not reckoned as factors at all when calculating the cost of a manufactured article," one official argued. But as another pointed out, "If all palm fruit were carried to a mill the women would escape their share of the work." So long as men could exploit the labor of wives, children, slaves, and other dependents, it was eminently sensible for them to keep making oil in the household. Women didn't climb trees to cut fruit, and by 1914, they had few other ways of earning a cash income in places like rural Sierra Leone. "I cannot imagine the Sierra Leone native taking kindly to a system whereby he did all the work and his wife none," an official noted.[62]

Land woes added to labor problems for concessionaires in British colonies. The POEM group found one of its Gold Coast palm leases enmeshed in a bitter dispute over the kingship of Ahanta. Partisans of one claimant inflicted "a series of outrages" on mill employees.[63] POEM muddled through the 1920s, importing laborers from Nigeria but finding itself unable to get rights to enough new groves to run an economical business. An anonymous Briton wrote about his own "bitter experience" with land and labor on a Sierra Leone concession.[64] By the 1920s, many Britons agreed that a successful palm oil concession would require "a monopoly of land and people, which amounts to robbing them of their birthright."[65] Colonial officials knew that their power was precarious, resting on the support of pliable chiefs and kings ("indirect rule," in colonial parlance). Seizing land would create political problems that might topple the whole system.

That didn't stop Lever Brothers from joining a committee with other big buyers of palm oil and kernels in a 1923 campaign aimed at opening up Britain's colonies for new concessions. Asserting that "the African of the farming class is not accustomed to regular work," the committee argued that only European capital could develop the oil palm industry.[66] They issued five demands to Britain's imperial government:

1. ninety-nine-year leases or freehold property for mills,
2. twenty-one-year monopolies on fruit buying and factory processing of fruit,
3. leases or freehold title for plantations up to 5,000 acres,
4. government coercion to force fruit deliveries under contract, and
5. monopolies on machine transport around mills.

The committee really presented only two paths: taking African lands for plantations or forcing Africans to work their own groves under monopoly conditions.[67]

Nigeria's Governor Hugh Clifford took the idea of colonial "trusteeship" seriously, believing it was his duty to promote the best interests of his African subjects (and to decide what those interests were). He balked at the proposals, preferring gradual evolution to radical change.[68] Other officials warned that the committee's proposals were merely "a renewed attempt to fix the shackles of monopoly on the West Africa palm oil and palm kernel trade," with no benefit to Africans.[69]

Clifford pointed out the "unpalatable fact" that companies like Lever Brothers needed Africans and their oil palms more than Africans needed Lever.[70] If British businesses were willing to put long-term industrial development ahead of profits, Clifford was willing to grant special privileges.[71] But British capitalists insisted that businesses had to make money quickly and that African farmers could not be trusted to deliver fruit at market rates. At a minimum, they sought a "nucleus" of plantation land to keep costly machines running.[72]

The Colonial Under-Secretary, Ormsby-Gore, eventually settled on a different idea: paying subsidies to mills to help them—and local farmers—adjust to a new way of working with oil palms. Alarmed by discontent over taxation in Nigeria (see chapter 6), Ormsby-Gore was eager "to show the Nigerian natives that the policy is not dictated by greedy concession hunters or European firms out for profit but to ensure the export of Nigeria's greatest natural product in the future." Ormbsy-Gore wrote, "I know this is rank socialism but so are other things in West Africa and we can't afford to be doctrinaire."[73]

The first subsidized mill opened in the Gold Coast, where the *konor* (king) of Krobo had long been agitating for machinery and investment in the oil palm industry.[74] The African & Eastern Co. (A&E) reported that villagers around its new mill at Bukonor were excited: "the women were always more eager than the men in their desire [for the mill]." Men had left to work cocoa farms, and women hoped the mill would lure them back to their palms and families. The *konor* promised to supply all the male harvesting labor needed.[75] He and his chiefs agreed to a fixed-price fruit-delivery contract, and despite reservations about the price and the size of the box farmers had to fill, the *konor* insisted he "would see that this people accepted it."[76] The colonial government promised the mill a rolling subsidy based on fruit purchases, recognizing that the machinery had to operate close to full capacity to repay the investment.[77]

When the mill opened, Krobo men found the prices too low. Women turned against the mill when they realized it would cost them their kernel-cracking earnings. The *konor* could not convince or compel his subjects to climb trees and sell fruit to the mill.[78] The mill worked intermittently through the 1930s, but it could never get enough fruit. The cost of the equipment and the salaries of the white managers who oversaw it meant that the mill needed to sell its oil in Liverpool at about £25 a ton. By 1933, the price was closer to £15 and dropping. The government gave up on the plan in 1937, shipping the machinery to Sierra Leone, where the next subsidized project was slowly unfolding.[79]

In Sierra Leone, the A&E agreed to invest £8,000 in a mill, but only if the colonial government put up an additional £16,000 and had a "nucleus" plantation ready when the machines arrived.[80] Surprisingly, the government agreed. In the depths of the Depression, it was desperate to modernize the oil palm sector. The nucleus plantation at Masanki was built with prison labor and was supposed to teach local farmers how to get the most out of their oil palms. Behind schedule and over budget, the plantation had only a few hundred acres ready when the machines from Bukunor arrived. The United Africa Company (UAC, which absorbed the A&E in 1929) offered to take over the project in 1939, but only if the government turned over 7,000 more acres. The government declined and abandoned the whole scheme in 1941 after a short experiment with hand-powered presses.

The governor explained, "the object of the undertaking is not primarily to enable this Government and the [UAC] to engage in a profitable trade, but to provide an 'object lesson in the application of modern scientific and commercial methods to the cultivation of the oil palm and the preparation of its products.' . . . It seems to me that the lesson has now gone on long enough to achieve its object."[81] Officials in Nigeria reached a similar conclusion after debating a series of UAC proposals, which included subsidies and requests for thousands of acres of land for plantations.[82]

The cumulative experience in British West Africa proved to UAC that "a palm plantation and mill should form a self-supporting unit, and should not in any way be dependent on supplies of fruit being brought by African farmers." A weary Nigerian government finally allowed UAC to lease land for two oil palm plantations in southeastern Nigeria in the mid-1930s, but UAC admitted that early results with seeds imported from Sumatra were "disappointing" and "unpromising."[83] In 1936, the government "decided that it had opened too far, and pushed [the door] back again," closing off access to land for foreigners.[84]

Belgian Congo

Sir William Lever failed to get monopoly concessions in Nigeria in 1911–14 and again in 1923, but he had more luck in Congo. The Congo basin hosted some of Africa's largest palm groves, but it was a minor palm oil exporter during the nineteenth century. That began to change with the 1885 Berlin Conference, which put a new power in charge of the region: Belgium's King Leopold II. Leopold claimed the region as a personal fiefdom through an entity called the "Congo Free State."[85] It was a "surrealistic oddity: a colony without a metropolis," ruled by Leopold, not Belgium, "in his imagined capacity as head of a federation of Congolese chiefs."[86] Leopold's brutal management of the Free State killed countless people through violence, overwork, starvation, and disease. Export duties on palm oil and kernels made up half of the Free State's revenue by 1890, but the trade was soon overshadowed by wild rubber.[87] Horrifying photographs of adults and children with limbs amputated as punishment for failing to meet rubber-collecting quotas helped galvanize opposition to the Free State in Europe. In 1908, the Belgian government forced Leopold to hand over Congo, making it a colony of Belgium rather than the king's personal property.[88]

Belgium was eager to remake Congo as a model colony. Capitalists willingly invested in Congo's copper mines, but they were skeptical about agriculture. In 1909, colonial minister Jules Renkin sent out agents to drum up interest from foreign investors. The government had its sights set on rubber, hoping to replace the collapsing wild rubber industry with plantation trees. Lever heard Renkin's pitch as he was negotiating with British officials for oil palm concessions. Lever sent two teams to survey Congo's palm groves. As historian D. K. Fieldhouse put it, "The future success of [Lever's project] largely turned on how well these men did their job." Unfortunately for Lever, "they did it badly."[89] One critic alleged that the surveys consisted of "a walk of a few miles into the interior" at three sites, supplemented by "unreliable and vague information" from traders and government officers.[90]

Lever wanted estimates of oil palm density: his intention, like other concessionaires, was to set up machinery in the middle of palm groves and start making oil immediately, saving the trouble of planting new trees.[91] There seemed to be "several million acres" of oil palms across Congo, and only a fraction of them appeared to be used by Africans. After hearing encouraging news, Lever told the Belgians, "Central Africa was not the country for rubber, but for oil."[92] Renkin enthusiastically agreed, boasting that Congo was "a land that literally drips with oil." He played up the idea of huge palm groves going to waste. "The natural product is badly exploited by the natives," he declared.[93]

Lever should have paid more attention when Albert Thys—a man with considerable experience exploiting Congo's people and forests—refused to join him in a new oil palm project. Thys wrote: "My dear Sir, form a company to exploit the wild oil palms in the Upper Congo, hopeless, there aren't any palms worth mentioning and I know my Congo." Thys urged Lever to invest closer to the mouth of Congo River, but Lever trusted his own experts and pressed ahead.[94]

In 1911, Lever formed a new company, Huileries Congo Belge (HCB), which then signed a treaty with the Belgian administration. Three things set HCB apart from other oil palm concessions. First, the concessions were enormous. HCB could pick out the most desirable land in five "circles," each covering more than a million hectares. If HCB met performance benchmarks, it could convert leases into freehold title in 1945. Colonial law limited African property rights to villages and farmland under active cultivation, making most palm groves fair game for HCB's surveyors. Second, the Belgian government used state power to provide HCB with labor that it could not otherwise recruit. Third and finally, Lever's own dedication to the project mattered a great deal. Lever poured money into HCB, and no one could convince him to abandon the project or reduce his ambitions as the project soured. In late 1912, he personally visited Congo, choosing sites for three oil mills in his "circles."[95] According to Lever's son, "Not a palm area was selected nor a site chosen, except on his authority; not a building was erected unless the plans had been passed by him, and very often these plans were largely his own."[96] Lever meddled in decisions as petty as what went into workers' "chop boxes" (they were horrified to find that he picked canned sheep's tongue).[97]

Lever's confidence in his own judgment and in his experts plagued HCB for years. To Lever's eyes, his Congo circles were full of "healthy and strong" palms, "fruiting as well as they can be expected to do in a wild tangle of bush with palms overcrowding each other."[98] Intending to "improve" the groves, HCB cut away young palms and brush. "It looked much nicer to have trees 30 feet high," remarked colonial agricultural director Edmond LePlae. This was a European vision of what a palm plantation should look like—tall trees, with bare ground between them, perhaps influenced by Lever's earlier foray into coconut plantations in the South Pacific.

Yet as Congolese farmers knew, old trees were too tall to climb and could be poor producers. HCB learned the hard way that the best trees were between ten and twenty years old.[99] HCB also discovered that brush and weeds were in fact an important part of fertility, protecting and regenerating the soil. "Clean-weeding" was an obsession of early-twentieth-century colonial plantation managers, who wanted to remove all plant competition from

FIGURE 5.2
A pruned oil palm in a clean-weeded palm grove in Congo, ca. 1912. Stereoscope slide in author's collection.

their land (figure 5.2). It took years for colonial experts to recognize the mistake.[100] As scientists working in Congo later discovered, exuberant palm groves didn't mark fertile soils, they built them. A 1942 report concluded that grove improvement and full-fledged plantation agriculture broke the nutrient cycle, while the "wastage side of the cycle, the rotting away and oxidation of the humus, continues—indeed, it is hastened by exposure to the tropical sun." The humus "very soon vanished into thin air."[101]

European scientists working in Congo eventually developed a good deal of respect for local agricultural practices. They praised Congolese farmers as "native Conservators," carefully managing the bush by felling undesirable trees and preserving valuable ones. "They will also never cut a palm tree," one official remarked.[102] A former HCB manager explained that although the land was "not 'cultivated' like a European coconut estate," it was nonetheless intensively managed. Farmers rotated crops and land through forest fallow cycles. Oil palms were so valued that communities held meetings to discuss the fates of diseased trees. Farmers were disgusted by the colonists' taste for palm "cabbage," which required the destruction of the tree. It was "about as economic as burning a house down to roast a pig."[103]

By 1915, HCB was already preparing to replace palm groves with plantations in the three northern circles where palms were much scarcer than anticipated.[104] But progress was slow. Clearing primary forest and palm bush for plantation agriculture was enormously expensive. A 1935 report estimated that it took fifty work-days to clear a hectare of grassland, 100 work-days for brush, 200 for light forest (which described many palm groves), and 400 for primary forest.[105]

The machines that were supposed to be at the very center of the HCB project also disappointed. Lever's steamship fleet suffered from constant breakdowns, and the first palm oil and kernel-cracking machines revealed serious design flaws. According to an HCB employee, sixteen of the units shipped to Congo were little more than "a lot of scrap knocking about" and were cannibalized to make a few working machines.[106] They were later replaced with new designs using centrifuges, rather than presses, to extract oil. By the 1920s, these new facilities were turning out top-quality oil, but like oil mills across western Africa, they were useless without labor to feed them with fruit.

HCB initially bought fruit from local people within the five "circles," working on the naïve assumption that they had left palm produce rotting in the groves for lack of a market. If this was where HCB had stopped, its experience would have differed little from other European concessions. Lever's treaty granted him the right to harvest trees in his concession zones, however. Company employees could cut fruit in these zones, regardless of whether local people claimed ownership of palms.[107] Officials were supposed to investigate the history and legal customs governing land before alienating it, but many simply ignored the law, declaring land outside actively cultivated village fields to be state property.[108] Company employees harvesting what were ostensibly wild trees provided most of HCB's palm produce through the 1940s.

The 1911 treaty required HCB to pay minimum wages and provide food, housing, and medical care to workers.[109] When Sidney Edkins arrived to work for HCB in 1911, however, he found few people eager to work for the company despite the amenities. Ivory hunters, rubber traders, slavers, and railroad builders had taken a horrific toll on the population during the Free State era. Edkins complained that in the Upper Congo, "hardly a village was to be seen" along the railway lines. Forced labor "had practically exterminated the existing population for a distance of 50 miles either side of the track, in addition to thousands of imported labourers."[110] All along his journey into Congo, he found abandoned villages and "miserable beings" suffering from sleeping sickness. "The remaining population was in such poor physical condition that they no longer had the energy to keep the larger wild animals at bay." Elephants and buffaloes raided farms and "appeared to have lost to great extent their fear of men." Edkins recalled a young European killed in an elephant raid on Leverville in 1913, found with a bent gun-barrel and a face smashed to a pulp.[111]

The Belgians—and some contemporary scholars—argued that apocalyptic descriptions like this exaggerated the situation.[112] Yet when Lever visited in 1912, he acknowledged that labor was scarce. Instead of raising prices paid for fruit, Lever asked for police patrols in the countryside, which could round up men to work to pay their tax bills.[113] Lever was convinced that Africans simply didn't want to work. He complained, "The Palm tree is in these parts the Banking account of the native . . . His bank is always open when he wants to draw on it."[114]

Colonial taxation was one of two tools used to find labor for HCB and other enterprises in Congo. The government charged men a head tax and levied taxes on wives, pressuring wealthier men to pay more. Notably, the taxes had to be paid in cash, not palm oil. This meant taxpayers had to sell produce to a trader or work for a foreign business.[115] In the Elisabetha circle, one of Lever's five concession zones, the tax was set so high that men would have to cut fruit for forty-eight days to pay it. Workers got food and housing while on duty, but they would pay everything they made in that term to the state. In the Barumbu circle, HCB's director asked the government to spread tax collection throughout the year, hoping to prevent workers from spending a few weeks on the job, earning their tax money, and absconding. He also asked for higher taxes in oil palm areas, and for "moral assistance" in persuading chiefs to send their subjects to work. "The natives in our region are apathetic and lazy," he complained. Forcing people to work was in fact "defending the best interests of the black man."[116] In the midst of the First World War, Belgium withdrew many of its troops, prompting some Congolese to merrily tell HCB recruiters "that they would no longer go to work, having no need of money now that *Bula Matari* (the government) no longer had soldiers to collect taxes."[117]

The other labor-finding tool was the use of traveling recruiters, who enlisted men for three-year contracts. Recruiters weren't a welcome sight: in 1914, HCB agents were "met with a volley of arrows" on at least four separate occasions.[118] Officials paid chiefs for delivering labor recruits, but chiefs could be deposed if they failed to meet quotas. Chiefs often compelled younger men to sign contracts. But recruitment patterns were uneven, leaving some villages "practically depleted of able-bodied men while others were largely ignored."[119] A Belgian official tasked with recruiting men for HCB complained in 1925 that he "believes himself daily becoming more and more a veritable merchant of men, when his villages become empty at his approach, as at the approach of a slave trader."[120]

At the Lusanga circle, an HCB official reported in 1915 that most of the workforce were in fact slaves, many of them children incapable of doing the work.[121] A 1931 report bluntly described the recruitment process in Kwilu: "The chiefs supplied slaves. The slaves, once designated, were resigned to their fate. They were given a work-book, a blanket and a machete. They were told what work was expected of them, and in general they do it fairly willingly." Slaves often had their three-year contracts automatically renewed. "In practice, a cutter is never set free but cuts fruit until he dies, or until his manager lays him off because of old age."[122] Climbing trees to cut fruit was considered slave's work in some parts of Congo; "few men would do it voluntarily."[123] A 1932 report claimed: "The native has a deep-rooted notion that once a cutter always a cutter. Even if the occupation were ten times better paid, it would continue to be despised."[124]

Climbing a palm trunk with a rope was arduous and potentially deadly work, and many men in Congo didn't even know how to do it.[125] One concessionaire in French West Africa declared that no tree over 35 feet tall was worth harvesting due the risk of deadly falls. In addition to the man falling off the tree, the tree itself might topple: the weight of a man at the top of a swaying palm could pull the shallow-rooted tree "right out of the socket."[126] Harvesters also risked encounters with venomous snakes in palm bush and at the top of palms. The greatest risk came from the tsetese fly, *Glossina palpalis*. The insect carries the *Trypanosoma* parasite that causes sleeping sickness, and it thrives in wet, shady palm groves. Villages weren't placed in the midst of "palm bush" for good reason.[127]

HCB learned that it could only expect six to nine fruit bunches per workday in typical groves. In Dahomey's intensively farmed palmeries, a cutter might get as many as fifty a day. The key factors were the condition and height of the trees, their distribution, the size of the bunches, and the distance from the collection point.[128] While HCB boasted about "how few minutes are needed to cut three or four of these clusters," other sources stressed that securing seven bunches in an entire day was "considered to be a difficult task to execute."[129] Men conscripted women as unpaid laborers, using wives and children to carry fruit from distant trees back to collection points. A doctor noted illnesses, deformities, and premature aging among women and girls subjected to this heavy work.[130] Men employed as cutters found it difficult to get married, since women knew they would have to carry fruit, adding to the unpopularity of fruit cutting.[131]

To earn the weekly ration, workers for HCB and other firms had to meet delivery quotas.[132] Managers showed no sensitivity to the seasonality of fruit yields or the distance and height of the palms. Instead, they employed sentries, one armed man "to harass 7 to 8 cutters," complained an official.[133] While HCB needed cutters to deliver fruit to make money, it faced additional pressure from the terms of the 1911 treaty. If HCB failed to meet export benchmarks, it could lose its leases. Belgian officials noted that company agents resorted to measures that were "hard to justify" when output fell. "[W]ithout any effort to find the causes of the drop in the production," agents used police and chiefs to whip and coerce workers. "We demand a constant return from them although palm kernel [*sic*] production is seasonal."[134]

At times, HCB failed to meet the standards set by the 1911 treaty, withholding clothing and rations. In a cruel irony, some workers were denied a regular supply of palm oil to dress their food.[135] Perhaps the most extreme penny-pinching involved prohibitions on collecting palm leaves: "We carefully measure out the leaves [from palms] necessary for their participation in tribal life, in agriculture, in the repair of their huts," one official stated. The company wanted to avoid damage to "its" palms from excessive leaf cutting.[136]

Conditions for Congolese farmers worsened after 1917, when Belgium instituted a policy of compulsory cultivation. Edmond LePlae, colonial director of agriculture, floated the idea in the Anglophone world in a 1914 article, noting the apparent success of forced labor in Germany's colonies. LePlae argued, "it would supply a fairly simple means of overcoming the apathy and heedlessness of the blacks."[137] He later declared, "It is deplorable to say to the natives 'Work if you want, you are free!' . . . It is necessary to engage the population in work, oblige them to produce and accumulate, and lead them by the hand."[138] The racist policy required Congolese families to plant crops dictated by officials. For some, this meant planting food crops to supply HCB and other firms. In other cases villagers planted new oil palms to renew and expand the groves for HCB. By the 1930s, two million Congolese were living under compulsory cultivation orders.[139] Failing to grow crops as ordered meant two months in jail. As much as one tenth of the entire adult male population of Congo was imprisoned at some point in the year for this and other offenses in the 1950s.[140]

Congolese men and women found ways to resist HCB's efforts to commandeer their labor and their palm trees. Edkins found a "secret village" hidden in the bush to escape recruiters, stumbling on a common survival strategy.[141] Theft and sabotage also plagued early HCB operations: Edkins noted that local people appropriated practically all movable metal objects from the first settlement at Leverville. Entire railways vanished at three other locations. At Brabanta, three out of the five Europeans stationed there were allegedly poisoned.[142]

The best-known episode of overt resistance came in 1931, when a group of Pende men refused to work as cutters. Angry at HCB and other companies, Pende communities flocked to a new sect dedicated to the removal of European influences. Sect members then killed and dismembered a Belgian officer, and they fought with armed patrols sent to arrest them. By the time the revolt was crushed, at least 550—and probably more—Pende men and women had been killed. Back in Belgium, one politician sympathized with the rebels: "The reasons [for their rebellion] are economic in nature," he insisted, pointing to low prices offered for palm fruit and the high rate of taxation. "The revolt is simply the logical and inevitable consequence of this oppression." The widow of the slain official agreed: she told the press that foreign companies "have mistreated the blacks, and exploited them."[143]

Did all this coercion and violence work? At least 26,000 people were directly employed by HCB by 1930, and by the end of the decade, the company could export more than 60,000 tons of palm oil.[144] Many of these employees worked voluntarily, earning enough money to buy amenities like bicycles and sewing machines. Life in the concessions offered relative freedom for those fleeing oppressive elders and chiefs or slave status. Men from diverse ethnic backgrounds turned to age sets as a way of organizing and supporting one another, prompting colonial ethnologists to fret about the breakdown of tribal authority and the "proletarianization" of workers.[145] But most Congolese still avoided work as fruit cutters, and wherever a good alternative presented itself—selling handmade oil and kernels to merchants, growing other crops, or gathering copal (a fossilized tree resin, like amber) from the forest—people chose the alternative.[146] Belgian officials gradually clamped down on these escapes, banning the sale of oil and kernels to petty traders. But no policy could hide the fact that climbing trees was tough and unpopular work. And without labor to cut fruit, Lever's machines were useless, no matter how many palm trees Congo held.

Understanding Palm Groves

Lever called the palm groves he saw around Leverville "The grandest sight I have seen in any part of the world."[147] But he was dead wrong in thinking, like most Europeans, that they were a "gift of nature."[148] Belgian missionary Hyacinthe Vanderyst was one of a few foreigners who recognized the anthropogenic nature of Congo's palm groves from an early date. Drawing on conversations with farmers in the Kwilu region of Congo, Vanderyst vigorously defended Congolese ownership of the palms. He heard oral traditions about the oil palm's ancient migration alongside humans, and he learned that farmers sowed palm seeds for the express purpose of building up groves. Scattered clumps of oil palms in grassland and scrub were signs of former villages rather than survivors of a destroyed rainforest.[149]

Many societies along the Congo River and its tributaries had clear rules about palm ownership: men who planted them and their heirs owned them, though a grove in a long-abandoned site might be open to anyone.[150] Earthen pits for pounding palm fruit could mark village and clan territory; a settlement without a historic pit "is subject to challenge on the grounds that the ancestor was incompetent or never existed in the first place."[151] Vanderyst stressed that many groves had been abandoned recently due to the sleeping-sickness epidemic (along with colonial depredations, though the priest avoided this subject).[152] This was true elsewhere in western Africa, where disease, war, and drought transformed large swaths of land in the palm belt, forcing farmers to abandon some areas and settle new ones.[153]

Yet for many Europeans, palm groves were synonymous with "forest" and were therefore wild. In nineteenth-century accounts, palms formed "the body of the forest" or were "intermingled more or less numerously among the common forest trees."[154] Accounts of Dahomey often described thick palm forest, when travelers were in fact moving along royal roads lined with plantations.[155] Some early works on oil palms were written by colonial foresters, working on the assumption that the palms were an integral part of the tropical forest.[156]

The French botanist Auguste Chevalier provided a rude awakening for Europeans, reinforcing Vanderyst's contemporaneous findings in Congo. Focusing on the Ivory Coast and Dahomey, Chevalier revealed that few, if any, palm groves were truly natural—in the sense that they nothing to do with humans. In Ivory Coast, Baoulé informants told him that their ancestors brought palm kernels with them when they migrated westward, recreating the palm groves of their homeland in a place that had previously been devoid of oil palms (N'Zi and Bandama districts).[157] Where others saw "forests" of palms, Chevalier learned to identify groves in varying stages of use and disuse.

Europeans were also slow to realize that *not* using palms was an important use. Fallowed oil palms fed animals, and scattered seeds for future palms; they also added significant amounts of biomass to the soil as fruit, leaves, and trunks fell to the ground and decayed. Palm roots and leaf cover conserved rainfall and checked soil erosion. Without humans to prune leaves and clear away brush, palms failed to reliably set fruit and were overgrown. Crowded palms grew tall and lanky in their battle for sunlight.[158] Chevalier argued that palm groves were better thought of as human-made orchards than as forests. He drew an analogy with the Mediterranean region: what European would claim that olive trees formed a forest?[159]

In a proper forest environment, deciduous giants eventually overtop the palms, dooming them to rot in the shade. Where such forest trees can't grow, old palms suppress younger palms, before succumbing to old age en masse. As a detailed study of one site in southeastern Nigeria in the 1950s and 1960s showed, a grove established on a village abandoned in 1870 was practically gone by 1950. What was left behind was open country suitable for farming, with a few surviving palms scattered around—a far cry from the thick forest scientists once thought was the natural home of the oil palm. Another village site abandoned in 1900 had thinning palms giving way to farmland, while a village abandoned in 1930 hosted a healthy, dense palm grove.[160]

The rivers of southeastern Nigeria and tributary rivers of the Congo were the only areas where palms grew wild in large concentrations, taking advantage of riverbanks and swamps that kept forest species at bay. Yet ethnographic evidence from Congo shows that swampy groves were economically marginal, with most farmers planting trees nearer to their farms.[161] In Nigeria, a colonial forester concluded that—despite thick stands of palms along waterways—the oil palm was "a semi-domesticated plant" for all intents and purposes, growing where people made room for it.[162]

Ideas like "forest" and "farm" were of course European ways of seeing the world. African societies recognized contrasts between cultivated and uncultivated land, but this wasn't the limit of their environmental understanding.[163] Tamara Giles-Vernick's study of the Mpiemu people's environmental knowledge offers one example. Living in today's Central African Republic, the Mpiemu describe the environment through relationships and narratives, not by rigid ecological typologies. One kind of forest is recognized as "uncultivated," another as a place to hunt game, and another as a place where the understory is so thin that a person can see another person a long way off. Such understandings extend across time, recognizing past uses and populations of species that might no longer exist.[164] In Nigeria, Igbo farmers similarly used at least seven distinct elements to classify farmland: geographical location, vegetation types, soil structure, soil color, historical uses, drainage characteristics, and ownership or tenure.[165]

Africa's palm groves reflected generations of farming and management of the landscape. They were neither completely natural nor entirely human built, but rather emerged as "a cultural creation and a lived environment."[166] It took a long time for colonial science to accept this fact, but by 1949, the leading Belgian oil palm researcher declared "that the oil palm did not occur in a truly natural state in the Belgian Congo. . . . Wherever it occurred it revealed the habitation of man."[167]

• • • • • •

The importance of humans in building, maintaining, and exploiting Africa's oil palms meant that any capitalists—no matter how impressive their machines—had to address the human factors in production. Belgian Congo was the only African colony in which the mechanized model succeeded in producing large amounts of palm oil and kernels for export markets. But as a British official concluded, there was "no doubt that the system existing in the Congo is accompanied not only by monopoly rights, but also by some elements of coercion." The model was "unthinkable in a British protectorate."[168] "You cannot do this, and thereby convert them into slaves of European capitalism," humanitarian activist E. D. Morel declared, "without the use of armed force, pitilessly, relentlessly, and above all, continuously applied."[169]

If forcing people to climb trees and feed oil mills was unthinkable, what else could be done? And how had a country like Nigeria—where concessionaires barely made a mark on the oil palm industry—managed to double its oil and kernel exports between 1910 and 1930, dwarfing HCB's accomplishments in Congo? Chapter 6 turns to the other side of the colonial story, examining how Africans worked with and against colonial governments to expand and transform the "traditional" oil palm industry.

6 African Smallholders under Colonial Rule

Africa's oil palms weren't as numerous or productive as Europeans imagined. The people who lived with them guarded their land as well as their labor, and the most successful concessionaires resorted to state-sanctioned violence to feed their machines. There was another side to the story of palm oil under colonialism, however. Across Africa, small-scale farmers ("smallholders" in colonial parlance) continued to make oil and kernels by hand. They fed a fast-growing domestic population and still managed to export much more oil than mechanized producers into the late 1930s.

But getting more oil wasn't enough: European consumers wanted *better* oil, no longer satisfied with rancid stuff for soap and candles. Much of the palm oil shipped out of Africa by 1900 had "a stink absolutely indescribable," coming in "every shade of color between golden yellow and black."[1] J. H. J. Farquhar recounted his own experience trying to get producers in Igboland to make better-quality soft oil around 1912: "their invariable reply is that they don't know the way," spoken with "an expression of mild sympathy and contempt for your suggestion." "This plea of ignorance," Farquhar insisted, was "untrue in most cases." At one market, he pointed to "some particularly unpleasant-looking and smelling hard oil, and on asking if they would be content to eat it they said certainly not, but 'them Aro-Chuku and Calabar man are fit to buy it for white man.'" When Farquhar asked a chief for some oil, "thinking it was for my personal consumption," the chief sent over "a bottle of perfectly sweet soft oil."[2]

Scientific studies concluded that the edible oil sold in African marketplaces was nearly as good as machine-made oil.[3] But export traders paid a miniscule premium for soft over hard oil.[4] Pounding out fresh fruit to make soft oil was, according to one observer, "about the most exhausting work on which I have ever seen men engaged." "[The] sooner this is done by machinery the better it will be for all concerned," he added.[5] Without labor-saving innovations, traders would have to pay much more for soft oil to make all the extra labor worthwhile. Producers interviewed by Farquhar also noted their "poverty and inability to buy the large iron cooking pots required" to thoroughly cook fruit.[6] Fuel and water were scarce in many oil-exporting districts, making hard oil a necessary adaptation to the environment. Hard-oil specialists also insisted that their methods extracted 5 or 10 percent more oil from fruit than did soft methods.[7]

Big, steam-driven machines could do away with these labor and material constraints, but as chapter 5 highlights, investors who could pay for such machines demanded sweeping concession rights. In West Africa, British and French officials were wary of giving away too much land and power to investors: the social and political disruptions that would ensue might alienate chiefs and topple the flimsy political framework of colonial rule. Yet they couldn't sit idle, either: new plantations emerging in Southeast Asia (see chapter 7) threatened to capture the market for palm oil, leaving millions in Africa without a viable cash crop. Officials and scientists in both empires sought out simple machines and other innovations that could give Europe the oil it craved, while causing "the least dislocation of labour [i.e., social] conditions."[8] Yet this desire to modernize, improve, standardize—in short, to control—the oil palm industry led colonial governments down coercive and counterproductive paths.

The business context for smallholders was also changing. Lever Brothers' purchase of the (formerly Royal) Niger Company in 1920 gave the firm incredible influence over oil and kernel markets in Africa. The Niger Company absorbed or destroyed rivals until it became the monopolistic United Africa Company (UAC) in 1929. The same year, Lever Brothers and its Dutch competitors became Unilever, a consumer-goods juggernaut that was the world's biggest buyer of fats. The new firm also happened to own the UAC. Farmers confronted this unprecedented monopoly in trying times. The Depression of 1929 sent prices plunging for palm oil and kernels. When Africans pushed back against low prices and colonial oppression, their resistance forced colonizers to reexamine the economic, social, and ecological limits of the smallholder oil palm sector.

Men, Women, and Machines

Small-scale palm oil machines weren't new—recall the oil press brought to Liberia in 1852. Few of these machines were ever sold, though. Inventors in Europe and Africa tackled small-scale palm oil extraction again as prices rose in the early twentieth century. One was John Buckman Esuman-Gwira, a Gold Coast engineer. The Gold Coast's importance as an oil and kernel exporter was waning, and that was precisely why the colonial government was so excited about Esuman-Gwira's 1907 design.[9] The machine was simple, costing £10. Workers put cooked palm fruit and hot water in a metal cylinder and turned a crank, which rotated a shaft studded with blades. After shredding the fruit, the workers opened a tap to drain off oily juices. John Mensah Sarbah, an eminent lawyer and the inventor's brother-in-law, declared that the machine met "all the requirements of the small producer." He stressed that African hands (and feet) never touched the cooked fruit and oil, addressing European anxieties about eating palm oil made by "primitive" methods.[10]

Tests with the machine proved disappointing, however. A team of two men, supported by one woman cooking fruit, could make eight or ten gallons of oil per day. The men thought the machine was small and awkward to operate. Tests with similar machines in Nigeria reached the same result.[11] These early designs sought to reduce human labor to mere motive power, endlessly turning cranks or wheels.[12] A Belgian official acknowledged that machinery like this "requires greater physical effort than indigenous methods, and for this reason alone it is improbable that farmers will ever adopt them."[13] Nigerian officials agreed: "True he often prefers a method that is slow rather than one that is quicker but involves harder work for a shorter time. This is due to the fact that he does the labour himself and therefore regards it from a different point of view from one who pays for it by the day or hour."[14]

Kernel cracking seemed like a more straightforward task for machinery. Yet when agriculture officials in the Gold Coast pitted a team of women against a kernel-cracking machine, they found that the women were more cost effective. Another test in Nigeria had a girl face off with a machine: the machine turned out sixteen pounds of clean kernels per man-day; the girl made four pounds in an *hour*.[15] Human hands could nimbly crack and sort every nut without smashing the kernel inside or mixing shell fragments with clean kernels. "Breaking by hand is tedious work," one official said, but he concluded it was "not so unprofitable as has been hitherto supposed."[16]

The most successful small-scale machines were screw presses that could wring out more oil from fruit pulp than bare hands or twisting nets could manage. When public tests at Calabar produced fifty percent more oil than manual methods, "a certain number of small capitalists" starting buying presses.[17] But these machines only touched one part of the labor process, requiring fruit to be cooked and pounded by hand. Moreover, at least a quarter of the labor needed to make palm oil went into fruit harvesting, and there was no way that machines would make this work any easier.[18]

Some communities made a virtue of the difficulty in harvesting: for the Ijaw of the Niger Delta or the Mende of Sierra Leone, cutting down a first fruit bunch was a rite of passage for young men.[19] Levi Uzozie recounted his own struggle with a palm in an Igbo community in the 1940s:

> To demonstrate to my father that I was going to prove a worthy successor, I decided one afternoon to climb and harvest from a young palm tree scarcely [!] 10 metres tall. The palm [fruit] head hit my climbing rope on its way down and nearly knocked me off balance. I remember throwing away the matchet, abandoning the rope to circle down, grabbing the tree trunk and rubbing my chest and abdomen on the rough stem all the way down, bleeding profusely. That was my first and last attempt at climbing the oil palm tree. The scars remain to remind me of the day I wanted to be a true Igbo.[20]

Uzozie wasn't considered a "true Igbo" by his extended family because of his missionary education. The more schooling boys received, the less likely they were to climb trees. John Mensah Sarbah complained about this in the Gold Coast; he thought machines would make the palm oil business more appealing to an educated man.[21]

The association between tree climbing and slave labor in many places didn't help. Slaveowners often maintained control over workers despite official decrees abolishing slavery, putting "dependents" to work as harvesters. The Nigerian "Native House Rule Ordinance" of 1901, for example, reclassified slaves as "domestics" and gave elite men sweeping powers to punish and control them.[22] In many regions, former slaves and their children "remained conscious of their socially inferior position," forced to defer to elites and pay tribute into the late twentieth century and beyond.[23]

The *pax colonia* put an end to wars and raiding that created most new captives, but it also meant that people could move without fear of enslavement. It trigged a continent-wide boom in labor migration. In Nigeria, workers from arid districts moved to the palm belt for work.[24] Aro men came to rely on hired "ropes" to climb palm trees, seeing the task—once done by slaves—as "beneath the dignity of the Aro people."[25] In the western Niger Delta, the Ikale had no experience with rope climbing and harvested only short palms. They started renting palm groves to Urhobo harvesters in the early twentieth century, sparking a major migration.[26]

Machines like Esuman-Gwira's were designed for the wealthy individual who could hire tree climbers and machine operators. But these were all male tasks: oil presses and kernel crackers were "too heavy to be worked by women and boys," and in any case, cultural norms meant that machines were strictly in the men's domain.[27] Replacing women's domestic work with masculine labor was costly, and women weren't enthusiastic about the prospect of losing earnings from selling oil and kernels. Some women earned kernels as payment for processing fruit; household heads kept proceeds from selling oil. In other communities, women sold the oil and kernels and split the proceeds with men. In still other cases, women bought fruit from their husbands with cash, and they sold the oil and kernels themselves.[28] The European notion of a patriarchal "family" or "household" oil producer was woefully inadequate in capturing these realities.

One merchant recalled how Nigerian women took "a kind of trade union view of things" when kernel crackers arrived, organizing to outbid merchants for kernels.[29] Women tried to block press operators from buying fruit in the markets, where women's organizations often had great power.[30] Moreover, the superiority of oil presses over women's methods was not always clear. Presses typically got 65 percent of the oil out of fruit pulp rather than the 85 percent promised by European officials, only marginally better than hand methods.[31] Men could machine press a gallon of oil in 1.5 hours, while a woman took 3.5 hours to make a gallon by hand. The press was faster, but it still needed women to cook the fruit.[32] And what else would women do? Their time was literally not worth as much as men's in cash terms. Women couldn't climb trees to harvest fruit, and the demands of food production and preparation, childcare, and trading meant that women couldn't easily work full-time as fruit carriers or cooks for a press operation. Cracking kernels and preparing oil could both be woven into household routines.

At least two innovations were designed specifically for women. The first was an iron kernel-cracking tool, intended for regions without good rocks (a surprisingly common situation in the Niger Delta). It was cheaper to buy a rock than the metal tool, however. The other device was a "rapid steam cooker," developed in 1924. Steaming fruit would sterilize it while saving fuel, water, and cooking time.[33] Unfortunately, the cooker proved too expensive for the average woman. Cheap cookware—notably the repurposed metal oil drum—didn't arrive until the post-1945 era.[34]

African women weren't Luddites. According to Chevalier, the introduction of a touring kernel-cracking car in Ivory Coast was a popular move: women "have welcomed this device eagerly," he asserted. The key was that women retained ownership of the cracked kernels.[35] Women in Igboland

FIGURE 6.1
A Duchscher-type palm oil press in operation. Two men turn a handle around a central screw, forcing the plate against fruit loaded in the wooden-slat cage. Oily juice runs out, and a man pours oil from a basin into repurposed four-gallon kerosene tins. Scene from Gold Coast Agricultural Development Corporation, *Mensah the Oil Palm Farmer*, a 1956 propaganda booklet aimed convincing Gold Coast farmers to plant oil palms, buy machines, and stop tapping trees for wine.

reportedly sought out men with kernel crackers as marriage partners, hoping to escape the drudgery of women's work.[36]

Meanwhile, many oil presses served as status symbols for chiefs and traders. Some didn't bother assembling the presses. One design did start to win working customers, however: the "Duchscher" screw press, a modified wine-press design that was more powerful and reliable than other models (figure 6.1).[37] More than 800 presses were in service in Nigeria by 1938.[38] French and British officials hoped that men would form cooperatives to buy presses, noting the importance of cooperative labor in traditional agriculture. Ijaw men in Nigeria, for example, formed oil-making clubs with up to sixty members. Men summoned colleagues with a drum to mash fruit in a communally owned trough, which could be nearly thirty feet long. Men shared the hard work, knowing they would call on their friends to do the same for them in due time.

Europeans imagined such clubs could easily become cooperatives. An anthropologist interviewing Ijaw men in the 1960s found that these associations were fragile, however. Men complained about lazy neighbors who "displayed concern with the job only when their own berries were being mashed." When an individualistic source of income arrived with palm wine distillation, men rushed to take it up.[39] Similar problems hampered French efforts to equip cooperatives with machines in the 1920s and 1930s. The presses chosen by officials were too big for one man to operate with his family, but sharing oil from fruit owned by multiple households was a recipe for conflict. Cooperative members also worried about breakdowns, which had to be fixed at members' expense. As a result, many machines "rusted away in the depots" or "were put to use only during official visits."[40]

British officials in Nigeria complained that simply organizing cooperatives was "uphill work." Men rightly feared that the more the state knew about their incomes, the more it would tax them.[41] Most presses wound up in the hands of wealthy traders, leading officials to despair about the creation of a "new class of middlemen (or middlewomen)" in an already-lengthy chain linking rural farmers to exporters on the coast.[42] Most of these palm oil capitalists were men: in one Nigerian district with 130 press owners, Mrs. Rose Uche was the sole woman proprietor. Women struggled to raise money to buy presses, a problem compounded by men choosing to give uncracked kernels to press owners in lieu of cash.[43]

Still, these entrepreneurs offered useful services to men with lots of fruit to press. The price for pressing a four-gallon tin of oil in Nigeria soon fell by more than half to 7d. and reached 2d. by 1946. This was a significant fraction of the export price of palm oil, but it was roughly equivalent to the cash value of labor invested by poorer families in hand-making oil.[44]

In the end, the most revolutionary machine in the oil palm belt was the bicycle. Bicycles proved highly efficient at moving oil and sacks of kernels, reaching villages that were miles from the nearest river or railway. Nigerians bought more than 100,000 bikes in the 1920s alone. Before the bike, oil sellers might travel up to ten miles with their calabashes to a riverbank, where traders filled puncheons. A bicycle carrying oil in recycled four-gallon kerosene tins doubled that range, with more than twice the capacity of a head-carrier.[45]

A typical bicycle trader could buy three tins of oil and sell them in a distant market for a 1-shilling profit. The proceeds bought kerosene and dried fish, which brought another shilling in profit.[46] With luck, a man could quickly pay off the cost of the bike and accumulate capital. Women tended to lose out due to cultural norms associating bicycles with men, as well as the cost of getting a bike. One Igbo woman saved for nine years to buy a £10 bicycle.[47] Women did preserve one niche in this changing palm oil economy: bulking oil. For decades, women sat outside the doors of European factories and bought oil from sellers, cleaning and consolidating it in larger containers before selling it inside the factory. Traders tolerated it because the women's profit margin was lower than the wages traders paid men to do the same work.[48]

The cumulative impact of hand-presses and bicycles brought the "traditional" palm oil export industry to new heights in the 1930s, even as prices for oil and kernels plummeted. Nigeria alone exported more than 140,000 tons of oil 1936, compared with roughly 50,000 tons in 1905. Palm kernel exports more than tripled over the same period, a feat matched by Sierra Leone and Dahomey.[49] It was an impressive rebuttal of industrialists like Lever, who insisted that only large machines and European capital could deliver the oil that European industry demanded.

Improving the Palmeries

Alongside small machines, colonial scientists tried to improve the trees Africans harvested. There was little scientific work to build on, however. As late as 1910, scientists acknowledged "a good deal of doubt as to the different species" of oil palm, along with great confusion over how palm types or even soil type might explain differences between hard and soft palm oil.[50] Palm fruit varied widely from region to region and even within groves. One tree might make thick-shelled fruit with little oil, while another produced fleshy, oil-rich specimens (figure 6.2). But studying oil palms was difficult. Except for the fused leaflets of the "fetish" palm, the fronds, trunks, and flowers of the other varieties all look alike. As Chevalier noted, "the natives themselves admit that they can only recognize [palm varieties] when they bear ripe fruit." This didn't mean that Africans lacked knowledge about palms: he reported that farmers could reliably identify palms around their homes, recalling who planted them and what they used them for.[51]

Chevalier recorded ten types in Ivory Coast, only two of which were routinely used for oil.[52] Gold Coast officials found eight varieties, though Mensah Sarbah complained that the list missed well-known types like the "porcupine oil-palm" with its extra-long spikes or the "broom oil-palm," whose leaf ribs made the best brooms.[53] Officials in Nigeria compared notes from Efik, Igbo, Ibibio, and Yoruba farmers, arriving at four main types.

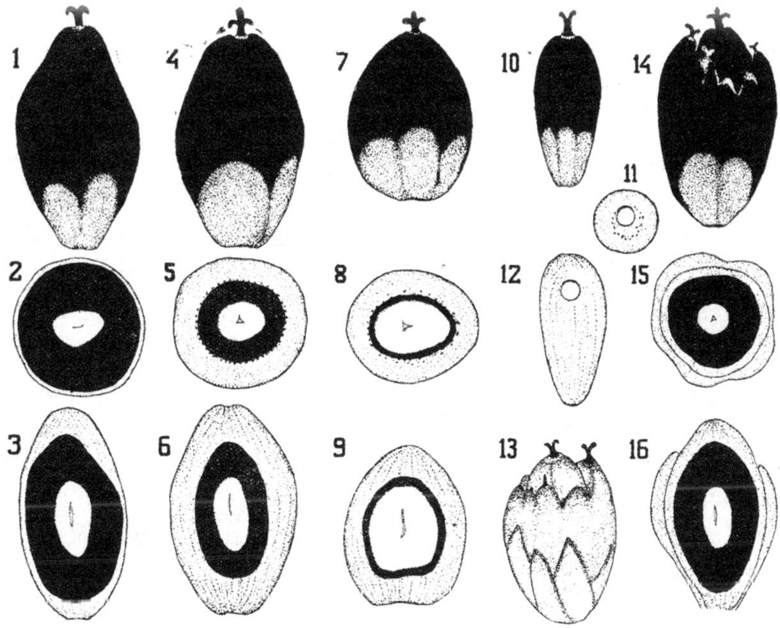

PLATE IV. Varieties of Elaeis guineensis. Explanation see text.

FIGURE 6.2 An early typology of oil palms based on fruit characteristics. In each cross-section, the light-shaded outer ring depicts the oil-rich flesh, followed by the shell (black) and the kernel (white). 1-3 show the thin-fleshed "Congo type" *dura*, 4-6 show "Deli-type" *dura*, 7-9 thin-shelled *tenera* type, and 10-13 shell-less *pisifera*. 13-16 depict a "mantled" or *diwakkawakka* palm fruit with unusual fleshy lobes. A. A. L. Rutgers et al., *Investigations on Oil Palms*, plate IV.

Farmers pointed out which were best for oil, and which—like the fetish palm, or those with thin pericarps—they left unharvested.[54] Gold Coast farmers warned officials not to bother with a palm that made fatty fruit without a kernel shell (likely *pisifera*). Animals ate most of the fruit, and most specimens were infertile.[55] Colonial researchers found that Mende men in Sierra Leone discussed palms with a nuanced vocabulary that described not only varieties but also stages of growth and the ripeness of fruit, points of practical interest to oil and kernel producers. French scientist A. Houard griped that these sorts of classifications were subjective and had no place in science. But without a viable alternative, he and others continued to use indigenous names and categories for years.[56] Western scientists eventually settled on three subjectively named types in their own vernacular, Latin: the shell-less *pisifera* (false), the common *dura* (hard), and the thin-shelled *tenera* (tender).

Paul Preuss generated excitement in colonial circles with a 1902 report on the "*lisombe*" palm, a thin-shelled type in Cameroon that gave more oil than common varieties.[57] Whereas typical *dura* palm fruit yielded 10 or 15 percent oil by weight, *lisombe* (*tenera* in modern classification) gave up to 50 percent. These palms form roughly 4 to 6 percent of oil palm populations in Africa, which are mostly *dura*.[58] Chevalier concluded that *lisombe* offered scientists "a magnificent subject for research."[59] Scientific breeding had turned plants like the sugar beet into economic powerhouses, and European industrialists salivated at the idea of an oil palm with double or triple the yields of "wild" palms.[60]

Yet as the botanists at the Royal Botanic Gardens at Kew noted, European palm-breeding enthusiasts had "interests at heart that are not necessarily coincident with those of the inhabitants [of the colonies]."[61] What Europeans wanted out of a palm wasn't always what Africans wanted. When colonial scientists bothered to ask farmers about it, they found that the *tenera* palm was well known.[62] Whether farmers used it for oil or not was "rather a matter of taste." (Oil from *tenera* fruit isn't as bright red or as flavorful as oil from *dura* fruit.)

It was also a matter of business. Was the farmer selling palm oil, or kernels, or tapping palms for wine? Thin-shell palm fruit has more oil but smaller kernels, with smaller fruit bunches.[63] In Congo, scientists found that *tenera* and *pisifera* palms were rare in some mature groves, despite being more common in young groves. This suggested that farmers actively culled these as less-desirable trees.[64] Conversely, farmers in the Kasai region of Congo favored thin-shelled palms, removing other varieties.[65] In several places in Nigeria, farmers traveled many miles to get thin-shelled kernels or seedlings, which they planted near villages.[66]

Importantly, farmers knew that the thin-shelled palm didn't reproduce true to type. A German scientist stumbled on this fact working with Preuss's palms. Only two of seventeen test palms produced *tenera* fruit; the rest were thick-shelled *dura*.[67] As Farquhar learned, "The natives do not consider this palm a true variety, as they know from experience it does not come true to seed."[68] Chevalier heard similar accounts in Ivory Coast and correctly theorized that *tenera* was a hybrid.[69]

By 1914, the Italian botanist Odoardo Beccari had grouped the oil palm into thirteen distinct varieties and numerous subtypes. These didn't agree with African typologies or classifications by other Europeans. A French botanist got up to nineteen varieties in 1917 before others started whittling down the list.[70] Colonial scientists embarked on a more rigorous study of oil palms in the 1920s. French and British researchers confirmed that rainfall explained a great deal about variable palm oil yields.[71] But breeding experiments faced challenges ranging from bad planning to rats stealing kernels from nurseries.[72] The Nigerian government couldn't convince farmers to sell enough land in the palm heartland for big experiments, forcing the agricultural department to settle for a subpar site near Benin City. French scientists had better results at stations in Ivory Coast and Dahomey, funded by a levy on French industries.[73]

Tenera turned out to be something of a white whale for scientists. Some lamented that breeding experiments "almost make it clear that it will never come up to expectations."[74] Researchers working in Congo only began to unravel *tenera*'s secrets in the 1930s.[75] Abiron Beirnaert's experiments at the Institut National pour l'Etude Agronomique du Congo Belge (INEAC) confirmed that *tenera* was a "genetically impure" palm, carrying genes for both thick-shelled *dura* and shell-less *pisifera* traits. Beirnaert and his collaborator Vanderweyen proved that crossing *dura* and *pisifera* always produced *tenera*. Their discovery came on the eve of the Second World War, and its real impact was not felt until the 1950s.[76]

Meanwhile, colonial scientists hoped to convince Africans to replace existing groves with higher-yielding varieties, grown under "scientific" conditions. These "improvement" or "*aménagement*" programs used a variety of methods. Belgian Congo fell at one end of the spectrum. Authorities made *aménagement* obligatory for farmers. Villagers had to cut trails and clear brush around the palms, and to carry out regular pruning. They became unwilling subjects in experiments with *tenera* palm breeding, which left them with many sterile *pisifera* palms.[77] French West Africa fell in the middle. As Chevalier argued in a 1931 article, farmers were capable of making rational choices about their land. He urged the state to provide as much aid as farmers wanted but not to force them into any particular steps.[78] Officials offered incentives for thinning and replanting groves, but farmers often chose to inter plant with food crops against expert advice, hedging their bets on oil palm.[79]

At the far end was Britain's low-budget, laissez faire approach. Pamphlets in local languages explained European notions of weeding and thinning, and encouraged farmers to plant thin-shelled palms.[80] Demonstration plots showed what "improved" groves could do, but officials could only offer modest subsidies and other incentives to convince farmers to take up new ideas. Chiefs in Sierra Leone "expressed amazement" at the high yields on short, easy-to-climb palms. But like all farmers in western Africa, they had to balance palm oil with food. In Sierra Leone, the farmer's "principal concern in life is his rice crop, and he knows that rice does not thrive on land well stocked with oil palms." The palm's dense root mat "robs other crops of their food."[81] Farmers sensibly chose rice over palm monoculture.[82] Chiefs were also wary of palm replanting, because it threatened land tenure arrangements: planting palms was a sign of ownership.[83] In addition, farmers and chiefs knew that governments rewarded owners of improved and registered plots with higher taxes.[84]

One official pondering the Masanki project (chapter 5) struck on the underlying problem with grove improvement efforts in a 1935 memo: "The curse of palm production in Sierra Leone is paradoxically, it may seem—the presence of the palm trees!" If he could wipe the landscape clean of oil palms, new palms "would have to be planted and tended," creating a plantation of short, high-yielding trees.[85] It was a difficult proposition to convince a landowner to cut down a mature tree and replace it with experimental seedlings, however. Nigerian officials doubted that owners "will be able to bring themselves to take the drastic step of cutting down good oil palms."[86]

Long-term studies showed that "improvement" strategies were counterproductive. HCB found that one hectare of "natural" palm grove in Belgian Congo produced up to 1,000 kg of fruit annually. Once pruned and weeded (see figure 5.2), it might make anywhere from 500 to 5,000 kg. Across 59,000 ha of HCB-managed groves, the average yield remained below 1,000 kg/ha—in other words, no better than a good "natural" grove.[87] This was largely because grove improvement broke rotational fallow cycles that kept soils fertile. Montague Dyke, a former HCB employee hired to study Nigeria's groves, found that farmers there "learnt long ago the lesson to be obtained from such experiments as these." The farmer ignored most of the grove, but lavished care on "what may be termed the 'Toilet' of his palms," cleaning a small circle around the base of the best fruit-bearing trees. Because a man could only climb a few trees a day, yield per tree rather than per hectare was what mattered most. Dyke refuted the common idea "that the natural palm stands are only very partially tapped," declaring that "the most accessible palmeries are already being cropped to the limit of their capacity."[88]

To protect groves that already existed, many governments banned palm felling and wine tapping.[89] Relatively few Africans killed palms for cabbage; most groups considered it either a delicacy or a famine food. They "understood very well that to destroy a palm tree is to diminish a source of wealth."[90] But felling palms to tap wine was more common in a few regions, and some palms existed solely for tapping. Oil palms around Conakry were "puny" because of regular tapping; in Togo and Dahomey, farmers maintained "vineyards" of oil palms. Some reportedly crammed 2,000 palms into a hectare (a modern plantation has 150).[91]

Tappers could climb and tap standing trees, gathering a gallon from six to ten trees. A single felled trunk produced eight gallons or more. If a farmer had palms to spare, the felling method made a great deal of sense.[92] Yet French authorities began restricting tapping and felling in 1906, hoping to boost palm oil exports.[93] The Germans shortly followed: a law in Togo required farmers to plant three new palms for every palm felled for wine. Officials thought the small population drank the equivalent of 300,000 palms every year.[94] British officials introduced bans at the district level, relying on chiefs to enforce them. One 1911 ordinance banned the destruction of oil palms "for any purpose" without permission.[95]

Palm wine was an important source of money for tree owners, however, and some officials stressed that tapping bans were "impossible and would indeed be unjust."[96] Wherever palm wine offered better returns than palm oil or kernels, farmers made wine. By the 1930s, a tree's worth of wine fetched two or three times the price of a tree's oil in Gold Coast.[97] One governor quipped that enforcement of a tapping ban would require "a policeman at the foot of every tree."[98]

Colonial efforts to stem wine tapping were complicated by new restrictions on liquor imports imposed in 1920. The bans inadvertently encouraged Africans to make even more palm wine and to distill it into harder stuff. Bonou Kiti Sodabi brought distillation technology back to Dahomey from France after fighting in the First World War. He and his brother distilled the first palm gin and were rewarded by the government with six months in jail. Beninese distillers call their palm gin "Sodabi" in their honor.[99] Stocky James Iso similarly learned about distilling while abroad, and he too served jail time for selling the secrets in Nigeria.[100] The popularity of palm gin drove a wave of palm felling in some colonies: a single trunk's wine distilled down to roughly a gallon of gin, meaning that each gallon of foreign spirits kept out by liquor restrictions caused the death of another oil palm. African elites and colonial officials despaired at the spread of distillation, claiming it undermined social order. Young men defied both colonial law and traditional norms by making and drinking palm gin.[101] In the words of one governor, they were "imbibing the lesson of defiance" to authority.[102]

The biggest problem for Europeans hoping to boost exports from Africa's palm groves wasn't wine-tapping or aggressive weeding, though. It was the basic fact that palm oil didn't pay as much as palm wine or other crops. Yoruba farmers recalling their switch from oil palm to cocoa or kola nuts told an interviewer, "the trouble [with palm oil] is too much!"[103] As one official put it, scientific demonstrations about palm improvement usually convinced farmers "that the process is not worth carrying out."[104]

Colonial Coercion and African Resistance

Colonial states balanced extension work—machines, cooperatives, and agronomic advice—with regulation, taxation, and coercion. Regulations were largely aimed at stamping out adulteration. As palm oil entered the industrial food system, quality became a vital characteristic of the commodity, and commodity markets relied on standard, universally recognized grades.[105] A bad calabash mixed with good soft oil could ruin a whole puncheon. The burden of producing and policing uniformity in the commodity fell on African producers, however.

Adulteration was rife in the "palm oil ruffian" era, as sellers mixed dirt, maize flour, or mashed plantains to pad out oil containers.[106] Cautious buyers checked for sediment and heated lumps of hard oil in pans to see what impurities were revealed as the fat melted.[107] Skilled coopers could take entire puncheons apart and shrink them, cheating Europeans who assumed they were the standard size.[108] At least one fraudster was caught nailing logs inside a puncheon before reassembling it.[109] It was tougher to adulterate palm kernels, but shell fragments and excess moisture could easily amount to a 15 percent loss in a bushel.[110] Europeans were the end targets of this fraud, but even African traders fell victim, like the early-morning buyer who found himself the owner of "two pots of muddy water with half an inch of oil on the surface."[111]

"Every conceivable mode of deception is resorted to, to cheat you," a British writer complained.[112] But Europeans were cheats, too. They bought oil with cotton cloth doped with excessive sizing to improve its feel and appearance; the sizing disappeared with the first washing, revealing "for its astonished owner a new and threadbare form." Merchants watered down liquor and poured cheap stuff into better bottles. "Commerce in West Africa had never ranked as one of the most honourable occupations," noted historian A. G. Hopkins, "and these fraudulent innovations lowered its reputation still further."[113]

Lagos was among the first colonial governments to introduce mandatory produce inspections in 1889, responding to "a wave of adulteration."[114] The first decree of Germany's administration in Togo forced traders to use standardized containers for palm oil.[115] French officials barred sellers from entering oil and kernel markets until their wares had been inspected and cleaned.[116] More intrusive regulations arrived across Africa in the 1920s. In Nigeria, government inspectors prowled inland markets, working on the theory that "the educative value of inspection is higher the nearer it is brought to the native producer."[117] A British official called for setting ever higher standards, arguing that "the sellers are past masters in presenting produce which is just on the border line of requirements."[118]

Northern manufacturers demanded quality control and standardization to facilitate the commodification of palm oil and kernels and their incorporation into new industrial products, but women in Africa felt the brunt of regulation. Traders forced them to supply firewood to clarify oil. Some women had to "crack the palm kernels with our teeth" to prove their quality. At one Nigerian market, an official instituted a "one-shilling test," rejecting oil if sediment in the bottom of the tin was thicker than a shilling coin. Furious women blocked the road and extorted tolls from passing traffic, forcing the inspector to rescind the test.[119]

Beyond regulation was taxation. Import and export duties were often quite high in colonial Africa; farmers never saw this tax collector, but they felt the sting in what they got for their oil and the prices of goods they bought.[120] Head and "hut" taxes were another matter. These taxes were imposed on households (and sometimes on livestock and trees) simply for existing. They served fiscal and ideological goals. As Nigeria's governor, F. D. Lugard, argued, paying taxes publically demonstrated submission to colonial rule.[121] Some administrations took—or required—taxes paid in kind, collecting palm oil, kernels, and other produce to shape economic activity.[122] But more often they insisted on cash. This forced people into the export economy, whether they liked it or not.[123] In French colonies, officials also levied the *prestation*, a labor tax, payable in service on public works or in cash.[124] When a British businessman asked a French official how they had pushed farmers to export so many kernels, the candid answer was: "we take them by the throat."[125]

All the colonial powers relied on chiefs to collect taxes from villagers; the chiefs knew better than visiting officials about the numbers and wealth of their people. States gave chiefs a powerful incentive to collect all the taxes they were owed (and then some) through rebates. In Dahomey, a chief could keep 3.5 percent of taxes collected.[126] In the Niger Delta region, where many groups never had hereditary or elected chiefs, colonial officials appointed new "warrant chiefs" to fill the role. The chiefs were unpopular, and they started collecting head taxes at the worst moment: on the eve of the 1929 Depression.

Igbo men reluctantly paid the 1928 tax, when the price of a single four-gallon tin could pay a man's liability.[127] Prices dropped in 1929, however, as the world plunged deep into the Depression. Edward Kanu Uku recalled: "The limits to personal power and influence were now in sight for those who had climbed to the top of the short ladder at their disposal and the steady fall in palm oil prices made for . . . widespread frustration."[128] An Igbo song captured the fears caused by new tax regime:

Tax time is running nigh!
Tax time is running nigh!
The man without money is crying.
We shall see the whiteman's war.[129]

Villagers heaped insults on the warrant chiefs tasked with collecting taxes, but most paid them peaceably.[130]

Nigeria's colonial overlords made a serious error in late 1929 by including women in tax assessments, however.[131] When one warrant chief's clerk tried to count households, a Ngwa Igbo woman named Nwanyeruwa challenged his authority and the very idea of taxing women. Thousands of women rallied to Nwanyeruwa's aid and besieged the chief's compound. A British officer handed the women the warrant chief's cap—his symbol of authority—and blamed him for inflaming the situation, but it was too late.[132] Protesters marched across Igboland and beyond. Women destroyed courthouses and the compounds of warrant chiefs and government clerks before moving on to European businesses. When Edward Kanu Uku tried to calm an angry crowd, he was "shouted down as a traitor in the pay of the government, a man who had turned against his own people."[133] At Aba, colonial troops fired on demonstrators, killing at least fifty-five women and wounding many others. The protests soon collapsed at the end of 1929. Vengeful colonial officials levied fines on villages linked with this "Women's War," burning villages that refused to comply.

Investigators probing the causes of the "Women's War" found many, including cultural distinctions between men's and women's spheres of activity and the impact of Christianity on family structure and social values.[134] But as one investigator noted, "Palm-oil dominated the situation, was the hub round which the whole economic and in consequence much of the social wheel revolved." As the Depression worsened, people asked: "How can we pay tax if Government will not order the traders to pay us more for our oil? We have no money, we have no food. We die of hunger."[135]

More protests broke out across Nigeria and western Africa in the 1930s, though never again on the scale of the Women's War. One strategy was the "hold-up," in which Africans refused to sell to European merchants. Urhobo producers organized two major palm oil hold-ups: the first, in 1934, targeted rock-bottom prices offered by UAC for hard oil. UAC acknowledged that an Urhobo family might spent "five to six days to produce a [4 gallon] tin of oil which attracted a 'Beach' price of only eleven pence."[136] One man was fined fifteen shillings by his compatriots for trying to covertly sell a few tins. British officials let this strike play out. The UAC could afford to wait, but Urhobo sellers who needed cash to buy food could not. A repeat hold-up in 1938 was equally unsuccessful.

Europeans complained that hold-ups were irrational outbursts, coming from people who couldn't grasp the economic concepts behind taxes and commodity prices.[137] UAC even carried out a propaganda campaign in Nigeria, stressing that bumper supplies of whale oil and other fats made the UAC powerless to raise palm oil prices.[138] Yet palm oil producers understood more than the UAC cared to admit. West Africans had written about the "menace of mergers" in the 1920s, as trading firms consolidated into the UAC monopoly.[139] Producers followed global markets, too: when Britain purchased a consignment of whale oil in 1934 to hold down domestic soap and margarine prices, the move "was known immediately throughout the country [Nigeria]." A researcher recalled: "in the remotest bush I have been asked to explain it and to give the reason why Government allowed it."[140] By the 1930s, it was abundantly clear to producers—and indeed, to colonial officials—that European firms like UAC and Unilever had no real commitment to the economic well-being of their African suppliers.[141]

"Food for Thought": Beyond the Export Market

The crisis of the 1930s highlighted an important, and often ignored, part of the oil palm story: the role of Africans as consumers. Colonial governments cared about exports, because they could measure and tax them. Most historical records on the oil palm industry left in archives deal with exports.[142] This was the entire point of colonialism: extracting resources for the benefit of the home country.

Many Africans were committed export producers. But a typical family making oil might sell it, consume it themselves, and buy oil to consume throughout the year. We can only guess at the size of the domestic market, but it shaped—and limited—the growth of the export sector. Food, soap, illumination, and cosmetic uses consumed vast quantities of palm oil. If the roughly 10 million people living in southern Nigeria in 1930 each used five pounds of soap annually (an estimate that is probably much too low), they would need at least 11,000 tons of oil.[143] Factory-made soap imported from abroad and made at a new Lever Brothers factory in Lagos only supplied enough for one pound per capita.[144]

Imported candles and lamps for illumination had an impact in the opposite direction, freeing up palm oil for export. Stearic and palmitic candles were popular across western Africa from the 1850s, replacing dim palm oil lamps for the affluent. The arrival of kerosene in the 1890s brought bright light to even more consumers, and constant demand for kerosene compelled families to sell oil and kernels.[145] If kerosene replaced palm oil on a one-for-one basis (an arbitrary estimate), Nigeria's kerosene imports in 1930 would have freed up nearly 20,000 tons of palm oil for other uses.[146]

The biggest African use for palm oil was in food, however. Historians have underestimated just how much palm oil Africans consumed, offering figures as low as ten grams per person per day. This was far too little for the palm belt. Prison inmates in the Gold Coast were entitled to seven to thirteen grams for food alone, while soldiers got at least twenty-six grams.[147] Historic records give figures ranging from twenty-two to ninety grams daily for food. One French report put the total figure, including soap, lighting, and cosmetic use at 180 grams, or sixty-six kilograms per year.[148] European observers concluded that Africans in Togo, Dahomey, and Cameroon used about four times as much palm oil as they sold to foreigners.[149]

Africa's booming ports and market towns created lots of new customers for palm oil. A bottle of edible palm oil sold for as little as three pence in Nigeria's palm belt, but it cost as much as fifteen pence in the northern city of Kano in the 1920s.[150] In Gold Coast, edible palm oil sold in Accra for "about three times what it is worth for export." The comparison "gives much food for thought," mused a colonial official. He estimated that a family used about half of a gin bottle of palm oil per week, and that "the quantity used is limited by the supplies obtainable." It was "not in the least curious," the official remarked, "that owners of oil palms, with this large and remunerative market open to them, do not exert themselves still more and manufacture a surplus for export."[151]

Domestic consumption worked against the export market. But this didn't mean that domestic consumption was an obstacle to mechanization. Lever's failure at Yonibana in 1914–15 was partly due to farmers' worries that the mill wouldn't return oil to them for their own use.[152] The Palm Oil Estates Managers mills in Gold Coast sold large amounts of oil locally, and the mill at Butre finally turned a profit in 1928 on account of local taste for its "pure" oil.[153] The failed Bukonor mill was supposed to start selling oil in tins to local buyers, but hadn't gotten its free fatty acid content low enough to meet local culinary standards by the time it closed.[154] It was no different in French colonies. Mills in Ivory Coast and Dahomey sold nearly all the oil they could make locally, leaving only kernels for export.[155] Southeastern Nigeria was one of a few regions that continued to export large amounts of oil, but even there, the domestic market consumed the biggest share. By the late 1930s, British officials estimated that about 65 percent of all palm oil made in Nigeria was consumed by Nigerians.[156]

The growth of palm oil exports also faced ecological limits, especially in the Niger Delta region. As farmers in many parts of western Africa turned to more lucrative cocoa trees, this region remained entrenched in the oil palm business. Soils in the region were poor and unsuitable for cocoa or rubber. The land was getting crowded, too: in Onitsha and Owerri districts, population densities reached 500 people per square mile, forcing farmers to keep half of the land under constant cultivation. In the 1930s, yams in Owerri were reportedly half the size they had been a decade earlier.[157] As early as 1912, Nigeria's director of agriculture warned about the Niger Delta's "poor acid soils" and low food yields. He thought doubling down on oil palm was the best solution, relying on other regions to supply foodstuffs.[158]

Archival and oral records attest to growing pressure on oil palms and the people who farmed them.[159] This leading palm-exporting region had surprisingly few oil palms to go around: some reports from the first decades of the twentieth century described an average of six palms per acre around Aba, and up to ten per acre farther south in Ibibio territory. Although these figures are averages across palm groves and cleared farmland, they are not very big numbers compared with the reports of hundreds of trees per acre that attracted concessionaires elsewhere.[160] As historian Chima Korieh has argued, the region entered an ecological crisis during the economic turmoil of the Depression. There was simply not enough land left to sustain both food and cash crop production.[161]

A 1938 survey of the Niger Delta region found between two and four acres per "taxable male," but split between as many as seven separate plots. The average plot was under half an acre, and some contained only two or three palms. One recorded plot was only six feet by twenty feet. By the 1950s, an average Igbo farmer was down to a single acre of land. Ibibio farmers had 1.4 acres spread across an average of four plots, requiring frequent trips between field and household.[162] This land hunger forced the West Africa Oil Palm Research Station (now the Nigerian Institute for Oil Palm Research) to locate in Benin in 1939, rather than in the heart of the palm belt to the east. The "population is so dense that there is very little land to spare," remarked the director.[163]

One researcher concluded that "wild" groves were rare in this region by the 1950s, and in densely populated areas, they "no longer exist." Farmers often ignored remaining groves, because the palms "grow so closely together that many never produce fruit."[164] Overall, the shortage of land—and therefore of palm trees—forced farmers to carefully manage what they had. Some Igbo farmers aggressively pruned palms to keep them from shading foodstuffs, trading slower palm growth for higher food yields.[165]

Tax pressures inspired many Nigerian communities to regulate harvesting of communally owned palms; the right to harvest was tied to "the inclusion of a man's name in the tax roll."[166] Harvesting out of season was punished with a fine. These rules ensured that wealthy men couldn't pick the groves clean with hired labor, leaving poor men unable to pay taxes. Even individually-owned palms might be fair game at tax time.[167] Still, court records show frequent complaints from elders about young men cutting fruit illicitly.[168] In one Igbo village, young men successfully overturned a new rule that would have restricted men to harvesting only the palms owned by their families, a move that would have prevented energetic youth from accumulating wealth, titles, and wives.[169]

Communal harvesting rules reflected the egalitarian politics of many communities, and they were adapted to do new things like paying for clinics and school fees. "Failure to harvest energetically" during communal collection periods was "a grave social offense," noted one researcher.[170] Over time, communal rights to village-owned palms often gave way to a model that required families to buy a "share" of the harvest. Women and the elderly could hire a man to cut their share of fruit. Even more changes came around the 1950s, when some leaders simply took over the communal harvest, hiring men and selling the fruit in the marketplace to pay for taxes and special projects.[171]

An "Eminently Suitable and Efficient" System

In 1939, Africans supplied the world with about 520,000 metric tons of palm and palm kernel oil, twice what they sold in the years leading up to the First World War.[172] Machines owned by HCB and others in Belgian Congo accounted for 110,000 tons, but the rest came from smallholders, most working without even hand presses. Looking at Nigeria, the single largest exporter, F. M. Dyke concluded: "the native farmers have evolved a system of palm culture which gives them the maximum output with the minimum labour expenditure, and must be considered eminently suitable and efficient for working natural palm forests."[173]

Some British officials griped that the empire sacrificed "economic progress to political security" by resisting plantations.[174] Lord Hailey warned in his 1939 *African Survey* that "natives must learn by painful experience how severely a competitive system punishes those who will not adopt the most efficient technique of production."[175] As chapter 7 shows, there was no denying that oil palm plantations could be efficient. One 4,000-hectare plantation with 1,200 workers could export more palm oil than the entire Ivory Coast managed in 1930.[176]

Yet most officials in British and French West Africa remained opposed to plantations. One British official declared: "It is not the consumers—the soap and margarine makers—who see 'menaces' [from plantations] in the air. It is the merchant, who is always trying to steal land which does not belong to him."[177] And African farming techniques weren't as "primitive" or "wasteful" as they appeared to foreign eyes. Europe's obsession with economy meant remaking every hectare to push yields as high as possible; but Africans adapted to the land to maximize the output of their labor.[178] By the 1930s, colonial administrators vowed "to look at everything from the native farmer's point of view."[179]

This meant acknowledging that many farmers were better off replacing oil palms with cocoa. It meant recognizing the importance of palm wine and palm oil in domestic consumption. It meant accepting that Europe's hunger for industrial fats didn't take precedence over Africans' rights to land and a livelihood. And unlike a plantation company beholden to shareholders, the smallholder farmer was flexible and resilient.[180] Chevalier, whose scientific work helped launch the concession boom, stressed the central importance of people over palm trees or machines in a 1931 article. He noted that in France itself, food production remained in the hands of the peasantry. They adopted "scientific" techniques and improved varieties, and he saw no reason why African farmers would not do the same in time. He wrote: "If I were a native, despite the enchanting picture . . . of the well-being of a Sumatran coolie, I would still prefer to be a simple Adikrou peasant . . . cultivating a few hectares of *Elaeis*, cocoa, vegetables, and fruit trees around my plot, living and working all day with my family." It might not be the most efficient use of land, but it was enough to get by in the tumultuous economic environment of the 1930s.[181]

7 The Oil Machine in Southeast Asia

Africa's palm groves tempted European industrialists with dreams of fast profits, but they left a string of disappointments. Despite the huge quantities of palm oil and kernels African smallholders produced for the world market, cheap edible oil remained elusive. The solution proposed by many Europeans was the plantation, an institution for controlling land, plants, and people as a single productive unit. But the plantation was a difficult proposition in colonial Africa: officials were wary of inciting conflicts over land, and companies learned hard lessons about the cost of labor. Africans simply weren't interested in a life of poorly paid wage labor when they had other options. Instead, the first successful oil palm plantations emerged in the 1920s halfway around the world in Southeast Asia.

Along the Strait of Malacca (map 7.1), investors found colonial states eager and willing to provide land and labor, along with a proven model of plantation production that would shape the future of oil palm cultivation. Both sides of the Strait bustled with trade and agriculture long before Europeans arrived in the sixteenth century. Nutmeg, cloves, sugar, coffee, pepper, gambier, and many other crops drove waves of cultivators—indigenous and foreign—to transform the region's landscapes with farms and larger plantations. Malay-speaking communities settled along coasts where rice could be cultivated, later joined by Chinese migrants and others. Forest peoples living in the interior participated in cash-crop booms, too, integrating new crops into forest fallow agriculture.[1] As Europeans expanded their grip over the area in the nineteenth century—the Dutch in Sumatra and the British in Malaya—European capitalists launched plantations of their own.

Oil palm was a latecomer to the region, following the paths carved into Southeast Asia's forests by another tree, *Hevea brasiliensis*. This South American rubber tree inspired an unprecedent plantation boom beginning in the 1890s. In many tellings, the stories of both *Hevea* and *Elaeis* plantation industries share a common fairy tale origin. Randomly chosen specimens arrived from distant lands. They thrived in rich Asian soils, free from the pests of their homelands. European capital turned the jungle into orderly plantations, where European science improved both the plants and the ways

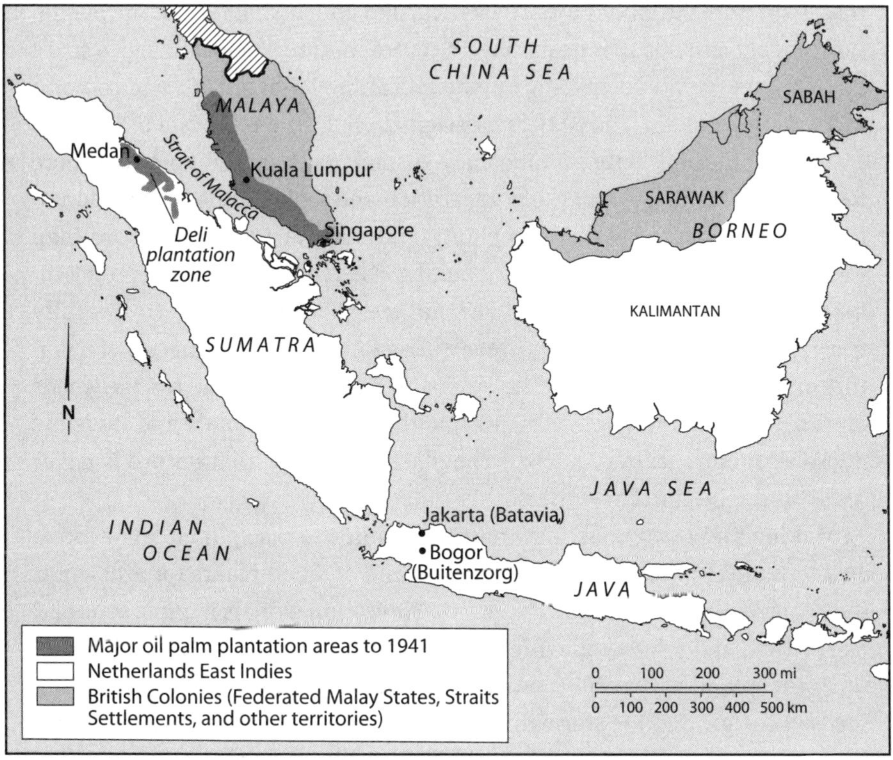

MAP 7.1 Colonial Southeast Asia, with major plantation areas hosting oil palm up to 1941 indicated. Most land in these areas was planted with rubber.

they were harvested. The arrival of modern plantations spelled doom for competitors in Africa and America relying on wild plants.[2]

Contrasting the Southeast Asian oil palm story with the African experience makes it clear that the rise of the plantation—whether populated with oil palm or rubber or anything else—wasn't down to a lucky plant, or scientific know-how, or entrepreneurship. The key was power over people and land. The ability to recruit and discipline labor and to transform land on scales that were impossible in most of colonial Africa was what made the Southeast Asian plantation a viable competitor to smallholder industries in the decades before 1941, laying the groundwork for an even larger plantation boom later in the century.

Land and Labor in Colonial Plantation Systems

A plantation is more than a big farm. It combines agronomic knowledge—how to grow crops—with political, economic, social, and even cultural tools that shape how the land is used, by whom, and under what terms. The plantation model demands that its managers control nature: "delocalizing" a space by clearing vegetation; evicting human and animal inhabitants; and contouring, draining, and irrigating the landscape until it fits a universal model.[3]

The plantation is built for efficiency. A small garden might produce more food per square meter, but the organization and technologies of the plantation allow a team of workers to plant, tend, and harvest much more than those workers could manage individually. The result is a sort of factory-in-nature, what some scholars call "new nature." The plantation is never fully free from the natural world, however.[4] The long, straight galleries of a mature oil palm plantation show the order and planning at work, but the weeds on the ground, epiphytes on trunks, and birds and animals and invisible microorganisms scattered across the place remind us that nature is never really under human control.

Making and using that "new nature" requires a disciplined labor force. Just as plantation managers select seeds and arrange plants for efficiency, they control the movements of workers. Plantation work is monotonous and exhausting, and few would willingly do it for the wages on offer. (Ironically, the term "planter" describes owners and managers, not the people putting seeds in the ground.) Historically, plantations have been "landscapes of oppression."[5] Some scholars whitewash the reliance on coerced labor, treating it as an unfortunate reflection of the mores of the era or a product of imperfect labor markets. They fail to see that some employers "prefer to recruit unfree labour."[6] A "plantation complex," as some historians call it, is not just the place where crops grow but also the political system that determines how land and labor are mobilized and exploited, and to what end.[7]

The American plantation complex, which first emerged to produce another crop requiring prompt, large-scale processing—sugar—relied on chattel slavery, that most unfree form of labor. In colonial Southeast Asia, plantations relied on hundreds of thousands of indentured "coolies" from Java, China, India, and other places. Unfree hands cleared forests, planted rubber and oil palms, and harvested trees. Recruiting and managing labor at this scale wasn't an option in much of Africa's oil palm belt, and it was never possible without coercive power, as Lever's experience in the Congo illustrated.

The rubber plantation complex matured a decade before the first oil palm plantations appeared, and it served as a template for planters turning to oil palm after 1910. Rubber arrived in Southeast Asia via the Royal Botanic Gardens at Kew, part of a coordinated effort to identify and distribute rubber plants across the British Empire (not a random consignment of seeds smuggled out of Brazil, as popular myth holds).[8] Remarkably, leaf blight and other diseases never followed *Hevea brasiliensis* out of South America. Freed from competing vegetation and its native plagues and pests, the tree enjoyed "ecological release" in Asian plantations. Europeans discovered that careful tapping methods could keep *Hevea* dripping latex for years, in contrast with the destructive "slaughter-tapping" inflicted on other latex-producing plants.[9] Meanwhile, competing South American plantations—including Henry Ford's massive "Fordlandia" project in Brazil—succumbed to disease.[10]

Hevea grew well in African environments, too, but rubber concessionaires faced the same struggles for land and labor that had plagued oil palm concessionaires.[11] In Southeast Asia, by contrast, colonial planters viewed land on both sides of the Strait of Malacca as practically vacant. Britons described the Malay Peninsula's interior as "virtually uninhabited," while the Dutch viewed eastern Sumatra as a "deserted corner" of their island empire.[12] The 1870 Agrarian Law of the Netherlands East Indies (NEI) put all uncultivated land under government control; by the late 1930s, nearly 700,000 acres of this "waste" land in Sumatra was in the hands of foreign plantation companies.[13]

Where colonial authorities nominally ruled in partnership with local kings, like the British-controlled Federated Malay States (FMS), the state held forest lands in trust for the rulers and their people.[14] Whether these rulers had rightful claims over forest-dwellers and their lands didn't matter.[15] Europeans could access these "vacant" and "unproductive" lands on generous terms: in Sumatra, planters could get leases lasting up to seventy years. In Perak, one could get a lease for "waste land"—defined as anything not part of a town, mining reserve, or occupied "under Malay tenure"—lasting 999 years. Perhaps concerned that a millennium wasn't enough, the British extended leases into perpetuity, though in 1911, new conditions allowed lease revisions after thirty years.[16]

Colonial officials did try to understand indigenous systems of land ownership in Asia, if only so that they could better regulate, tax, and expropriate land. British officials acknowledged the importance of forest fallow cultivation for forest peoples, but the land titling system they imposed on Malaya was totally incompatible with it. If land wasn't actively cultivated, it couldn't be owned.[17] While farmland surrounding Malay villages received special protection in 1913, the spaces used by groups practicing forest fallow agriculture became crown lands.[18]

Dutch officials had a similar interest in customary law and took measures to separate what they saw as the "modern" plantation sector from the "traditional" sector, with its diverse tenure systems.[19] On both sides of the Straits, concessions often had to set aside land for village (*kampong*) cultivation.[20] The work of recognizing local land rights and defining boundaries was simultaneously the work of exclusion, however.[21] Governments and concessionary companies took landscapes full of movement—of plants, animals, and people—and chopped them into static mining parcels, timber tracts, and plantation zones, restricting forest dwellers to "native reserves."[22]

To colonial eyes, forest fallow farming was primitive at best, environmentally destructive at worst. Officials were also critical of Chinese entrepreneurs, whose small plantations drove much of the nineteenth-century cash crop boom: after growing crops like pepper and gambier, they abandoned worn-out fields. These fields became "lalang wastes," colonized by *Imperata cylindrica*, a tenacious grass detested by planters and agricultural scientists.[23] In some of the Malay states, governments insisted that all new concessions be planted with "permanent" crops.[24] By the early twentieth century, officials offered European planters land seized from Chinese planters on steeply discounted "lalang terms."[25]

In eastern Sumatra, the European planters themselves were responsible for destructive soil-mining practices, thanks to their choice of crop: tobacco. Tobacco planters in the Deli district took up shifting cultivation on an industrial scale, rotating huge tracts through tobacco and fallow cycles.[26] When it arrived, rubber "was a godsend for many bankrupt tobacco and coffee plantations" who had exhausted all the easily cleared land.[27]

Chinese planters had in fact been the first to try rubber in the region, developing intercropping and sharecropping strategies in and around Singapore.[28] But the colonizers had no interest in building up a locally owned, small-scale plantation industry: they wanted European capital to take the lead. One FMS official called European planters "the backbone of the country . . . deserving of the utmost encouragement which Government can give them."[29] And planters got more than encouragement: officials gave them reduced rent, licenses to import indentured labor, and even subsidized loans.[30] Planters got privileged access to land along new roads and drainage canals, and they used state-built roads, railroads, and ports to ship their products to the world.[31] The planters did pay for agricultural research through privately funded bodies like the General Association of Rubber Planters on the East Coast of Sumatra (Algemeene Vereeniging van Rubberplanters ter Oostkust van Sumatra, AVROS) and the Rubber Growers' Association (RGA) in Malaya, but government scientists helped out, too. Government research stations imported and studied rubber and oil palm before they were taken up by planters.[32]

Although planters might occasionally regret the devastation they brought to tropical forests, they shared no sense that plantations posed an existential challenge to forest ecosystems.[33] In some places, land seemed infinite: in Pahang, only 2.5 percent of the land was occupied by 1921, at least in the sense of "occupation" meant by colonial record-keepers.[34] Capital and labor, what one planter called the "sinews of war" in the planter's fight against the forest, were the real limits to plantation expansion in this era.[35]

Neither *H. brasiliensis* nor *E. guineensis* was capable of invading Southeast Asian forests on its own. Human labor cleared the forest, drained swamps, planted trees, and harvested produce. Plantation methods demanded gangs of laborers working under strict discipline to achieve maximum efficiency. As one British businessman candidly remarked: "You cannot be efficient without compulsion."[36] Like their African counterparts, in Southeast Asia, "no locals were mad enough to work on plantations on the terms of the planter whilst they retained ownership of their own land."[37] The key, then, was finding people who weren't free to choose. The legal and logistical machinery for supplying such unfree labor in Sumatra and Malaya was already in place: "coolies" from China, Java, and India had been at work across the region for generations, and they had been laboring as far away as the Caribbean since the early nineteenth century.

Coolies were indentured, though the legal contracts that bound them were so flagrantly unfair and frequently violated that many historians call the institution "neo-slavery."[38] Some workers signed up voluntarily, believing the lies told by recruiters or perhaps hoping conditions wouldn't be as bad as returning survivors described. Some were tricked by recruiters or were forced to enlist by creditors in a form of debt slavery. Others were plucked from famine conditions. An unlucky number were simply kidnapped.[39]

Planters regularly withheld wages from indentured workers for food, housing, clothing, medical care, and eventual repatriation. As a Javanese man named Roosman (known to his employer as "Coolie 394") complained, the $2.31 per week he was promised was a fiction, because his employer deducted $1.52 for expenses. Workers who tried to escape were jailed by the colonial government or captured by trackers and punished by the planter himself. Adding insult to injury, runaways had to repay the cost of their capture.[40] Punishments for defiance or shirking could be brutal and sometimes deadly.[41] A bit of doggerel titled "The Labour Question Solved" conveyed the callous attitude of some planters—most of them young European men looking to make a fortune in a hurry—toward their workforce:

> If a cooly gives you cheek, And you can catch the giver,
> To make him civil heave a brick, And dislocate his liver.[42]

Signed "Shakespeare Junior," the piece went on to discuss stomping on genitals and hurling coolies into boiling water.

Even well-treated workers faced dangerous conditions in the tropical environment. Straits Plantation Ltd. reported that fifty-four workers out of 516 on one estate died in the second half of 1911, with another nineteen perishing in the first half of 1912. The firm told shareholders this mortality was "unprecedented in the history of the estate," but the figures weren't extreme outliers.[43] In Malaya, indentured Indians suffered much higher mortality than did the general population, even though most were "in the prime of life."[44] The 20,000 deaths recorded among Indians in the FMS in 1918 (the year influenza struck) amounted to 14 percent of the entire Indian labor force.[45] Across the Straits in Sumatra, conditions could be just as poor or worse. As many as one in four workers died in servitude in the last decades of the nineteenth century.[46] Mortality rates did decline as work shifted away from land clearing, and companies often used promises of medical care to attract and retain workers. But the Dutch didn't abolish the "penal sanction"—the laws that forcibly kept indentured workers on the job—until the 1930s.[47]

The FMS abolished formal indenture for Indians in 1910 and for Chinese in 1914, in part because of harrowing reports of abuse and in part because fast-growing rubber plantations were illegally poaching workers from one another with ever-higher wages. After 1910, many planters turned to contractors known as *kangani* to recruit and discipline Tamil workers from India.[48] More than a million *kangani* recruits arrived in Malaya between 1900 and 1940.[49] The *kangani* paid for workers' passage, creating a debt obligation. Social hierarchies of gender, class, caste, and age from Tamil villages reappeared on the plantation, helping the *kangani* control workers without an indenture contract. The *kangani* might have also purchased debts owed by the recruit, transferring debt bondage from India to Malaya. Chinese recruiters and "lodging house keepers" kept Chinese workers in a similar state of unfreedom, controlling their wages to repay transport, housing, and food costs, often topped up with debts from opium and gambling.[50]

Workers did push back against this abuse. They ran off to rival plantations offering better food or pay, or they stayed in place and resorted to theft, sabotage, and shirking.[51] In 1913, a group of 350 men and women marched forty miles in Malaya to complain to authorities after their manager docked their entire paycheck for food debts. A commentator wrote: "With any class of coolies less tractable and peaceable than Tamils such action . . . would lead to rioting."[52] And coolies—including Tamils—did riot. They protested more boldly when states abolished indenture and penal sanctions on labor. But those with debts to employers often had "little choice but to remain as nominally 'free' labourers on plantations" after their contracts expired. Off the plantation, law and custom segregated people along racial lines.[53] After the end of indenture, many workers had only the freedom to sell their labor to a new employer, not to acquire land and plant crops of their own.[54]

Unfree labor migration had a long tail, leaving plantations with what one official called an "efficient, economical and docile" labor pool long after the most overt forms of coercion ended.[55] As one plantation director noted in 1949, his workforce was "old established—the children and grandchildren of [the founder's] old coolies—and field work is done well and cheaply."[56] Off the plantation, migrants and their descendants who did manage to amass capital provided valuable services to planters as contractors. Contractors could clear forest at much lower costs than could plantation companies, due largely to the inflated salaries paid to European overseers, whose main tasks consisted of surveilling and disciplining laborers (figure 7.1).[57]

By the 1930s, most estates did provide decent medical care and schooling for children.[58] Managers recognized that mistreating workers or firing them when commodity prices slumped was counterproductive, given the difficulty of training for skilled tasks like rubber-tapping and oil-palm harvesting. As one company chairman noted, planters who treated their "coolies" callously risked "losing touch with, and perhaps seriously offending, a clan whose members had been servants of our Estate for many years."[59] And migrants who left India, China, or Java did sometimes find that material conditions on the plantations were an improvement over the places they had left. Yet relatively good treatment didn't erase the power disparities inherent in the labor system. Plantation development hinged on the fact that cheap labor was, in the words of industry experts, "available to almost any extent."[60]

"History Repeats Itself": The Oil Palm Arrives in Southeast Asia

By 1900, the working parts of the plantation complex in Southeast Asia—land, labor, capital, and the colonial state—were ready for the oil palm.

FIGURE 7.1
This cover illustration for *The Planter* (1931), the trade magazine for the Incorporated Society of Planters, depicts a manager posed in a field of tree stumps while "coolies" labor in the background.

Adrien Hallet, a Belgian planter, is usually credited with discovering the plantation potential of the oil palm in 1911. According to the conventional narrative, Hallet discovered decorative "Deli palms" when his car broke down along a palm-lined avenue. Impressed by their large fruit bunches (figure 7.2), he immediately launched a plantation company, and his friend Henri Fauconnier brought the first Deli palms to Malaya shortly thereafter.[61]

The Deli palm acquired a reputation as an exceptional variety. Botanically, the Deli palm is just a *dura* palm, with a thinner-than-average kernel shell and big fruit bunches. When grown in Sumatra, it produced fruit with large, oil-rich pericarps. This was "not the result of deliberate selection, but simply the good luck of the first planters of oil palms in Sumatra," wrote

FIGURE 7.2
Two "coolies" carrying a seventy-six kilogram Deli palm fruit bunch on a Sumatra estate. Photo by J. W. Meijster, ca. 1921–1922, KITLV 107672, Royal Netherlands Institute of Southeast Asian and Caribbean studies, reproduced under CC-BY 4.0 license.

A. A. L. Rutgers, the head of AVROS.[62] Tobacco, tea, sugar, rubber, cinchona, coffee, and other foreign plants had all grown exceptionally well in Sumatra, thanks to "fertile soil, cheap labour and a thorough application of modern science." Rutgers was confident that oil palm would be no different: "History repeats itself," he declared.[63]

The story of the oil palm in Southeast Asia began with four seedlings that came to Java's Buitenzorg (now Bogor) botanical gardens in 1848.[64] Johannes Teysmann, the director of the garden, tersely noted the plant's arrival in 1850: "Elaeis guineensis, received from the Hortus Botanicus from Amsterdam and D.T. Pryce from Bourbon. This palm produces an oil, which is a major commodity on the coast of Guinea."[65] Nothing else is known for certain about the provenance of these four *dura* palms, whose genetic material still dominates the world's oil palm breeding stock. Teysmann added to the uncertainty in 1858 when he described the palms again and said that two came from Amsterdam and two from "Bourbon (or Mauritius)." He thought they might be different varieties, or pairs of male and female trees.[66]

In the twentieth century, scientists like Rutgers puzzled through Teysmann's cryptic notes, concluding that all four palms came from Mauritius or Reunion (the *île de Bourbon*): two straight to Java, arriving in February, and two via Amsterdam, arriving in March. The progeny of the four palms were so similar—and appeared so distinct from African varieties—that many scientists thought all were a special breed, perhaps not native to mainland Africa. A half-century later, scientists argued that Mauritius and "Bourbon" were red herrings, merely indicating a stopover at Reunion for shipments sent from a single site in western Africa via Amsterdam.[67]

These revisionist accounts ignored the man Teysmann credited with the "Bourbon" palms, David Tannatt Pryce.[68] Pryce was a Batavia merchant with a penchant for economic botany. Among his many business interests, Pryce experimented with peanut oil manufacturing and shipped some samples of palm oil to an Australian exhibition in 1862. Notably, Pryce is credited with shipping other species to Buitenzorg from "Bourbon," alongside the oil palms.[69] Oil palms didn't grow wild on either Mauritius or Reunion, but the Mauritius "king's garden" and several private houses in Grand Port (formerly Port Bourbon) had collections of oil palms dating to at least 1837.[70] It's not clear where these palms came from, but the British were enthusiastically distributing seeds for economic plants in this period.[71] It seems likely that Pryce shipped two palms from Mauritius. The origin of the two "Amsterdam" palms remains an open question. It makes little sense for them to have been shipped from Mauritius via Amsterdam; they may well have come from Dutch outposts in the Gold Coast.[72]

None of the four original palms survive today, and genetic analysis of the modern Deli population has so far provided only vague clues about an African origin. The original quartet were prolific parents, however.[73] Thousands of their seeds were planted across Java and outlying islands. Teysmann rejected plans to import more kernels from Africa, noting the abundant fruit resulting from his own uncontrolled hand-pollination experiments.[74] Keenly aware of the growing market for palm oil in Europe's soap and candle industries, Dutch authorities tried to get planters and peasants alike to grow oil palms.[75] While some hoped the oil palm would outproduce the native coconut palm, Teysmann thought it "doubtful" that the newcomer could "stand up to the coconut palm."[76] Coconut oil was a staple food, and those accustomed to eating it thought that palm oil added "an unpleasant taint" to food.[77] Furthermore, farmers and would-be planters struggled to make palm oil economically without machines. Local soap manufacturers bought whatever oil farmers did make, but the NEI government called for an end to new plantings around 1865, and coconut remained king in the NEI.[78]

A series of British experiments with oil palms didn't fare any better. Singapore's Botanic Gardens (founded in 1859) received oil palms in 1870, possibly—but not certainly—from Buitenzorg.[79] Another consignment arrived from Ceylon in 1875; the Ceylon population dated back to at least 1850.[80] In Sarawak, Lady Burdett Coutts had her Quop estate planted with oil palms—probably sourced from Africa—in the 1860s. The "very poor" land was sold in 1872 to a Chinese planter, who grew pepper.[81] A few years later, the governor of Labuan arranged for Kew to send a consignment of kernels from West Africa to a planter on the small island of Daat. The governor thought palm oil "might prove an industry well calculated" for the region's "rough unskilled labor."[82] Planted in 1877, the palms did well, but the owner lost interest and turned cattle loose on the fallen fruit. He eventually replaced the oil palms with coconuts.[83]

A fresh experiment at the Quop estate in the 1880s proved the futility of competing with African palm oil. British officials bought the estate from its Chinese owner, put in 40,000 new seedlings, and started making oil. As an 1891 report noted, "The difficulty now is to find a market for this [oil], as in Singapore it is not enquired for and appears to be almost unknown."[84] Samples sent to Britain were "not particularly well reported on, when compared with Lagos oil."[85] The oil cost so much to make, to say nothing of shipping it halfway around the globe, that it could never compete with African produce.

By 1900, the most notable use of oil palms in Singapore's botanic garden was as a host for ferns, orchids, and other epiphytes.[86] Across the region, planters valued the tree solely for its looks. "Along all the roads I planted the African palm oil tree, which will in time give this clearing a very pretty appearance," wrote one Dutch planter in 1884.[87] In the booming Deli tobacco belt in eastern Sumatra, planters created long avenues shaded by oil palms. They left the fruit to rot and used kernels to pave roads.[88]

The First Oil Palm Boom

Ironically, it was a colonial official in Nigeria who renewed interest in the oil palm in Southeast Asia. Walter Egerton worked in Singapore and Malaya before taking up the governorship of Lagos in 1903. He sent a consignment of palm kernels to Singapore and "suggested to the Governor there planting up some acres." Apparently unaware of the history of oil palm experiments, he told his friend Henry Ridley, head of the Singapore Botanic Gardens, that the oil palm was "a more valuable tree than the coconut—and that is saying a great deal."[89] Soon afterward, the German scientist Paul Preuss paid Ridley a visit. Preuss "was surprised on seeing how well and quickly this plant grew in Singapore." Like Egerton, he assured Ridley "it was a more valuable palm than even the coconut . . . it is the only plant in the world which can with the least possible care, and without diminution of crop furnish a rich harvest for many decades."[90] Egerton wrote another letter in 1908, urging Britain to promote oil palm across the Empire. Even as applications for oil palm concessions arrived on his desk in Lagos, Egerton wrote: "The great product of West Africa is the Oil Palm; that palm should grow well all round the equatorial belt of the world."[91]

Not everyone was enthused about oil palms. The FMS agriculture department accepted more African seeds, but its experts were "a little doubtful as to the suitability of this cultivation for the Native or for the Chinaman," given the "tiring" work of harvesting and extracting oil. They were certain that rubber-crazed planters would only plant oil palms if offered "special rates for land."[92] Scientists at Kew were also pessimistic, arguing that "the West African Palm Oil Industry is capable of a much greater development."[93]

Still, the concession boom in Africa kept palm oil in the headlines. Malayan planters soon asked for their own seeds, and by 1909, the FMS agriculture department was distributing samples from Africa.[94] Harry Darby, a planter in Perak, imported seeds from his brother in Nigeria around 1909. Darby claimed the palms "thrived as if in their own native soil" and "were the means of supplying many nurseries for other ventures" in Malaya.[95] Another planter put in oil palms in Johor in 1910.[96] Across the Strait in Sumatra, Karl Schadt launched a plantation in 1910, following the lead of fellow Germans in Cameroon and Togo.[97] A Chinese planter, Cheong Ah Fie [Tjong A Fie], put in his own oil palms in 1912 at the Rambong Sialang estate in Sumatra.[98] But this trend wasn't a boom yet: far more planters were diving into the coconut business. The same industries driving up the price of palm oil also wanted coconut oil, and the coconut tree was a proven crop.[99]

Though he wasn't the first to attempt commercial cultivation of *Elaeis guineensis* in Asia, Adrien Hallet brought skill and capital to bear when he joined the fray in 1911. Trained as an agricultural engineer, Hallet started his career in the Congo Free State in 1889. He abandoned his employer's plans for tobacco on the island of Mateba and put in oil palms. Hallet managed the small plantation and its oil-making factory for five years before moving on to other work, most successfully with *Hevea* plantations in French Congo.[100] Frustrated by difficulties acquiring land and labor in Africa, he turned to Southeast Asia. He found his footing in the region while helping Belgian and French investors navigate the rubber and coconut booms, expanding his operations from Malaya to Sumatra and French Indochina.[101]

Contrary to legend, Hallet didn't discover the Deli palm by accident. He had intended to cultivate oil palms from the beginning of his pivot to Asia, and his car trouble on a palm-lined avenue in Deli occurred on one of his investigative trips. In June 1911, he toured Sumatra with his friend Henri Fauconnier with the explicit aim of collecting palm fruit, extracting oil, and identifying the most promising candidates for plantation work.[102] Fauconnier recalled: "The results exceeded our expectations. We acquired proof that the *Elaeis* in Sumatra gave much higher yields than on the coast of Africa."[103]

Hallet struck a deal to harvest "avenue palms" on Deli tobacco estates, and he put 50,000 seeds in a nursery at his own Soengei Lipoet rubber estate farther west in Aceh.[104] Before returning to Europe, Hallet bought the Poeloe Radja estate in Deli for a new oil palm company, Huileries de Sumatra.[105] Hallet poached Ernst Fickendey, a leading German scientist in Cameroon, to serve as scientific adviser for the firm. His company directors complained that oil palms were a distraction from rubber, but Hallet forged ahead.[106] Fauconnier brought some of the Deli seeds gathered in 1911 to Malaya and planted them at his own coconut plantation, Rantau Panjang, which was now part of Hallet's plantation empire.

At this stage, far more money was heading into rubber and coconut plantations, and planters knew next to nothing about oil palms and palm oil.[107] Notably, the big European soap and margarine firms declined to join Hallet's venture or launch their own. Lever had already settled on Congo. Jurgens turned down a proposal to invest in what might have been the biggest oil palm plantation ever seen, on up to 30,000 hectares in Sumatra.[108]

The First World War interrupted planting and prevented Hallet and others from buying oil-making machines. The plantations and their scientists nonetheless used the time to study their palms. Fickendey found that oil palms enjoyed unusually good conditions in Sumatra. There was plenty of rain year-round but also plenty of sunshine. In West Africa, the rainy season is cloudy, while the dry season slows growth and fruiting.[109] AVROS published a series of notices in Dutch and English, calling attention to the impressive yields of young plantation palms.[110] One planter shipped Deli fruit to

FIGURE 7.3 Intact and bisected samples of Deli *dura* palm fruit. These samples were shipped to Britain ca. 1922 by Lewis Smart, who hoped to attract investors for mechanized plantations in western Africa. Author photo. EBC355558, Economic Botany Collection, Royal Botanic Gardens, Kew.

Europe (figure 7.3), claiming they were "so magnificent they must be seen to be believed."[111] By the end of the war, high palm oil prices and eye-catching claims about Deli palm yields set the stage for an oil palm boom.

As one handbook for planters put it, oil palms represented "A vast industry; Vast possibilities; Vast opportunities for Error."[112] Investors would have been wise to heed the warning. The postwar oil palm boom (1919–21) was feverish and speculative. Reports from Malaya showed "better returns under expert treatment than any records so far published in Sumatra could show," wildly inflating the prospects of the tree.[113] Plantations selling their first consignments of oil boasted of profits that had "swollen beyond comparison since the outbreak of war."[114] When rubber prices slumped in 1920, investors turned to oil palms for relief. Guthrie & Co., a trading house with big stakes in rubber, tried to coax Lever Brothers into buying new Malayan oil palm estates. The firm assured Lever Brothers that oil palm plantations were "most desirable nowadays—especially if it is undertaken by a concern of the standing of Lever Bros."[115]

Lever Brothers didn't take the bait—and wisely. Palm oil prices crashed from £55 per ton to £30 in 1921.[116] To prop up interest in oil palms, FMS officials offered land on "pioneer" terms, with cheap rent.[117] Though they had reservations about oil palm, British officials worried about Malaya's over-reliance on rubber. They joined officials in Sumatra in welcoming oil palm as new industry. But FMS officials were also wary of speculators. One official warned in 1920, "these applications for large areas for the cultivation of the Palm Oil Tree are really only a cover to obtain more Rubber land," citing the case of a firm in Johor that had gotten a large piece for prime land for oil palms—"granted practically for nothing"—and turned it over to another company for rubber.[118]

Fauconnier and other planters initially tried to hedge their bet on oil palm, asking officials for permission to interplant rubber or coconuts with oil palms, cutting out whichever tree proved least profitable. Officials usually answered with a firm "no," only allowing less permanent catch-crops like tapioca or coffee on land paying discounted rent.[119] Eager to get land near Kuala Lumpur, Fauconnier pledged to only plant oil palms under pain of forfeiture, a condition the FMS imposed on future oil palm concessions.[120] In contrast, interplanting was standard practice on Sumatra. New oil palm plantations like Hallet's Poeloe Radja estate mixed coffee with oil palms, just as the early rubber plantations had done.[121]

Not content with cheap rent and catch crops, planters asked for other subsidies. Hallet's group had the audacity to ask for an interest-free loan from the FMS, matching private capital dollar-for-dollar.[122] Other companies asked for soil surveys, agronomic advice, and free seeds.[123] One manager balked at having to pay a seed bill for $27, claiming that he was doing officials a favor: "we are embarking on an experiment which may be of considerable value to the State."[124] But planters saw plenty of value for themselves in oil palms: the sorry state of the rubber market inspired many to diversify, and oil palm had appealing features. It was widely seen as a "forest" tree that would need little care, based on the assumption that it grew wild in Africa. Workers and neighboring farmers were unlikely to steal palm fruit to eat or sell in the market, as they could with coconuts or rubber. (Latex theft was a surprisingly common complaint.)[125] Applications for land continued to reach British and Dutch officials after 1921, despite lower prices for palm oil. By 1923, FMS agriculture officials griped that the industry no longer deserved "pioneer" rates.[126]

Importantly, planters thought oil palm was immune to competition from smallholders.[127] Smallholder competition plagued the rubber plantations: experts thought their untidy plots "constitute a menace to the industry," harboring disease and flooding the market with low-quality rubber. By the

1920s, a third of Malaya's rubber came from smallholders.[128] Latex could be handled with simple tools, but making edible palm oil seemed to require big machines and big plantations to feed them.[129] One writer crowed, "There can be no native small holding competition, such as that which has swamped rubber, because the small holder can make nothing of it here."[130]

The cost of machinery, in turn, was used to justify big land acquisitions, bigger than officials would have liked in some cases. Hallet's managers argued that one 2,000-acre tract they were offered in Malaya was too small "to be worked economically." Their expenditures on machines and salaries of European staff could "only be justified by a larger acreage."[131] There were efficiencies in scale, but this didn't mean smaller plantations were impossible. The FMS granted several concessions to planters using simpler machines, working a few hundred acres. A 1926 report on Sumatra's plantations stressed that there was no reason that "a plantation industry must oust a more primitive method of production." The author rejected planters' claims that markets wanted only high-quality oil: "the question of quality is of secondary importance; the market wants a great bulk of low or medium-grade produce."[132]

Soils, Seeds, and Science

By the late 1920s, the first wave of oil palm planting slowed as the supply of cheap land in prime areas disappeared. No one was certain about where to turn next: some claimed oil palms would grow anywhere, while others insisted that the tree required "first rate quality" forest land. Many planters began to target logged-over areas for their concessions. Logging provided a double subsidy, as it shifted land from "forest" to "degraded" status, making it suitable for alienation while also leaving behind roads and removing trees that planters would pay to clear anyway.[133]

The apparent versatility of the oil palm—along with notions that some land was unsuitable for rubber—helped companies acquire more land. When Brooklands Rubber Co. applied for 200 acres in Selangor to plant oil palms in 1924, the firm attested that "the land is under virgin jungle and the soil is mainly peat: it is unsuitable for rubber cultivation."[134] Shand Halldane and Co. also saw oil palm as a use for otherwise useless land, getting a 3,000 acre concession in Selangor that was unsuitable for rubber "owing to its hilly nature." The firm later changed its mind, though, and petitioned to plant rubber instead.[135] The giant Harrisons & Crosfield plantation group held only one oil palm estate in Malaya before 1945, on land that turned out to be unsuitable for rubber.[136] Hallet's 1913 application for 2,000 acres in Selangor was denied at first because of labor trouble and denied again on appeal on grounds that the land was too good for oil palm and should be used for rubber.[137] (The "labor trouble" might have been an inability to recruit labor. It might also refer to a strike by 1,500 workers at Fauconnier's Rantau Panjang estate in 1912.[138] Or perhaps it was a 1913 incident on Hallet's Poeloe Radja Estate in Sumatra, where European managers shot two men dead and wounded twenty-two others.[139] "Labor trouble" wasn't uncommon in the plantation zone.)

Before the 1920–21 rubber crash, planters could access credit solely on the basis of their forest leases: forest land was seen as inherently fertile and valuable. As commodity prices slumped, banks withdrew easy credit and forced planters to reevaluate land. Like their Malayan counterparts, planters in Sumatra kept putting in rubber on the best soils, such as the prized *terre rouge*. Replacing rubber with oil palms was out of the question; planters instead used oil palms to fill out undeveloped "reserve" lands.[140]

Much of what Sumatran plantations held in reserve was lalang grassland, the legacy of tobacco cultivation. Initially experts thought it was worthless; the "mere presence [of lalang] testified to the sterility of a landscape."[141] It might take 250 years for primary forest to reclaim lalang fields.[142] Once planters demonstrated that oil palms could grow on lalang grassland, however, the pace of expansion surged. Sumatran planters could clear grassland and scrub much faster than their Malayan counterparts could hack into the primary forest.[143] In the long term, fighting lalang grass would not be cheap, requiring labor, tractors, and chemical herbicides. Chevalier correctly warned that palms planted on grassland would run out of nutrients much faster than palms planted on forest soils.[144]

Peatlands—a source of huge controversy today—presented different challenges to colonial planters. Peat, an agglomeration of wet, slowly decomposing plant matter, could be meters deep in tropical forests. When peat swamps are drained, most crops, including rubber, can't tolerate the loose, acidic soil left behind. Unfortunately for planters in Malaya, much of what was left in prime agricultural districts by the late 1920s was "peaty swamp, not yet drained."[145] Around 1920, researchers found a group of oil palms thriving on peatland more than 1.5 meters deep in a corner of the Bukit Rata plantation in Sumatra. The shallow-rooted palms toppled over, but they could still grow upward from the prone trunk as the peat soil settled.[146] The findings helped convince experts in Sumatra and Malaya that shallow peat, once drained, was suitable for oil palm.[147] Planters turned to peat out of necessity, using oil palms to develop swampy corners of their concessions.[148]

In these conversations about land, Europeans treated forests as if they were vacant. As in Africa, European experts were slow to recognize the ecological dynamics of forest fallow, seeing it as a menace to valuable timber. Under colonial law, forest peoples in Malaya and Sumatra had no way of claiming land for future use in shifting cultivation. FMS officials blamed these communities for creating big patches of lalang grass; officials in Sumatra made similar complaints. One report on a Pahang oil palm estate alleged that the soil was "not very fertile and has in the past been subjected to ill treatment by the Sakais [an indigenous group]." Only the "good management" of planters had made the palms thrive.[149] In another case, the United Sua Betong Rubber Co. walked away from 1,000 acres in Selangor over the issue of indigenous occupancy. Officials hoped to relocate the twenty-six–person community to the Kelambu Forest Reserve, but the group was reluctant to abandon its five-year-old rubber trees, along with fruit and coconut trees. The district officer offered to carve out fifteen acres occupied by the community from the concession, but the firm wasn't interested in sharing.[150]

Though the rights of forest peoples didn't come up at all in most concession applications, this didn't mean that they weren't affected by plantations. Instead of staying and contesting claims to their land, many groups retreated deeper into forests, often taking up less-productive land to avoid conflicts with planters, along with tin miners and Malay and Chinese smallholders who were pressing into interior regions.[151]

Choosing the Right Seeds

Contrary to the popular narrative, Hallet and the rest of the planting class weren't confident in the Deli palm, whose plump fruits had sent investors rushing to Sumatra and Malaya in 1920. Experts cautioned that while the Deli palms were good *dura* specimens, their best virtue was uniformity, thanks to their limited parentage.[152] Still, that uniformity was exaggerated. As an AVROS scientist noted, visitors to plantations were "immediately struck with the great amount of variation in the trees," including malformed and sterile specimens.[153] The fabulous Deli palm yields that Rutgers reported from Sumatra in 1918–19 came from a group of trees ten to fifteen years old, "and were obtained the first year after cleaning [pruning] the trees and for six months under the influence of artificial pollination." "These figures cannot be taken as normal," Rutgers warned.[154]

Yet Rutgers and others promoted these numbers as if they were normal for the Deli *dura*.[155] A French expert, Yves Henry, complained that Rutgers's estimates "remained an article of faith in Sumatra; they are cited in all places, they are provided for information purposes and serve as the base for calculating the production of mature plantations." Actual plantation yields were lower than Rutgers's figures and showed worrying declines over time; to Henry, they seemed no better than a "good African palm grove in adulthood."[156]

As the excitement of the first years of large-scale planting wore off, critics complained that the Deli palm's yield had been wildly overstated. "Interested experimenters, with the aid of a pocket knife, a pair of scales and a few selected fruits have persuaded themselves into believing that Malayan grown oil palms produce fruit yielding double the amount of oil given by oil palms grown in West Africa," scoffed one critic.[157] Deli palms did yield more oil than did African palms on average, but they made smaller kernels. Much of the difference in oil yield was down to rainfall and the fresh nutrients available in virgin soils, rather than to genetics.[158]

And the genetics of the Deli *dura* worried Rutgers, the palm's leading publicist. Were the 1848 palms really *E. guineensis* at all? Were they randomly chosen, or a hand-selected variety? What valuable traits might the palms be lacking? And what problems might lurk in their inbred gene pool? Even the origins of the St. Cyr tobacco estate "avenue palms," which provided much of the seed for Deli plantings, were uncertain. The St. Cyr manager said they came from Singapore (and therefore from either Buitenzorg or Ceylon), though several Dutch experts insisted they came from Java.[159] Despite all these questions, Rutgers and AVROS urged companies to avoid importing seeds from Africa on their own. The Deli palm was "something of more than average value" and was good enough, until researchers could identify a better palm.[160]

Deli *dura* promoters sought to breed an even better palm out of the Buitenzorg quartet's offspring. Palm no. 249 at Medang-Ara was one early star, making eleven heavy bunches per year. If it bred true, it promised six tons of oil per hectare.[161] British experts in Malaya were initially skeptical of Deli *dura* breeding, however. They warned against "allowing estates to plant up with the local immature seed from Sumatra," preferring to source seed from Nigeria instead. These scientists argued that the longevity of oil palm rewarded caution: "A palm oil tree is at full maturity at 80 years old and has a life of 120 years so it would pay to be careful when starting such an industry."[162] They did not yet understand that harvesting, rather than yield per tree, was the main factor in profitability. The FMS ordered a million Nigerian seeds in 1920, but poor germination rates forced the administration to supply plantations with Sumatran seeds, restricting new African imports to experimental work.[163]

Fauconnier took seeds from Deli to plant at his Rantau Panjang plantation in 1912.[164] But his company reported in 1915 that it had "obtained from the Congo 26 bags of oil palm seeds which we are now putting down in nurseries," probably at what was to become the Tennamaram Estate.[165] Tennamaram Estate's manager told colonial officials that "as far as I know," his trees came from Sumatra,[166] but another manager contradicted this. He wrote: "We received the seeds from the Congo (Belgian)."[167] When an official toured Fauconnier's estates in November 1920, he gave a glowing report of palms up to seven years old, "grown from seed imported from the Congo." He observed that "they bear quite as heavily as palms of similar age in Sumatra."[168] This decision to plant African seed alongside Deli seed was consistent with what Hallet was doing on Sumatra. On what were to become the estates of the plantation conglomerate SOCFIN (the Société Financière des Caoutchoucs), Hallet had huge consignments of Congolese seed planted. By the late 1920s, SOCFIN boasted of having "thousands of hectares established with seeds imported directly from Africa."[169] African palms were grown in "mixed" plantations with Deli palms, and it was likely that the Deli population was "contaminated with characters inherited from this material."[170]

SOCFIN's experts knew it was absurd to base a breeding program on four specimens. They happily sold selected Deli seed to other planters—becoming the largest distributor of oil palm seeds—but their goal was the oil-rich, thin-shelled *tenera* palm, which Hallet knew from experience in Africa.[171] Rutgers also saw the appeal of *tenera*. He wrote to Kew in 1921 asking for "some less common varieties" of oil palm and emphasized that he wanted to work with *tenera*.[172] But *tenera* proved elusive (see chapter 5), and it wasn't successfully bred for commercial plantations until the late 1940s.

In the end, more money could have been made if SOCFIN and other firms had stuck with the Deli palm.[173] However, British experts complained that the early Sumatra data were "practically valueless."[174] There wasn't enough knowledge (at least among Europeans) about oil palms for the kind of scientific planting advocated by both colonial officials and plantation companies. It didn't help that AVROS carried out its tests in secret on private estates, solely for the benefit of Sumatran planters.[175] Anecdotal reports and rumors fueled enthusiasm as well as skepticism throughout the 1920s.[176] Eminent Dutch researcher P. J. S. Cramer complained about the secrecy of AVROS, noting that in Africa, oil palm research was "more multilateral" and widely shared, giving much more information than reports on yield per hectare.[177] Outside of Hallet's SOCFIN group, the big plantation companies remained wary of shifting too many resources from rubber into oil palm, reflecting this broader uncertainty about the viability of the oil palm as a plantation crop and the future of palm oil as a commodity.

"A Machine for Converting Fertiliser into Oil": The Maturing Plantation Complex

The key claim underpinning the spread of plantations across Sumatra and Malaya in the 1920s wasn't that they could make more or even cheaper palm oil than Africans could: it was that they could make *better* oil. Like the concessionaires in Africa, the planters were sure that modern machines were the key to making cheap, high-quality oil.

The early results did not impress. The first Sumatran plantations made rather mediocre oil and left nearly 25 percent of it unextracted, marginally better than African manual methods.[178] Plantations were secretive about their machinery, but a British visitor concluded in 1921 that it was "so unsatisfactory that there is no point in making any recommendations based on what was seen there."[179] But as the plantings of early 1920s boom matured, new hydraulic presses and centrifuges arrived. Only the best-capitalized estates could afford them: Sumatran planters were "keenly jealous" of British capitalists in Malaya, who spent lavishly on the latest novelties.[180]

The planters' preoccupation with palm oil meant that kernels sometimes went uncracked.[181] The U.S. market, which became the main outlet for Southeast Asian palm oil, had little interest in palm kernels. Coconut from the Philippines enjoyed preferential tariffs in the United States, and shipping costs from Southeast Asia to Europe were much higher than from Africa. Deli fruit was approximately 33 percent oil and 8 percent kernel by weight, but Asian kernel exports always fell short of this ratio. On top of the shipping disadvantage, the machines used for cooking palm fruit stained the kernels, lowering their value. Some Sumatran estates shoveled kernels into their boilers as fuel into the 1930s.[182]

If Sumatran firms rushed to plant trees without getting the machinery right, Malayan companies had the opposite problem: British experts insisted that efficient machinery was the sine qua non of an oil palm plantation. One columnist attacked this "Gilbertian notion," comparing it to "[spending] all one's money on a churn without the wherewithal to make butter."[183] Yet FMS officials had a history of favoring high-tech solutions for rubber processing, and they applied the same thinking to oil palm. They encouraged planters to buy complex centrifugal oil mills, a technology demonstrated at the 1924 Wembley Exhibition with Deli fruit shipped in the for the occasion.[184]

Most of the Sumatran estates opted for hydraulic presses built by Krupp or Stork, squeezing out oil with brute force rather than the clever physics of the centrifuge. Fickendey worked with Krupp to develop a process that wrung 99 percent of the oil out of palm fruit with two pressings and a bath in benzene solvent.[185] Yet planters were happy to get oil with 7 percent free fatty acid (FFA) out of this process, no better than African soft-oil techniques.[186] The cost of working with solvents—and workers's fears of suffocation and explosions—meant that most firms abandoned the method.[187]

Plantation managers did gradually drive down FFA levels, revealing the true advantage of the plantation method: speed and efficiency in harvesting fruit. Oil palms grown in open fields develop thick, squat trunks instead of sending their stems skyward to fight for sunlight. Plantation palms can be harvested from the ground for up to twenty years, eliminating the slow and dangerous climbing process. Just as important, coolie labor gave planters a cheap and disciplinable workforce. A Javanese coolie would cut anywhere from thirty to seventy bunches a day, his labor costing the same as an African palm-climber delivering twelve bunches.[188]

Harvested fruit reached the factory faster on a plantation, avoiding the fermentation that caused high FFA levels. Roads and tramways swiftly carried fruit to central factories, where steamers—and later, pressurized sterilizers—cooked the fruit to perfection.[189] SOCFIN saw its oil FFA content fall to 5 percent after the first sterilizers got to work in 1930, and by 1939, it was down to 2.8 percent, comparable to values for today's plantations.[190]

Machines couldn't keep oil palms healthy, though. Unkempt Deli palms on poor soil weren't better than any other variety. As one critic noted, "Such palms are to be seen almost everywhere in Sumatra . . . bearing no fruit at all worth mentioning and presenting not even an ornamental appearance."[191] Some planters believed that aggressive pruning stimulated fruiting; others thought pruning harmed the tree. Malayan planters had poor results with their early palms because of overly cautious pruning. A visitor taught them Nigerian methods in the 1920s, which spread by word of mouth.[192] Hallet recognized that pruning and cultivation practices were key to good yields. He noted that "Asia has not yet learned how to cultivate the tree as it is known in Nigeria, Dahomey and on the Ivory Coast." Hallet pointed to Africa, not Sumatra, as a guide to best practices, praising "the care that the natives of Africa consecrate to the Elaeis."[193]

The more fundamental health problem was with the soil. Early planters held the old idea that the oil palm was a "forest" tree, not recognizing the importance of rotational fallow in building soil fertility in Africa.[194] Planters hoped that green manures and compost from leaves and stems would be enough to maintain fertility.[195] But their trees quickly exhausted what geographer François Ruf has termed the "forest rent": the nutrients accumulated over hundreds of years in forest soils.[196] Plantations exported these nutrients halfway around the world in the form of palm oil and kernels, and more disappeared as smoke up the chimneys of oil factory boilers burning fruit bunch stems and kernel shells.[197] Sun exposure and erosion took even more nutrients away. As early as 1920, the FMS Agriculture Department warned that formerly-forested soil was "losing its virginal fertility with such rapidity that artificial compensating elements must shortly be resorted to on a large scale."[198] Within a decade, all oil palm planters turned to artificial fertilizers to make up for the loss of "forest rent."[199] A 1943 report warned that even "on the richer soils of Sumatra[,] fertilizers are now in general use." In the view of many planters, the oil palm was now merely "a machine for converting fertilisers into oil."[200]

"Clean weeding," an obsession of plantation managers, didn't help the nutrient situation. Topsoil ran off weeded fields, clogging rivers across Malaya and Sumatra. Herbicides used to wipe out lalang grass—which often popped up on clean-weeded fields—killed off useful plants as well as people, cattle, and just about anything else exposed to it.[201] While many rubber planters stubbornly ordered coolies to clean-weed into the 1930s, oil palm planters realized the mistake and turned to leguminous cover crops. An early favorite was *Mimosa invisa*, a spiny plant that shielded the soil while replenishing nitrogen. AVROS and the FMS followed *Mimosa* experiments with studies on leguminous genera like *Centrosema* and *Pueraria*.[202] They conclusively proved that clean weeding, however aesthetically appealing, was ecologically unsound, a lesson demonstrated in untidy but healthy rubber plots managed by Malay and Chinese smallholders.[203]

No longer drawing on the fertility of Southeast Asia's forests, the plantation complex consumed material from far-flung ecosystems. The "guano islands" of the Pacific provided much of the fertilizer, notably phosphates.[204]

When those islands were stripped bare and abandoned as desolate moonscapes, miners turned to rock phosphates in northern Africa and other parts of the world, further expanding the ecological footprint of the plantation.

High oil yields also hinged on keeping trees free from natural threats. Rutgers cheerily asserted that "very few tropical plants are so free of pests and diseases as the oil palm," encouraging the early planters. Yet he understood that plantation monocultures were "unnatural crowds," creating risks.[205] No plague like rubber leaf blight hung over the oil palm in Africa or beyond, but the tree did encounter all sorts of problems in its new Asian environment. Wild pigs, rats, squirrels, and porcupines developed a taste for young palms and their fruit. One Sumatran estate lost 10,000 seedlings to porcupine raids over ten days; pigs ate another 40,000. Historian Susan Martin documented a long struggle against rats on one Malayan plantation that included a failed cat-breeding project, imported Highland terriers, and a bounty for rat tails (which inspired an enterprising contractor to breed the rodents in cages).[206] Cobras were unwelcome surprises for workers in the crowns of palms, but they were "voracious eaters of rats," and planters often left them alone.[207]

Elephants were a more imposing threat, smashing through electrified fences and even filling trenches with dirt to get at tender palm stems.[208] Elephants destroyed 2,137 palms in a single raid of a Pahang estate in 1932.[209] Tigers also stalked plantations, targeting fruit cutters working alone in long, shady rows of palms. The cats also attacked transport animals like mules and bullocks. One Kelantan estate reported in 1939 that most of its animals had been killed by tigers, the few survivors bearing scars from tiger claws.[210] No mammal posed an existential threat to the plantation complex, however: humans could trap, poison, or shoot their way out these problems.[211]

Smaller creatures were more challenging foes. By 1928, one expert reported that the "list of insect pests in Sumatra has already passed that of the insect pests of oil palms in West Africa."[212] Insects preying on the native coconut palm found the oil palm equally tasty, and a few insects that ignored coconuts attacked oil palms.[213] Planters resorted to chemical warfare, dousing the palms—and their workers and everything else in the area—with toxic pesticides.

Ironically, a lack of insects working as pollinators also caused problems for plantations. Poor fruiting rates had been noted as early as 1858 in Java. Teysmann successfully hand-pollinated his first four palms, but the practice was forgotten: ornamental palms didn't need to bear fruit. The idea was "rediscovered" in the 1900s "and spoken of as an entirely new thing."[214]

Fickendey, working for Hallet in Sumatra, was aware of the importance of a weevil (*Elaeidobius kamerunicus*) in pollinating palms in Africa. He found "no insect to take its place in the East." French researchers confirmed that "pollination is almost entirely by insects" in African palm groves.[215] Yet experts warned against introducing the weevil to Asia: it would "be extremely dangerous . . . owing to its habit of destroying the flowers: no insect should be imported until the fullest investigation had been made into its intimate life-history."[216] The African weevil would have to wait until the early 1980s for its first foray into Southeast Asia.

Rutgers saw no alternative to hand-pollination. "It is obvious that the natural pollination of oil palms [in Asia] is far from satisfactory and is indicated by the low percentage of fruits on the bunches," he declared.[217] Workers shook anise-scented pollen from male flowers into bags, and then squeezed a rubber bulb to dust receptive female flowers (figure 7.4). They marked flowers with a small dot of paint to prove to overseers that the task had been done.[218] Hand pollination was so successful and pervasive that some experts later insisted that oil palms had "undoubtedly" evolved for wind pollination, with insects playing an incidental role.[219]

Like harvesting, pollination relied on cheap labor and the "new nature" created by the plantation: short palm trunks. As palms grew taller, hand-pollination and harvesting became increasingly costly and dangerous.[220] In 1954, a quarter of all injuries on Malayan oil palm estates were caused by falls from trees. Artificial pollination had another cost: it stressed palms and accelerated the loss of soil nutrients. Both consequences caused disease. By the 1920s, plantations catalogued "stem rot," "bud rot," and "crown disease," among other ailments. The toll taken by diseases steadily grew: in 1933, one Malayan estate lost 40 percent of its crop to "fruit rot."[221] "Crown disease" (probably today's Common Spear Rot) was widespread in Sumatra as early as 1913 and plagued estates across the region. The cause is still debated, but the inbred Deli population is particularly susceptible to it.[222]

Fungal infections, especially by *Ganoderma* species, emerged as a long-term threat to oil palms in Southeast Asia. Once established in the soil, fungal diseases proved impossible to eradicate. Coconut stands often hosted *Ganoderma*, and replanting with oil palms led to new attacks.[223] The fungus also grew in Africa, where Lever Brothers gave early warning about deadly fungal attacks on its "improved" Congo palm groves in 1915.[224] Fungal outbreaks rarely threatened more biodiverse palm groves in Africa, and African farmers valued fungal species growing on living and dead oil palms as foods.[225]

FIGURE 7.4 Left: two men collect pollen from a Deli *dura* palm, shaking a male flower over a funnel. *Oils and Fats*, British Empire Exhibition Malayan series no. VI (1924), facing 18. Upper right: Photo of male (left) and female (right) oil palm flowers. Lower right: a pollen-collecting funnel (right), and a tool for spraying pollen (left). When squeezed, a rubber bulb fitted over the glass bottle gently puffs pollen out of a narrow tube. Rutgers et al., *Investigations on Oilpalms*, plate XII and fig. 3.

Humans fell victim to new diseases created by plantation agriculture, too. Beyond the usual threats from malaria and other mosquito-borne diseases, palm plantations created outbreaks of "scrub typhus." Rats carrying *Orientia tsutsugamushi* bacteria colonized the margins of new plantations, feasting on palm fruit and workers' rations. They passed the bacteria to a species of chigger, whose population exploded in grasses that grew around young palms. The chiggers infected workers carrying out weeding, pruning, and harvesting with the bacteria. One Malayan estate recorded sixty-eight cases and ten deaths in 1929. This ailment disappeared with time, however: maturing palms cast more shade, suppressing the grasses favored by chiggers.[226]

Despite a growing catalogue of problems, the acreage covered by oil palm plantations steadily grew in the years up to 1941. Planters could solve or abate nearly any problem with more land, more labor, more chemicals, and machines.[227] Yet oil palm plantations were far from inevitable or unstoppable. Planters put far more land under rubber than they put under oil palm, and investors generally had little enthusiasm for oil palm in both Malaya and Sumatra up to the Second World War.

The Sumatra Menace?

Starting from nothing, exports of palm oil from Sumatra and Malaya did grow quickly throughout the 1920s and 1930s. Alex Cowan, who later ran UAC's plantations in Africa, warned that "the West African forest palm as it is to-day cannot hope to hold its own" against this threat.[228] But the "Sumatra menace," as some officials took to calling the Southeast Asian oil palm plantation industry, stalled in the midst of the Depression. The region overtook Africa in palm oil exports by the end of the 1930s, but this was because Africans responded to pitifully low prices by selling less oil.[229]

Capitalist planters had no choice but to make oil. They had to generate returns on their investment, and oversupply sent prices downward as more and more hectares of palms matured. By 1937, palm oil sold for less than 40 percent of what it had fetched in 1929 (graph 7.1). Several Sumatran estates went bankrupt. A Johor estate idled its workers and machines for the entire 1934–35 season.[230] Hallet's Huileries de Sumatra lost more than a million francs in 1931.[231] Even a well-managed plantation group like United Plantations was lucky to stay afloat, paying tiny dividends to angry shareholders.[232]

Planters also faced the same new monopoly as African smallholders did in 1929: Unilever. "I'm very much afraid that you are all now at their mercy," warned one writer. "If I had the slightest idea that such a combine was possible, I would never have advocated oil palm planting," he wrote.[233] The chairman of the Harrisons & Crosfield plantation group agreed: he "reportedly declared that only a fool would go in for vegetable oils while Unilever dominated the market."[234]

If Unilever wasn't enough of a threat on its own, planters also had to contend with a chaotic market full of tariffs, quotas, and new entrants like hydrogenated whale oil.[235] British firms, including Unilever, bought six times more whale oil and ten times more soybeans in 1934 than they did in the late 1920s, taking advantage of a "race to the bottom" in prices for oils and fats. The plantations got some relief from the United States, whose manufacturers bought up most Sumatran oil when it arrived on the market. But the United States imposed harsh tariffs in 1934 to protect domestic producers. At the prices prevailing by the late 1930s, these tariffs amount to nearly

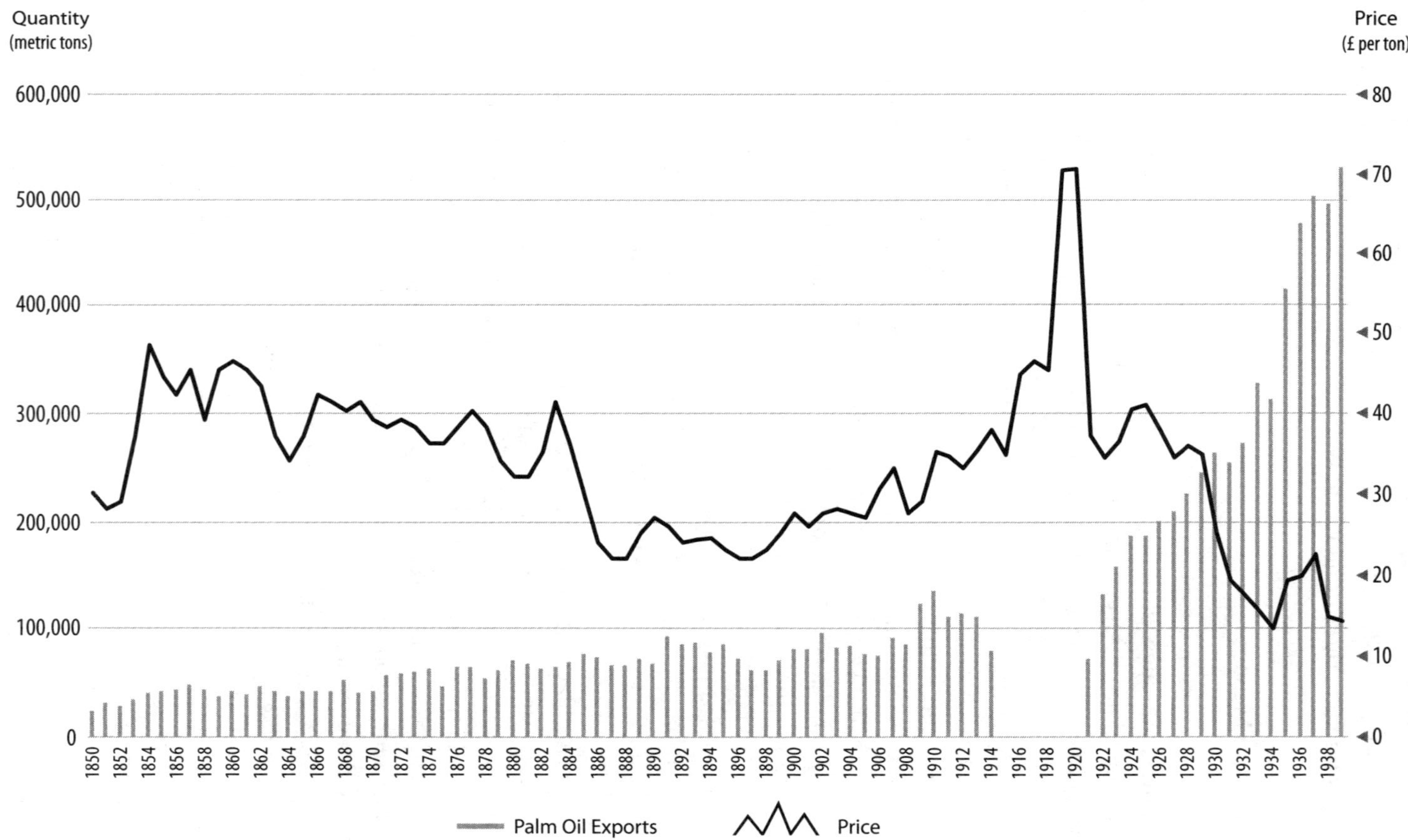

GRAPH 7.1 Global trade in palm oil and prices, 1850–1939. Data to 1914 show imports (metric tons) to Europe only; U.S. and Brazilian imports not shown. Data after 1921 shows exports (metric tons) from producing countries. Prices are UK averages (£ per ton). Data from Jacks, O'Rourke, and Williamson, "Commodity Price Volatility"; League of Nations, *Statistical Yearbook* (1922–1939), and graph 4.1

100 percent of the market value of palm produce.[236] On top of this, drought in the American Midwest pushed farmers to carry out "the wholesale execution of starving cattle," gutting worldwide markets for palm kernels and other oilseeds.[237] More than 60 percent of the Sumatran palm oil sold in America wound up in soap or tinplate, with the rest going to shortening and baking fats, and this only because drought had sent prices for cottonseed oil and other domestic fats skyrocketing. Edible, low-FFA palm oil was the cheap substitute, not the premier commodity.[238]

Nevertheless, the plantation system survived. By 1937, Hallet's *Huileries de Sumatra* was paying handsome dividends again.[239] Though firms benefited from a modest recovery in palm oil prices, ferocious cost cutting ultimately made the difference. The Depression gave firms an excuse to slash wages and employment, including overpaid European managers. But personnel cuts fell hardest on the coolie class. Malayan plantations "repatriated" a third of the entire Indian workforce. The eastern Sumatra coolie workforce fell from 336,000 in 1930 to 160,000 by 1934. By this point, the Dutch had abolished indenture and all the paternalistic trappings associated with it, freeing planters to push penniless workers back to Java. A "careful selection process assured that only the most trusted, docile, and hardworking married men and women were retained." Lucky to have any work at all, the remaining laborers submitted to strict discipline.[240]

Malayan planters formed a Palm Oil Producers' Association in 1932 to press their interests with the colonial government. By 1934 it was asking for low-cost or zero-interest loans and export bounties. Some companies pleaded that they "planted no other crop." The imperial government agreed to encourage palm oil consumption and cut land rents, but it warned that planters would "ultimately be required to contribute to the cost of research" to sustain the industry.[241]

Some partisans of the African industry predicted the demise of their plantation competitors, noting that in Africa, "the native farmer . . . from empirical experience alone, has evolved in the course of generations a very efficient system of palm culture, and the yields of fruit which they can obtain are comparable with, if not quite equal to, those of the Sumatra plantations."[242] Frank Stockdale, Britain's top colonial agricultural expert, didn't bother visiting an oil palm estate during his 1938 tour of the FMS. He argued that plantation oil palm was a poor choice, given declining soil fertility and the success of smallholders with rubber. Oil palm was among several crops he urged smallholders to adopt.[243] The official *Malayan Agricultural Journal*, which had trumpeted the arrival of state-of-the-art centrifugal extraction plants in the 1920s, began carrying advertisements for the humble Duchscher screw-press.[244]

Though local disgust at the idea of eating palm oil held back smallholder adoption of oil palm, by the late 1930s, officials began to see oil palm as a viable smallholder crop for the region. A few plantations invited smallholders—who had taken up oil palm on their own initiative—to sell fruit to their mills.[245] The most important justification of the plantation system, that only big oil mills supplied by big plantations could give the world market the edible fat it needed, was falling apart.

8 From Colonialism to Development

It was the aftermath of the Second World War, rather than the war itself, that brought the next wave of momentous changes to the oil palm industry. By 1945, it was clear that the world needed much more food.[1] The situation for fats was especially dire, a "world famine position" in the view of one official.[2] Privately owned plantations in Asia had a lot of land under oil palms, but the Japanese occupation devastated infrastructure across the region. Although the sub-Saharan African colonies didn't suffer directly from enemy action, officials in Britain's Ministry of Food fumed that colonial administrations hadn't squeezed more palm oil out of Africa. One declared his conviction that colonial officers "will be unable to prove that they have done anything at all."[3] The charge ignored the strenuous efforts of African men, women, and children to make oil and "crack kernels, crack Hitler," as one propaganda campaign urged.[4] But he had a point with regard to colonial policy: mechanization and "grove improvement" projects across Africa had accomplished little in the British Empire. The French record wasn't much better. Africans sold huge amounts of palm kernels and oil, but the quantities didn't seem likely to grow fast enough to meet postwar demand.

Before the war, most colonial policymakers held that the best policy allowed "the native 'to develop on his own lines.'" Lever's right-hand man in Africa, T. M. Knox, argued that these policies were rooted in "nostalgia for the supposed simplicity of native life." In his view, they "condemn the African to live for ever in the comparative poverty which is still his lot."[5] Some officials took blatantly racist positions, accusing Africans of "unwillingness to adopt improved agricultural methods."[6] Sir Alan Pim held up Belgian Congo and Huileries Congo Belge (HCB) as exemplars, down to the policy of compulsory cultivation. "It is for us to take [the African farmer] by the hand, to say 'take your hoe' and 'work your field,'" Pim wrote.[7] He criticized colonial policies on land tenure and plantations, declaring that "native institutions must be reformed, not preserved like a holy of holies."[8] Nigeria's wartime governor made a similar complaint, writing that the "unimpeachable principle" of respect for local land rights had "been diverted into almost fanatical fetish worship."[9] Officials more concerned with European margarine prices than colonial welfare demanded "bold experiments" with plantations and more mechanization.[10]

Colonial administrators pushed back, insisting that new projects deliver benefits to farmers, in addition to meeting export demand.[11] The Nigerian palm producer "never had a price which did not condemn him to a very low standard of living," complained one officer. "[It] cannot seriously be suggested that the peasants of Eastern Nigeria, whose standard of living is one of the lowest in the Empire, should subsidise foodstuffs in the United Kingdom," he wrote.[12] African elites joined the chorus warning that colonized subjects needed to see real economic improvements. Returning servicemen were a particular concern. Disillusioned soldiers could—and as it turned out, did—foment anticolonial politics.[13] The need for change in Asian colonies like Malaya was just as clear, a fact reinforced by a communist insurgency ("The Emergency") which lasted from 1948 to 1960.

Britain and France adopted a try-everything approach to oil palm after the war, supporting plantations where it was feasible and working with farmers where it wasn't. New institutions like Britain's Colonial Development Corporation (CDC) and France's Fonds d'Investissements pour le Developpement Économique et Social (FIDES) promised to invest in the colonies, setting aside huge sums for edible fat projects.[14] As the names of these organizations suggested, postwar policymakers rebranded and repackaged colonialism as a new thing: development.[15] The tools and policies they created to revitalize the oil palm industry shaped oil palm landscapes across three continents for the next half-century.

The Postwar Plantation in Southeast Asia

The Southeast Asian plantations were in a sorry state by 1945. Most Malayan planters joined British troops retreating from the Japanese invasion in 1942, destroying equipment as they fled. Many of their counterparts in Sumatra stayed put and found themselves interned in camps for the remainder of the war. In both places, Japanese managers forced "coolies" back to work. Rubber was their biggest concern, but the occupiers did see uses for palm oil. They urged local people to use palm oil in soap and grease, and as a substitute for diesel fuel. One of the occupation government's least popular moves was its promotion of "Red Medical Palm Oil," unrefined oil rich in vitamin A. Everyone from schoolchildren to prisoners of war were forced to take a spoonful of this stuff once a day, inspiring lifelong revulsion toward red palm oil.[16]

Japan's surrender in August 1945 spared Malaya and Sumatra from what was sure to be a destructive Allied invasion. The British quickly reoccupied Malaya, and planters returned to their estates. Most expected the government to compensate them for destroyed machinery, and companies asked for reconstruction grants and the use of surrendered Japanese troops as forced labor. They also expected to be rewarded with high prices for their oil and kernels when production resumed.[17]

They got none of these things. Britain's Ministry of Food commandeered every bit of fat the plantations could produce, paying a fixed price below market rates. The only concession from the government came in the form of low-interest loans.[18] Colonial administrators didn't want the plantations to fail, but they weren't keen on helping them expand, either. Officials denied new land applications, and the CDC flatly refused to lend to planters. Officials "served British capitalism badly," noted one historian, but that was because their main concerns were Malay peasants and a growing nationalist movement.[19]

The plantations did recover. Unilever bought a large Malayan estate in 1947, noting that the Ministry of Food wanted "to see British [companies] exerting their influence in promoting an increased production of oils and fats."[20] Other estates planned for gradual expansion into their reserve lands into the 1950s, undeterred by the start of a communist insurgency in 1948.[21] Planters complained about low fixed prices offered by the Ministry of Food, but they were nonetheless caught off guard when Britain ended its monopoly buying system in 1953. A cooperative pool for selling oil, formed before the war and used to manage sales to the Ministry of Food, collapsed in 1957 as prices fluctuated.[22]

Sumatran estates were in worse condition than their Malayan counterparts. As the war progressed, Japanese authorities ordered workers to plant food crops, cutting down older trees to make room. Hungry workers tore out young trees on their own initiative. About 16 percent of Sumatra's oil palms had been cleared by the end of the war.[23] Japan's surrender prompted Indonesian nationalists to declare independence, sparking a bloody conflict with Dutch forces trying to reimpose colonial rule. When Europeans returned to their estates, they found that the "old social order was destroyed." Nationalist and communist organizers had built new "labor unions, peasant organizations, and political parties, all prepared to challenge the rights of the planters."[24]

When the Netherlands conceded that Indonesia had won its fight for independence in 1949, the planters found their position even weaker. Politicians on the right and left called for the expulsion of foreign capitalists and the expropriation of plantations. Moderates in the new government took a careful position, hoping to appease peasants and workers without scaring away the foreign capital needed for reconstruction and development.[25]

Sumatran labor unions organized slowdowns and strikes, letting palm fruit ferment and spoil on many estates.[26] In 1952, the government tried to remove workers squatting on the Tanjung Morawa tobacco estate (the site of Hallet's famous car trouble), provoking protests and riots. People continued seizing land throughout the 1950s, even diverting irrigation channels and flooding oil palm estates to water their food crops. Ann Stoler's interviews with former estate workers show that many turned to squatting "to break their ties altogether with the [plantation] companies."[27] Official patience with strikes and squatting dwindled with the nationalization of Dutch properties in 1958, however. Now the government was a major owner of plantations—and thus became an opponent of organized labor and squatters.[28] Moreover, many unions and peasant activists were aligned with the Communist Party, attracting the suspicion of centrists and right-wingers in the government. The bloody 1965–66 anti-communist purge wiped out much of what was left of Sumatra's militant labor organizations.[29]

If political woes were not enough for plantations in Asia, experts raised alarms about spreading fungal disease. One planter admitted in private that "the biggest problem" at his estates was *Ganoderma,* which "cannot be eliminated from the ground and is carried over to replantings."[30] A 1946 report warned that there was no cure for *Ganoderma* and that "the land is rendered unsuitable for growing oil palms in the years ahead." In this new environmental reality, the economic life of an oil palm plantation might only be twenty years, forcing planters to rethink their methods. Even worse for planters, experts started to identify more fungal threats from *Fusarium* species.[31] One type of *Fusarium* caused "Panama disease," a blight that was steadily wiping out Latin America's banana plantations and had jumped to Africa.[32] The most practical solution for both fungi was to destroy the trees and plant something else.

To escape disease, planters needed more land. Malayan estates continued planting up their reserves, but they found that the government (especially after Malaysia's independence in 1957) wasn't eager to alienate new tracts for foreign companies; it wanted to preserve timber and save good agricultural land for Malay smallholders.[33] Many companies also found new oil palm land in their existing plantations. During the earlier rubber boom, planters hungry for land cut out coconut and oil palms in favor of rubber. By the 1950s, it was rubber's turn to fall, a decision helped by the growing threat of synthetic rubber.[34] Plantation managers also put more peat soil under oil palm cultivation. By the 1970s, planters armed with new techniques and chemical inputs could plant in deep peat soils previously considered unfit for any crops.[35] The shift to deep peat opened up a new frontier of swampy forestland across Southeast Asia and beyond.

Oil palms themselves posed the most serious natural challenge to the plantation complex, however. Most Malayan and Sumatran palms were more than twenty years old by the end of the war. Even with squat trunks produced in plantation conditions, these palms had grown too tall to harvest from the ground. Workers often fell from ladders as they pruned, pollinated, and harvested. The average Malayan oil palm plantation suffered thirty-two serious accidents annually for every 1,000 workers, compared with eight in the rubber sector. The owners of Elmina Estate (a much-trumpeted modern plantation of the 1920s) told officials in 1946 that they would harvest their palms for two more years due to "the serious shortage of fats and oils," but they pleaded that the trees were too tall to harvest indefinitely. They asked permission to plant rubber between the palms, noting "impoverished soil conditions" due to a lack of wartime fertilizer treatments. "The oil palm takes so much out of the ground," the company lamented.[36] But despite soil exhaustion, aging trees, and an increasingly challenging political situation with regard to cheap land and labor, the postwar plantation complex left the 1950s much stronger than it had entered the decade, thanks largely to new technologies that made palm oil cheaper than ever before.

Late Colonial Science and the Future of the Oil Palm

A wave of technological change swept the plantation industry in the 1950s and 1960s. Some innovations were relatively simple, leading to incremental cost-savings. Others were revolutionary, reshaping the very trees at the heart of the plantation sector. Unlike the secrecy of the first plantation boom, the postwar era was marked by remarkably free exchanges across colonial and private research networks, spreading plants, tools, and "best practices" across the industry.[37]

The origins of one of the simplest innovations, a pole-knife for pruning and cutting fruit, isn't known. It probably emerged out of similar tools used to harvest coconuts and cacao. The pole-knife spread across the Malaysian plantation sector in the 1950s, though using the long, wobbly pole takes strength and "an acquired skill." The pole-knife made harvesting a skilled task, improving the status and wages of those who wielded it—and it firmly marked harvesting as a man's job.[38]

Chemical inputs also helped firms squeeze profits out of older palm groves. Planters dumped on more and more fertilizer to keep up fruit yields, adding micronutrients like boron to the usual nitrogen, phosporhus, and potassium blends as scientists learned more about the oil palm's needs. Herbicides were particularly important in slashing wage bills. Weeding, most of it performed by women, was the average plantation's single biggest payroll item. A smaller workforce armed with herbicide sprayers could keep an estate "free of lalang at minimum cost." "Not a blade of grass left standing," promised one company.[39] All told, new chemicals, pole-knives, and "rationalized" labor routines helped plantations cut their workforce by a third.[40]

The most revolutionary changes were in the oil palm itself. The postwar era was an exciting time for scientists: as one researcher put it, the goal was "to undertake a fundamental study of the oil palm as a plant," rather than focusing on specific problems in plantation or smallholder cultivation.[41] F. W. Toovey, head of Britain's Oil Palm Research Station in Nigeria, argued that "it was clearly not in the best interests of Nigeria that production should be so largely dependent on an unselected wild plant."[42] Influenced by ideas learned at the Institut national pour l'étude agronomique du Congo belge (INEAC) in Belgian Congo, Toovey expanded the Oil Palm Research Station with government funding, turning it into the West African Institute for Oil Palm Research in 1952 (WAIFOR, now NIFOR).[43] France created a new Institut de recherches pour les huiles et oléagineux (IRHO) to research oil palms and other crops. Private companies with deep pockets, like the Société financière des Caoutchoucs (SOCFIN), Guthries, United Fruit, and Unilever, expanded their own research programs as well.

INEAC researchers kicked off the first wave of innovation as news of their *tenera* research spread. By crossing *dura* with *pisifera*, they got *tenera* every time, avoiding earlier problems with sterile *pisifera* and thick-shelled *dura* offspring that resulted from *tenera* crosses. The new trees promised to boost yields to four tons of oil per hectare.[44] One planter told a thrilling tale of smuggling *pisifera* pollen out of Sumatra disguised as face powder, but most exchanges were more mundane.[45] This was an era of transcolonial—and soon postcolonial—science, no longer dominated by top-down relations between European institutes and colonial outposts. Ideas, specimens, and scientists flowed between and across national and imperial borders.[46] IRHO led a grand "international experiment" in breeding beginning in 1946, drawing on expertise and plant specimens from INEAC and SOCFIN's Malaysian estates.[47] As Susan Martin has stressed, the success of the scientific enterprise hinged on "the willingness of planters to exchange research results and to sell or even give away their vital supplies of pollen and seedlings" in cooperation with state-backed researchers.[48]

Deli *dura* and its monstrous bunches of fruit were no longer the star of the show.[49] Scientists recognized the importance of expanding the plantation gene pool, searching for new African specimens and exploring cross-breeding with the American oil palm (*E. melanococca*, now *E. oleifera*). High yield was only part of the goal for planters: palms also had to be cheaper to harvest. Scientists in Malaya addressed that problem with a shorter palm.[50] Dwarf palms had first been identified in the 1920s, but by the late 1930s, researchers focused on one specimen on the Elmina Estate. Known as "Dumpy E206," the palm had an unusually thick trunk, much shorter than its Deli *dura* peers. Dumpy E206 didn't survive the war, but its progeny did. Researchers crossed Dumpy palms with high-yielding *duras*, and they crossed those trees with *pisifera* to get a short, high-yielding *tenera* by the mid-1950s.[51]

The physical properties of *tenera* fruit stimulated changes in processing technology that made planters even more money in the long run. Southeast Asian plantations used hydraulic presses or centrifuges to process *dura* fruit. In the press, fruit often had to be crushed two or three times to get all the oil out, a slow process. Centrifuges were more effective, but they were precision machines that needed frequent maintenance. When an eleven-year-old centrifuge at the Mayak Oil Palm Estate wore out in 1941, it exploded and killed two men.[52] Centrifuges struggled with the clumps of fiber found in *tenera* fruit. The expeller press proved to be the solution to problems with both centrifuges and conventional presses. Used for other oilseeds since the late nineteenth century, the machine forces materials against metal screens with a powerful screw. It continually loads and discharges, saving labor, and could extract 96 percent of the oil out of *tenera* fruit without damaging the kernels.[53]

While *tenera* research proved most productive, researchers also investigated genetic oddities. "Mantled" fruit's additional fatty lobes turned out to be a damaging genetic defect, but palms with colorless oil promised to save money for food companies. Bleaching was expensive and trapped a significant amount of oil in bleaching media that could only be recovered with costly solvents.[54] Officials ultimately decided that white palm oil was ill suited for Africa on customary as well as nutritional grounds, since the color in red oil comes from vitamin A.[55]

British officials weren't thrilled to hear that Unilever and other groups were pursuing "a strain with the minimum size kernel necessary for reproduction" alongside colorless oil. Unilever insisted that the kernel "cannot compare with the pericarp oil and is in fact somewhat of a nuisance in estate oil palm practice."[56] Vanderwyen similarly hoped his *dura* × *pisifera* breeding program at INEAC would lead to "the disappearance of palm kernels as a commercial by-product."[57] The machinery for separating, drying, and cracking kernels was disproportionately expensive, given that most of a mill's earnings came from pericarp oil.[58]

But other scientists defended the kernel. Toovey noted the sheer scale of African kernel exports and their significance to women. "If this most important trade were lost," he warned, "there would be a serious social upheaval."[59] By 1950, African exports topped 770,000 metric tons of palm kernels, compared with a mere 32,000 from Asia. While African exports fell to 679,000 tons a decade later, Asian exports had only reached a tenth of that figure.[60] Under Toovey's direction, WAIFOR bred *tenera* palms as well as new *dura* × *dura* crosses in the hopes of finding a palm that balanced oil and kernel output. IRHO also took a "cautious attitude" toward the kernel in its breeding programs.[61] After all, in the African smallholder industry, most kernels were by-products of domestic oil consumption. The kernel trade tapped women's labor that couldn't easily be diverted elsewhere.[62] Hand-cracked kernels also fetched higher prices than plantation kernels: the latter were often discolored or contaminated with shells and dirt. And finally, kernels were a critical source of lauric acid, a necessary ingredient in the fast-growing market for detergents, for which coconut oil was the only viable alternative (see chapter 9).[63]

Aside from the defense of the kernel, what good were these innovations for smallholders? When Nnamdi Azikiwe (later Nigeria's first president) talked to scientists about reforming the smallholder sector in 1955, they delivered unwelcome news: hard oil was finished, with lower grades of tallow already being "almost unsaleable." If smallholders wanted to compete with the plantations, they had to improve both the quality and quantity of oil they produced.[64] Across western Africa, colonial officials launched new drives for mechanization, grove improvement, and entirely new ways of working with oil palms. The most important strategy was the "Nucleus

Estate-Smallholder" (NES) model, which combined "the virtues of the plantation system of management and consistent standards and practice" with the "social attractions" of peasant agriculture.[65] It promised to keep farmers on the land, while delivering the high yields and oil quality demanded by export markets. It was a model that shaped the oil palm industry in Africa, and even in Asia and Latin America, for the rest of the century.

State-Directed Development in Late Colonial Western Africa

The NES model was a climax of colonial development thinking: it imposed new crops, new ways of working, and even new people on landscapes with little regard for local knowledge and ecologies.[66] At the core of the NES was a "nucleus" plantation, run like any other, but surrounded by smallholders—often settlers—growing the same crop. Like similar arrangements in bananas and sugarcane, the nucleus processed fruit from the plantation and smallholders in the same factory. The nucleus plantation served as insurance, keeping the oil mill running if smallholders failed to deliver. In addition to buying fruit, the plantation dictated how smallholders grew their crops, giving them instructions, seedlings, and chemical inputs.

Early iterations of NES model in oil palm dated back to the 1920s, but the first full-fledged experiment at Masanki in Sierra Leone had been canceled in 1941. The United Africa Company (UAC) revived the idea in a 1944 memo, proposing a "tripartite syndicate" between the state, UAC, and smallholders. The state would provide land and money. UAC would develop and manage the estate, hiring settler-laborers (ideally demobilizing soldiers and emigrants from overcrowded regions), who would earn rights to a portion of the land. UAC declared: "It is absolutely certain that if the plantation is to be a success, it cannot be run by the Africans themselves." The only company capable of running such a project was, of course, the UAC. A managerial fee would pay the company's costs, while UAC's owner, Unilever, would benefit by having more oil and kernels on the market. The settler would be "virtually the sole owner and draw the entire profit [after UAC's fees] for himself from the enterprise."[67]

UAC's plantation director, Alex Cowan, had long argued that plantations and smallholders could coexist. Nobody "who knows the African should doubt that he will take up the running once it has been made plain to him the direction in which his real interest lies," Cowan argued. He added: "It is a clear duty however, to show him the way."[68] UAC imagined that settlers would carry "the gospel of plantation methods and sound agricultural practice" across Africa, achieving more than four decades of colonial experts had ever done.[69]

Britain's Colonial Office rejected the proposal, insisting that "the natural palm" sector was its priority.[70] The UAC didn't drop the idea, however.[71] They revised the memo and published it in *African Affairs* in 1948. In private, UAC told colonial officials they had no choice. If Nigeria (by far the biggest exporter in Africa) lost its oil and kernel markets to competition from rebounding Asian plantations, it would "lead to such a collapse of the local economy as to result in three or four million people being reduced to famine conditions."[72] Some administrators agreed, seeing a choice between modernization or "starvation and famine."[73]

This wasn't hyperbole. In southeastern Nigeria, population pressure forced farmers to dramatically change practices, including shortened fallow and new tenure arrangements. Communal access to palms withered as wealthier landowners planted new palms as private property.[74] Poorer families used cash from oil and kernels to buy food: "The degree of reliance upon this tree is a reliable index of the breakdown of traditional crop farming in the high density zones," noted one geographer.[75] WAIFOR's director warned that "intensive food cropping" in groves had already done the work of "grove improvement" by ruthlessly eradicating weeds and thinning low-yielding palms.[76]

While Britain's Colonial Office rejected UAC's proposal, the company had already won over an important ally: the Colonial Development Corporation. The 1947 "Bamenda-Cross River-Calabar" (BCC) scheme was not an auspicious start in the oil palm business for the CDC or the NES model, however.[77] BCC was supposed to ease "land hunger" among demobilizing veterans and landless farmers in Nigeria.[78] Each male settler got five acres of oil palms and another twelve acres for food crops and housing. Settlers chafed under the discipline imposed by CDC managers, however, preferring to secure their food before working on palms. Out of 126 families, the CDC only found eleven "satisfactory in all respects."[79] The CDC shortly abandoned the whole project, handing it over to the regional government to develop as a standard plantation.[80]

The CDC blamed the participants rather than the NES model for the failure. It had a "shopping list" of NES projects for Nigeria, but it faced a new political reality by 1950: Nigerians were now in charge, not British officials.[81] Politicians in Nigeria's Eastern Region scoffed at the NES model as a step backward. They wanted state-of-the-art plantations and new machinery to fuel an industrial revolution, and they swept in with cash as BCC and several other CDC projects fell apart.[82] They had money to spend because of a 1949 decision by the imperial government to keep wartime "marketing boards." These boards monopolized produce sales, pocketing the difference between fixed prices paid to farmers and the world market price. The proceeds were supposed to stabilize prices and finance development projects.[83] Nigerian politicians "correctly saw the potential [of this money] . . . in the struggle for greater political power."[84] Authorities spent more than £8 million raised from oil and kernel sales on plantations and other new projects, and almost £2.5 million on research between 1947 and 1954—far more than the government had ever spent on oil palms in the prewar period.[85] In effect, politicians taxed smallholders to finance plantations, a strategy warmly endorsed by the World Bank in 1955.[86]

Nationalist politicians also embraced a new "Pioneer Oil Mill" program. The Pioneer mill was a UAC-designed centrifugal plant, powered by kernel shells and other by-products. Responding to the stand-alone mill failures of the 1930s, it was supposed to be cheap to run in smallholder districts, managed by government officials or farmers' cooperatives.[87] Nigerian politicians saw the machines as another path toward modernization, commissioning a film titled *Invitation to Wealth* to entice Nigerian capitalists to buy them.[88] It was part of a broader strategy aimed at replacing household oil production with machines at all scales, ostensibly to boost yields and export earnings, and also to free up labor for new industries.

Under colonial supervision, the first Pioneer mills shipped without kernel crackers. Managers let women buy kernels for hand-cracking in proportion to the amount of fruit their husbands delivered.[89] They rightly worried that the Pioneer mill "would strike at the roots of the family economy."[90] Yet women ferociously protested when men sold fruit to the mills.[91] Nwga women told officials a husband's failure to give his wife her share of kernels and unrefined oil was "a good ground for divorce."[92] Some protests focused on the threat to water posed by mill effluent. The biggest protest began in 1948 with the destruction of a woman's cassava field by mill workers; hundreds of women destroyed the mill building before burning down the Native Court and storming the police station. The government backed down, relocating mills to communities that actually wanted them.[93]

Protesting women weren't stuck in the past, however. One group of women proposed that a Pioneer mill "be owned and run by them, while men served as paid laborers." The government ignored their proposal.[94] Many protests were aimed against "government control [rather] than against mechanization as such."[95] Others—joined by men—were in defense of hand-press owners, who feared that their investments were about to become obsolete.[96] The Eastern Region administration rejected more than 3,500 applications for loans to buy hand-presses in 1950–51, insisting that money was better spent on Pioneer mills: a tool for the "benefit of the entire society" rather than a lone press-owner.[97]

The protest movement gradually fizzled out, helped by the fact that mills were increasingly owned by local entrepreneurs or managed by government boards.[98] One lawyer defending women protestors, Jaja Wachukwu, bought the very mill his clients opposed. In another community, men organized to cooperatively run a mill they had opposed alongside their wives.[99] By early 1953, seventy-eight Pioneer mills were operational in Nigeria, with another eighty-six on order.

Yet official enthusiasm for plantations and machines was misplaced. The "traditional" industry, with the addition of hand-presses, proved remarkably adaptable. By the mid-1960s, some 40 percent of Nigerian palm oil was still extracted without machinery. Hand-operated presses made up another 50 percent, and the remainder came from Pioneer mills. The £2.3 million spent on these mills could have bought a lot of hand-presses and cooking pots.[100] And contrary to the predictions of European experts and Nigerian politicians, smallholders did improve the quality of their oil. The share of export oil classed as "special grade" (good enough for margarine) rose from 60 percent in 1954 to nearly 90 percent in 1963, by which point Pioneer mills only accounted for 13 percent of "special grade" oil. All this came as Nigerians doubled the total amount of oil and kernels they exported, on top of feeding a burgeoning domestic population.[101]

The domestic market was at the center of this story. Hand-press operators, making oil for both local and export use, paid more than Pioneer mills did for fruit. As a result, they got plenty of it.[102] The French found similar results in Cameroon, where farmers "reserved the best bunches to treat themselves, while selling the worst to the oil mill."[103] Many farmers reported that they "never had it so good," thanks to high local and global prices for soft oil. Hard oil vanished from the market, partly in response to punitive prices for low-quality oil set by marketing boards.[104] Local traders carefully tested oil quality, even buying laboratory kits to measure FFA content in remote villages.[105] Despite serious shortages of land, the smallholder oil palm sector in Nigeria outperformed expectations, with only marginal contributions from NES projects and Pioneer mills.

Oil palm farmers in French West Africa had a similar experience, though with one key difference: more continuity in strategies and institutions from the late colonial era to independence. The focus was squarely on West Africa, in part because oil palm projects in French Equatorial Africa had proved rather disappointing: thousands of hectares of plantation palms planted in the 1930s had been abandoned or destroyed by fires by 1950, and smallholders were not enthusiastic about either "grove improvement" or delivering fruit to new steam- and hand-powered oil mills.[106]

"The oil palm is undoubtedly a source of wealth for Africans," one French expert noted. Yet he asked: "Why does the native lose the habit of climbing and only sporadically collects bunches? Simply because he doesn't think the work pays." The answer, in his view, was a "coordinated policy" to boost oil quality and yields, and therefore increase earnings.[107] That policy looked much like Britain's NES model. Modern mills ringed by plantations and "improved" smallholdings formed the center of the IRHO's *Plan "Huile de palme."* Launched in 1948, *Plan "Huile de palme"* used FIDES money to construct eight new mills in Ivory Coast, Dahomey, Togo, and Cameroon.[108]

IRHO spent a disproportionate amount of money in Dahomey, which along with Togo had some of the least hospitable conditions for oil palms in western Africa's palm belt. The reason was simple: for most families, "Palm oil [and kernels] remains the main source of revenue."[109] Dahomey had no other major exports. Dahomey's elites welcomed the mills, which promised to stimulate replanting and new agricultural methods, but farmers—especially women—were less enthusiastic.[110] The mills deprived women of kernels as well as valuable palm fiber and kernel shells, both sold as fuel. Women protested, and men withheld fruit from the mills, delivering a "psychological shock" to French planners. Some officials debated turning the facilities into fruit juice factories or drying sheds for coconuts.[111]

The mill managers eventually offered women some shells and fiber, easing tensions.[112] When roads feeding the mills finally opened, new truck services helped win over farmers. The trucks collected fruit and paid farmers in cash, saving them the trouble of hauling fruit long distances. The innovation cut women's labor out of the industry altogether, however. By 1954, the Dahomey mills and several others in French West Africa were working at maximum capacity.[113] But the machines couldn't fix Dahomey's dry climate, which kept oil yields relatively low per tree. French experts were also infuriated to see farmers tapping their palms for wine instead of harvesting fruit. The "Palm grove renovation department" began cracking down on tapping in 1954; patrols combined repression with education as they punished tappers and lectured them about the benefits of scientific agriculture.[114]

If French officials worried that tappers were destroying trees that should be making oil in some regions, they found the opposite problem in other areas: farmers refused to kill *dura* palms to plant *tenera*.[115] These older palms often went unharvested because of a lack of climbers. The shortage of climbers was accentuated by France's abolition of compulsory labor in 1946, but it also reflected a broader trend across western Africa. Young men with access to education and more lucrative job prospects saw no reason to take on the hard, low-status work of tree climbing.[116] Women also refused to carry fruit back to collection areas without wages: many had historically received kernels in exchange for their labor but lost these to mill operators.[117]

Despite labor problems, French colonial administrators were successful in forging ties between colonial organizations—IRHO, FIDES, and European companies—and nationalist politicians. Ivory Coast was the best example. A cohort of Ivorian planters had amassed considerable fortunes by the end of the war, mostly from cocoa and coffee. They bought tractors and other machinery, acquired more land, and hired immigrant labor, ultimately winning political battles against European planters.[118] When these elites led Ivory Coast to independence in 1960, they maintained close ties with the former colonial power. With the help of the IRHO, the European Development Fund (a successor to FIDES), and the World Bank, the Ivoirian government launched an ambitious NES-type program for oil palm in 1963.[119] By this point, Ivory Coast was importing palm oil, something that the *Plan palmier* quickly reversed. The country exported 100,000 tons of palm oil and kernels in 1970.[120] Critics accused President Houphouët-Boigny and the Ivoirian elite of supporting "neo-colonialism," but the partnership delivered real economic results—though not necessarily for the poorest farmers.[121]

The NES Model at Kulai

By the time the Ivory Coast inaugurated its grand NES-style program, the model was already recognized as a resounding success: not in Africa, but in Asia. After the CDC's debacle in Nigeria, a new project fell into the CDC's lap in 1950: the Kulai Oil Palm Estate (KOPE). KOPE was a small Johor plantation of 1,772 acres, abandoned during the war by its Chinese owner. It was a far cry from the 10,000 acres that the CDC was initially looking for in Malaya, but it had two important attributes: a large block of mature oil palms and a bargain price. The CDC imagined it as a ready-made "nucleus" to settle smallholders around.

In their enthusiasm for getting the "nucleus" running, the CDC postponed inviting smallholders to the estate, focusing on expanding the plantation and building a state-of-the-art oil mill.[122] Colonial officials were skeptical. They liked the idea of smallholder participation, but they wanted it sooner, not later. In 1950, the CDC had no plan for how smallholders would be integrated around the "nucleus," or how much land would be set aside for food crops and housing, or even what the CDC was willing to finance. The CDC seemed to lack "any idea of the way in which a long-term scheme for the development of the oil palm industry can be worked out."[123]

Fortunately for the CDC, they didn't have to know what they were doing. Under their charter, neither the British Colonial Office nor the Treasury "would query the commercial aspects of the project which would be regarded as within CDC competence."[124] If the CDC said it was a good idea, the Colonial Office had to give them the benefit of the doubt. And by 1950, the CDC had learned a few lessons from failures like BCC, most important of which was hiring professional managers. For Kulai, CDC chose Guthrie & Co., subcontracting management to the long-established firm in the same way that investors had used "agency houses" like Guthrie to run plantations in earlier decades.[125]

When the CDC requested more money from Britain's treasury in 1952 to expand KOPE, "Guthrie's steadying influence" was a deciding factor in getting the money. Colonial officials were still skeptical: Guthrie was "an efficient firm" but was "not concerned with the overall economics of the scheme and is not at risk." Guthrie refused to buy shares in the CDC project, and had in fact turned down an offer to buy KOPE outright. A pack of major firms—Guthrie, Harrisons &Crosfield (H&C), Unilever, and SOCFIN—also turned down the CDC's earlier offer of a joint venture at KOPE. Still, officials insisted that Guthrie was "one of the leading people in the Far East, and can be trusted to run a venture of this kind at maximum efficiency."[126]

Colonial officials continued to press for smallholders at KOPE. In their view, helping smallholders was "the only justification" for the CDC "entering on a commercial venture which might well have been carried out by private enterprise."[127] But Guthrie's manager, W. A. Gibson, was in no hurry for settlers to arrive. By 1954, he had 1,283 acres in production, with another 3,083 planted. The estate shipped more than 650 tons of oil that year. Gibson estimated that the plantation would be turning a profit of £156,000 by 1963. He ran a tight operation. In 1953, for instance, the estate had two local women prosecuted and fined $100 for stealing $3 worth of palm fruit. The women "claimed the fruit was worthless and they had collected it for fuel."[128] Overall, the CDC thought Gibson's work was a great accomplishment, especially "against the background of constant terrorist activity."[129]

It was that "terrorist activity" which gave the CDC an excuse to continue delaying the smallholder phase of the project. The "terrorists" were communist rebels. Malaya's Communist Party fought against the Japanese occupation and—like their counterparts in Vietnam—launched a new fight against returning Europeans when it was clear the colonizers were not going to leave any time soon. The British called the war, which lasted from 1948 to 1960, "The Emergency." The first victims were European planters, and rebels routinely attacked plantation staff, sabotaged equipment, and destroyed plantation trees.[130] Britain responded by flooding the country with troops. Colonial forces issued identity cards, set up road checkpoints and curfews, and forced "squatters" to move out of the forests into fortified villages. These settlements, surrounded by barbed wire and armed guards, kept insurgents out and villagers in. Planes dropped bombs and sprayed defoliant on forest hideouts, killing off forest cover and food crops. Hungry rebels reportedly resorted to pounding stolen palm fruit by hand to make cooking oil during the conflict.[131]

At KOPE, Gibson survived a series of attacks by "bandits." He managed to find enough workers to keep the plantation operating, though as the CDC explained, "for security purposes it is necessary to make arrangements for all labour to be concentrated at a central point in a barbed wire enclosure."[132] This was no place for a family resettlement project. But British officials insisted that the CDC had the whole idea backwards: settling smallholders would prove that the colonial state wasn't simply exploiting Malaya but was giving the people land and a stake in a modern enterprise. It would weaken the appeal of communist propaganda. A colonial official wrote, "it is the type of development . . . which is needed to bring stability in Malaya just BECAUSE of the present political conditions."[133]

That lack of stability was made painfully clear in July 1954, when rebels killed Gibson and another CDC employee in a targeted ambush.[134] They had already killed a foreman on the estate in March.[135] Subsequent raids killed a clerk in 1956, while an assistant manager survived a gun attack in 1957.[136] The plantation faced internal problems as well. Workers went on strike for two months in 1956 and again in 1959 over the firing of union organizers accused of "misconduct."[137] Malaya's regional labor groups formed a national union in 1954, but they did not head down the more militant road taken by their counterparts in Sumatra.[138]

Meanwhile, the CDC's 1950 proposal for Kulai inspired discussions in other parts of Malaya. Officials working on what became Malaysia's Rural Industrial Development Authority (RIDA) examined oil palm as a smallholder crop in 1952. A major draw was the fact that oil palm was a food crop—potentially for Malaysians—and that it grew on less desirable soils. The problem, as always, was harvesting and processing the fruit. If smallholders sold to a plantation mill, they were "in the hands of the estate."[139] Conversely, officials sympathetic to planters argued that smallholders were "a burden" to industry. Before the war, the government had in fact discouraged smallholders from planting and selling to nearby plantations. One plantation-smallholder arrangement in Pahang caused "constant trouble," though another in Selangor "had done quite well."[140] European planters accused smallholders of poor quality control, overproduction, and environmental mismanagement. They had levied the same charges against rubber smallholders before the war. Moreover, they insisted that oil palm—because of the need for fast harvesting and processing—was uniquely suited to plantation methods.[141] Planters also complained about farmers helping themselves to seedlings from plantation nurseries and stealing fruit to sell back to the estates.[142]

As late as 1962, Guthrie's chairman urged Malaysians to trust the plantation model.[143] But the planters didn't get their way. Malay leaders in Selangor were enthusiastic about the NES model for oil palm: one said it was "the best means to improve the economic and social well-being of the Malays."[144] A 1955 World Bank report endorsed the NES approach, providing Malaya's nationalist politicians with more encouragement.[145] Later that year, the United Malay National Organisation agreed to pursue a strategy of coordinated, state-backed land development focused on small farmers. The next year, the Federal Land Development Authority (FELDA) was founded.[146]

By the time Malaya declared its independence from Britain in 1957, FELDA and other state and federal agencies were preparing big plans for smallholder settlements across the country. British and Malayan forces had nearly wiped out the Communists in Johor, and the CDC was finishing up its new palm oil factory at KOPE. Kulai was, at last, ready for smallholders. First, though, CDC had to decide who counted. Initially they had hoped that settlers would own fifty or even 250 acres. These wealthy farmers would become shareholders in the factory, using their clout to enforce quality standards among their neighbors. Officials hoped they would also better understand their position in the world market, resisting the urge to abandon oil palm when prices slumped.[147]

The CDC ultimately settled on a more modest limit to "smallholding": twelve acres. "The occupier of anything over this will have to be an employer of hired labour," wrote one CDC employee.[148] As Johor's chief minister Haji Hassan bin Yunus put it in a 1959 speech opening KOPE's smallholder phase, the goal was to "satisfy the cry of the 'small man' for land." Following the minister's speech, settlers wielding machetes and axes "began attacking the jungle . . . felling and burning trees." A newspaper boasted that the settlers were transforming 5,000 acres of what was "once the worst terrorist black spot in the State" into a model for development.[149]

A total of 455 families applied for the first 300 allotments at Kulai; most of those chosen were Malay, despite government promises to include Chinese and Indian families. Managers chose settlers "on the basis of their age, the size of their families, physical fitness and agricultural background." The government promised settlers that they would soon be self-sufficient, noting "The market for palm oil . . . has always been steady."[150] Each family received twelve acres for oil palm, and more land for a house and garden (later reduced to 5 hectares total). Settlers also got cash loans and a chance to buy $50 in factory shares for every acre of oil palms. On top of this, the government guaranteed that "a school, water supplies, a health centre, shops and a mosque" would soon be built. CDC launched a 5,000-acre extension in 1962 as Kulai Two, with plans to expand to 20,000 acres.[151]

A visiting planter admitted that labor conditions and housing at KOPE were "better than those of neighbouring estates." He thought the whole project "sounds a little out of this world," but concluded that CDC was willing to pay whatever it took make the project work.[152] And it did work: by 1969, the project had repaid the CDC's investment and was home to hundreds of Malaysian families.[153] KOPE served as a nursery for FELDA's new palm plantings, and it trained a generation of engineers and agricultural staff in the arts of managing oil palms and people.[154]

Oil Palm at the End of Empire

If oil palm sectors in Malaysia and Ivory Coast saw relatively smooth transitions from colonial to postcolonial development through NES-style projects, the same could not be said for the rest of the tropics. Indonesia was the first country with a significant oil palm industry to gain independence from colonial rule. As discussed at the beginning of this chapter, nationalists and communists organized strikes and land occupations against foreign-owned plantations. Sukarno's government seized Dutch properties and forced all other estates to surrender a third of their land to the Ministry of Agriculture.[155] The government urged surviving firms to reinvest profits in the country and to replace foreign staff with Indonesians.[156] After surviving a foreign-backed coup attempt in 1958, Sukarno pressed foreign companies harder, seizing British and other foreign assets.[157] His policy of *Konfrontasi* ("confrontation") with Malaysia and its British backers escalated into armed conflict in the 1960s.[158] Suharto's coup against Sukarno in 1965 didn't halt the confrontation, but it pushed the government hard to the right. At least half a million suspected communists were massacred in the ensuing purge. The U.S. government, now seeing Indonesia on the "right" side of the Cold War, arranged for International Monetary Fund and World Bank loans to redevelop the country.[159] Unilever, H&C, and several other big companies struck deals with Suharto's government to get their businesses back.[160] The oil palm boom that ensued is addressed in chapters 10 and 11.

Although some of Malaysia's politicians talked about nationalization, the government allowed big companies to keep their properties.[161] An H&C chairman noted that his decision to replace old rubber with oil palms was helped by the fact that independent Malaysia was "a friendly country anxious to have foreign investment."[162] H&C and other companies adopted a policy of gradual Malaysianization-by-attrition, hiring Malaysians as European staff retired.[163]

But H&C wasn't entirely convinced that its plantations were safe; one executive warned, "it is not inconceivable to think that a situation similar to Indonesia will some time arise in that country [Malaysia]." H&C pondered growing oil palms in Queensland, noting that such a plantation could evade Australia's "savage rate of duty" on imported fats. The company ultimately concluded that labor costs would be too high.[164] H&C settled on a joint venture with Australia's colonial government in Papua New Guinea in 1967.[165]

Unilever also sought out safe havens for plantations, proposing an NES project on Kolombangara Island in the Solomons in 1950. British colonial officials tore into the plan, which threatened to revive the "coolie trade" and trap workers in debt bondage. One official warned against letting Unilever get its "tentacles" any deeper in the Pacific, noting the "strangle-hold" the company held over African trade through the UAC.[166]

But Unilever's strong position in Africa was no shield against nationalization. Under Kwame Nkrumah, Ghana (independent in 1957) moved aggressively against foreign holdings, promoting new state-owned plantations instead.[167] These "state farms" included several oil palm projects, but they were poorly executed. Unilever's own estates in Ghana, inherited from various companies bought by UAC, were barely profitable. One plantation was still using a 1918-vintage centrifuge. Frustrated with fees, title disputes, and conflicts with local farmers, Unilever sold its plantations to Ghana.[168] The state farms couldn't make up the difference in oil output, though, and by the early 1960s, Ghana joined Sierra Leone and Ivory Coast among the ranks of Africa's palm oil importers.

Nigeria pressed ahead with its own plans for state-led development after independence in 1961. Eastern Region authorities promised nothing less than "an agricultural revolution."[169] It was clear that traditional agriculture was under strain: fallow cycles that had once taken ten years were down to four years or less. "[The] point is rapidly approaching in some places where no fallow is possible and the soil is being completely worked out," warned researchers. Nigeria's plans called for "modern scientific agriculture" in all its forms. Nigerians wanted CDC money, but they also looked to new sources of aid from the World Bank, the Ford Foundation, and the U.S. Peace Corps.[170]

Foreign experts warned that new state-backed plantations would "create far reaching changes in the social and economic systems," shattering communal landholding.[171] For Nigerian politicians, this result was a necessary evil. Critics in the CDC complained about the "heady confidence" of the nationalists and their disdain for the gradualism of the colonial NES model.[172] Planners pushed aside farmers to secure large blocks of land, but the plantation projects failed to generate anything near the number of jobs predicted, thanks to a heavy reliance on tractors and other machinery. Plantations disproportionately employed men, hiring women only for low-paid seasonal work like weeding.[173]

Foreign experts did back Nigerian plans to replant palm groves with short *tenera* palms, boosting yields while eliminating the need to find skilled tree-climbers.[174] Researchers compared *tenera* with the "miracle" rice strains that were transforming Asian agriculture, predicting similar benefits for African farmers.[175] And importantly, Nigeria's agricultural officers had learned from colonial failures: while they insisted that farmers cut down old palms, they offered stipends to make up for the lost income while young trees matured. They also allowed farmers to grow food crops between rows of palms.[176] Most aid was only available to farmers with at least five acres, however. Thus, the main beneficiaries were not poor farmers but chiefs and town-dwelling landlords. Replanted groves offered little work for the growing numbers of farmers without enough land of their own.[177]

At least 50,000 acres of "semi-wild" palm in eastern Nigeria had been replanted along "scientific" lines by the end of 1966. The government hoped to replant 590,000 acres and put in 315,000 acres of new oil palms by the end of the decade, covering an area more than twice as big as Malaysia's oil palm plantations.[178] Farmers got access to high-yielding seedlings, fertilizers and pesticides, and cash. Agricultural expertise was in short supply, however: Nigeria had only one extension worker for every 50,000 people. In the United States, the ratio was one for every 300 farmers.[179]

The World Bank financed much of the replanting, but it warned that the new palms wouldn't do much for Nigeria's palm produce exports: increasing domestic consumption was projected to consume nearly all the new oil.[180] In any case, the experiment crashed to a halt in 1967 when the Eastern Region seceded from Nigeria, starting a bloody civil war. Fighting raged across the oil palm belt until early 1970, destroying infrastructure and palm groves. Among the hundreds of thousands of victims of the war was S. C. Nwanze, the first Nigerian director of NIFOR.[181] The Federal blockade of the Eastern Region caused famine, forcing refugees to cut down palms to make way for food crops.[182] By the time World Bank–sponsored projects resumed in the mid-1970s, Nigeria's oil palm industry was in shambles, and the country was a net importer of edible oil.

Conflict also shattered Africa's other big exporter of palm produce, Congo. Unilever's Huileries Congo Belge and other European-owned concessions worked throughout the 1950s, replacing "natural" palm groves with *tenera* plantations.[183] The colonial government also intensified its efforts to extract palm produce from farmers, resettling nearly 10 percent of the entire population on smallholdings near roads and market centers. Families in these *paysannats indigènes* received training, seeds, chemicals, and tools, but most families went only "because the colonial state wanted them to." The project "was utterly disliked" and collapsed with independence in 1960.[184]

Still, between small farmers and the big companies, Congo's palm oil and kernel exports surged after 1945. The situation was a welcome change for HCB, which had been a "disastrous failure" as a business up to that point. The company hadn't paid a single dividend. HCB's treaties with the colonial administration forced the company to assume the costs of social reproduction—ensuring that workers and their families had good food, housing, education, and medical care—on top of the costs of maintaining and harvesting palm groves. A 1933 study showed that Hallet's Sumatran estates invested £32 for every ton of palm oil produced annually, compared with £58 for HCB.[185] After the war, HCB started to behave more like a typical plantation company, scaling back benefits for workers. It also diversified into rubber and cocoa.[186]

HCB and the Belgian administration were complacent about African nationalism. When HCB named its first Congolese manager in 1958, it was thinking about money rather than politics: the company paid Africans much less than it did Europeans.[187] Congo's rush to independence in 1960 took HCB and many other firms by surprise. Congo entered a period of political chaos following the overthrow and murder of Congo's first prime minister, Patrice Lumumba. With racist overtones, Europeans sniffed "that all these fruitful territories would once again steadily revert to the jungle."[188] Yet the 1960s were surprisingly good for HCB, even as other concessionary companies collapsed. HCB was more popular than its Belgian-owned competitors, thanks to the services and pensions it provided. One company official noted that Pierre Mulele, who took up arms on a Lumumbist platform in 1964, "declared that as far as he could influence matters the installations of HCB were not to be damaged." Mulele had once been HCB's "brightest pupil" as a trainee before turning to politics and guerilla warfare.[189] No matter how nationalist their rhetoric, Congo's warring politicians had "no wish to kill this golden goose."[190]

With or without war, the future of the plantation model for oil palm remained an open question at the end of empire. Malaysian plantations were poised for a massive expansion in output, thanks to *tenera* plantings. Unilever's plantations in Cameroon grew after independence, making money for the company and keeping employees satisfied with good wages and services.[191] But everywhere the plantation model relied on cheap land and labor, conditions first made possible under colonial rule. Indonesia's decision to nationalize plantations raised fears across the industry that other countries would follow. In Congo (renamed Zaire), Mobutu seized foreign plantations in 1973, marking the beginning of a precipitous decline for Congo's oil palm industry. Even in business-friendly Malaysia, the government made it clear that it was going to set aside more land for smallholders and indigenize plantation firms through share purchases.

As chapter 9 shows, these growing plantations and smallholder settlements enjoyed access to a growing world market after 1950. Palm oil remained the cheap substitute, but its physical properties became important once again as controversies over the health impacts of different fats raged into the late twentieth century. And its cheapness inspired chemists to find new ways of using it, transforming industries and creating new markets with seemingly limitless potential.

9 Industrial Frontiers

"Golden Gifts," Gas Tanks, and Great Apes: Palm Oil's Industrial Frontiers

The oil palm's first commodity career began in earnest in the nineteenth century, as African farmers tapped their trees to feed a burgeoning world market. But as important as this trade was in the story of western Africa, it was a marginal part of the story of industrialization in Europe. No one *needed* palm oil soap or candles. The color and odor of palm oil—qualities once prized in Europe—disappeared in industrial products, and occasional appeals to the fat's African origins were little more than marketing ploys. Palm oil was the cheap replacement for other substances, subject to substitution itself.

The oil palm's second career as a supplier of industrial foodstuffs was more a matter of necessity. The world's population skyrocketed from around 1.6 billion people in 1900 to three billion by 1960, reaching six billion by 1999. Accounts of the "Green Revolution" that kept this growing population fed rightly center on grain crops like rice and wheat, but getting more fat was an important and often unappreciated part of the story. Since 1970, three times as much new cropland has been put under oil-producing plants than under grains.[1] With per hectare yields ten times higher than for other sources of fat, the oil palm has played a vital role in feeding the world.

The world's population got richer, too, and the richer people got, the more fat they consumed. Average daily consumption of dietary fat jumped twenty grams to fifty-three grams per person between 1967 and 1999. North America and western Europe were global gluttons, adding more than 20 percent to an already-high level of consumption. But China's fat consumption tripled over that period, and across Asia and Latin America, people ate far more fat than their grandparents did. Africa was the lone exception to this trend; per capita fat consumption stagnated and even declined.[2]

Faced with these facts, the oil palm—and more specifically, the oil palm plantation complex that delivers high yields of edible-grade oil—seems necessary, even inevitable. But was palm oil a "golden gift to the world," as some apologists claim, or an example of what sociologist Jason Moore calls "cheap food," extracted at great ecological and human costs from marginal regions of the world economy?[3] The previous section of this book examined the changing ways people lived with oil palms around the world. This chapter returns to the consumer side of the story, tracing palm oil's sometimes uneven path through a globalizing world economy. It sets the stage for following chapters that explore the costs and benefits on the production side.

Palm oil met many needs in the twentieth century, but not everyone wanted it. The oil palm's place in the global economy and food system ultimately hinged on a series of political and cultural choices. U.S. policies—which encouraged oil palm cultivation in the 1960s and 1970s—moved against palm oil, before indirectly boosting it again in the 1990s and 2000s. Industrial consumers in rich countries embraced palm oil as an invisible replacement for animal fats that end consumers were increasingly rejecting, but many companies dropped palm oil in a controversy over heart disease in the 1980s. A decade later, a new panic over trans fats sent manufacturers racing back to palm oil. Outside food industries, palm and palm kernel oils seized back market share from petroleum products and moved into new areas like fuel, as industries built up their "green" credentials to appease consumers and government regulators.

As the last section of this chapter emphasizes, palm oil remains a controversial commodity. Critics of the plantation complex emphasize the invisibility of palm oil in food and industrial products, intimating that manufacturers tricked consumers into participating in deforestation and habitat destruction in the tropics. Meanwhile, industry-led efforts to recast palm oil as a sustainable commodity have been hampered by the very thing that made palm oil so attractive: its cheapness.

Fats in the Industrial Food System

An integrated, industrial food system emerged on a global scale in the 1950s. Heavily mechanized farms fed industries churning out processed foods, which began to reach consumers even outside the industrialized North.[4] Palm oil served key roles in this system beginning with the earlier margarine boom, but postwar technological developments secured palm oil's place. Solvents, centrifuges, vacuum pumps, and a range of other machines and techniques transformed palm oil into finely tuned products. Stiff palm stearin might wind up as candle wax or part of "vanaspati" ghee. Fluid palm olein might become a golden-hued oil for stir-fries or salad dressings. Harder "palm mid-fractions" could replace cocoa butter in chocolates.[5] End consumers never saw, smelled, or tasted these fats. Manufacturers chose them because they were cheap and effective in replacing more expensive fats (cocoa) or culturally undesirable ones (lard, tallow, whale oils).

As a Unilever chairman explained, his company was "trying all the time to put ourselves in a position to use less of the oils and fats which are in short supply, and more of those which are easier to get." Yet full interchangeability was never realistic. Consumers wouldn't readily accept a new texture in margarine or different lathering qualities in soap, even if it meant a cheaper product.[6] Manufacturers could always use hydrogenation to turn liquid oils into fats with any plasticity they wanted, but this required specialized machinery. The cost of hydrogenation motivated many manufacturers in Europe to adopt palm oil in the 1950s, even as American firms relied more and more on hydrogenated cottonseed and soybean oils.[7]

Refined, bleached, and deodorized (RBD) palm oil became, as historian Susan Martin put it, a "blank canvas upon which manufacturers could create the desired impression."[8] Food technologists created flaky pastries that stayed tender and crisp for days; powders that stirred into coffee as "cream"; boxed biscuit and cake mixes that lasted for months on shelves—just add milk and eggs. Instant noodles ready with a cup of hot water; peanut butter that you never, ever have to stir oil back into: these were the miracles of modern convenience food, made possible by refining, fractionating, and hydrogenating fats like palm oil.[9]

Anthropologist Claude Lévi-Strauss theorized that food has to be "good to think" before it can be good to eat. Yet the invisibility of palm oil in industrial foods meant that most consumers never had a chance to think about it. Manufacturers liked it that way. As one margarine promoter put it, margarine "looks good, tastes good, and is a good economical food. . . . That is about all they need to know about the product."[10] If a list of ingredients appeared on a package at all, palm oil appeared as "vegetable oil." It might turn up in margarine, chips, or even ice cream.[11] Palm oil took new, unnatural forms in many of these foods, blurring the line between "food" and "chemical."[12] No one ever bought a stick of palmitin or a jar of diglycerides in a supermarket, but they ate them nonetheless in packaged foods.

Palm oil became especially important for manufacturers trying to replace animal fats in the postwar decades. Although animal fats were just as invisible as plant fats in industrial food, advertisers often reminded consumers of the purity and cleanliness of all-vegetable products, a pattern dating back to the first all-plant shortenings and "nut butters" of the 1890s. "Oleo oil" practically disappeared from oleomargarine by 1950 (taking the "oleo" prefix with it), relegated to cheap brands often made in the same factory as all-vegetable products. Consumers in the West did not want to be reminded that they were eating rendered cows, pigs, and whales, no matter how tasteless and odorless the end product was.[13] In margarine and baked goods, palm oil had the virtue of mimicking the plasticity of customary fats used in the North, like lard and butter, with little or no hydrogenation needed. Coconut and palm kernel oils are too stiff and brittle, and most other plant oils are too fluid.[14] Food manufacturers also recognized that they could sell new margarine and soap blends made without animal fats to consumers observing Jewish and Muslim dietary laws. Animal fats lurked in all sorts of supermarket products, and replacing them with palm oil (or hydrogenated oils) was an easy fix.[15]

Palm oil remained invisible as it permeated food markets in the global South, especially in Asia.[16] Yet the biggest growth came late, beginning in the 1980s. It wasn't a direct product of population growth but rather was associated with growing wealth. Consumers with money to spend embraced industrial foods, and palm oil was the perfect fat for making instant noodles and deep-fried snacks. Its triglyceride composition is well suited to resisting oxidation, and it contains very little linolenic acid, a major culprit behind odors and bad tastes in heated oil. These same attributes give products made from palm oil a longer shelf life, staving off rancidity. Meanwhile, refined palm olein swept away coconut oil and other local favorites as an all-purpose household oil. One researcher described it as nothing less than a "palm oil revolution" in Asian factories and kitchen pantries.[17]

But this revolution demanded palm oil's invisibility. Unrefined palm oil was, like all foods, an acquired taste. European colonizers initially assumed "coolie" workers would take to it as a replacement for coconut oil, but workers and managers alike found the stuff unappetizing. One colonial-era experiment with eggs fried in palm oil resulted in a dish covered in "a deep orange film, suggestive of varnish or floor polish," accompanied by a flavor that was "not quite pleasant." Malaysians only started eating palm oil from the country's fast-growing plantations in the 1960s, thanks to domestic refineries that stripped the color and taste from the oil. RBD palm oil slipped into Malaysia's food industries like it had done before in the North, with most food labels not bothering to identify the stuff.[18] One businessman explaining his company's shift from coconut to palm oil said, "We just did it quietly . . . and the consumer not only had no objection, but as a matter of fact, grew to like it."[19] A similar story played out in Indonesia, where palm oil accounted for 94 percent of domestic fat consumption by 2010.[20] The skyrocketing cost of coconut oil—mostly grown by smallholders—meant that people had little choice but to eat palm oil products in packaged foods and cooking oil. A state-backed advertising campaign stressed the merits of "golden" palm oil (refined palm olein), pitching it as an economical and healthy choice. Out of the public eye, RBD palm oil seamlessly replaced coconut oil and other fats in bakeries and instant-noodle factories.[21]

The place of palm oil in everyday diets was less certain outside the oil palm's growing plantation empire, however. A Unilever expert warned in 1963 that "The future lies, to a large extent, in the hands of the U.S. Government." The Americans didn't produce a drop of palm oil, but their government's policy of distributing cheap soybean and cottonseed oil in developing countries had an enormous influence on markets and foodways. On the one hand, dumping American fats relieved overproduction in the United States and crushed foreign competitors, like palm oil. On the other hand, the abundance of cheap fat created new habits and import dependencies. People eating curries and noodles and chips made with subsidized soybean oil in the 1960s wouldn't give those foods up when American oil disappeared. As the next-cheapest thing, plantation palm oil went on to capture this "new and permanent market for oils and fats" across the globe (graph 9.1).[22]

China and India offered the two biggest emerging markets for oil palm growers. Convoluted ecological links across the global food system meant that India, not China, became the world's largest single consumer of palm oil by 2015. Chinese consumers did eat more fat as they got richer, but they ate a *lot* more pork. From 7.6 kilograms per capita in 1975, Chinese pork consumption jumped to 29 kilograms in 1995 and topped 40 kilograms in 2015.[23] Those pigs ate staggering amounts of soybean meal. Because oil is a by-product of crushing soybeans for meal, China has had a cheap and abundant supply of oil since the economic reforms of the 1980s.[24] Chinese firms nonetheless began importing large quantities of palm oil in the 1990s for manufacturing and to replace hydrogenated oils in prepared foods. A swine flu epidemic in 2019 led to soaring Chinese palm oil imports, illustrating the close link between soybean oil, pork, and oil palms.

Palm oil first reached Indian markets in the 1920s and 1930s, much earlier than in China. Some wound up in soap, but the most prominent use was in vanaspati (vegetable ghee). Indian dairy farmers and campaigners for "pure" food vigorously opposed the stuff—made from all sorts of hydrogenated fat, including whale oil and tallow—mirroring fights against margarine in

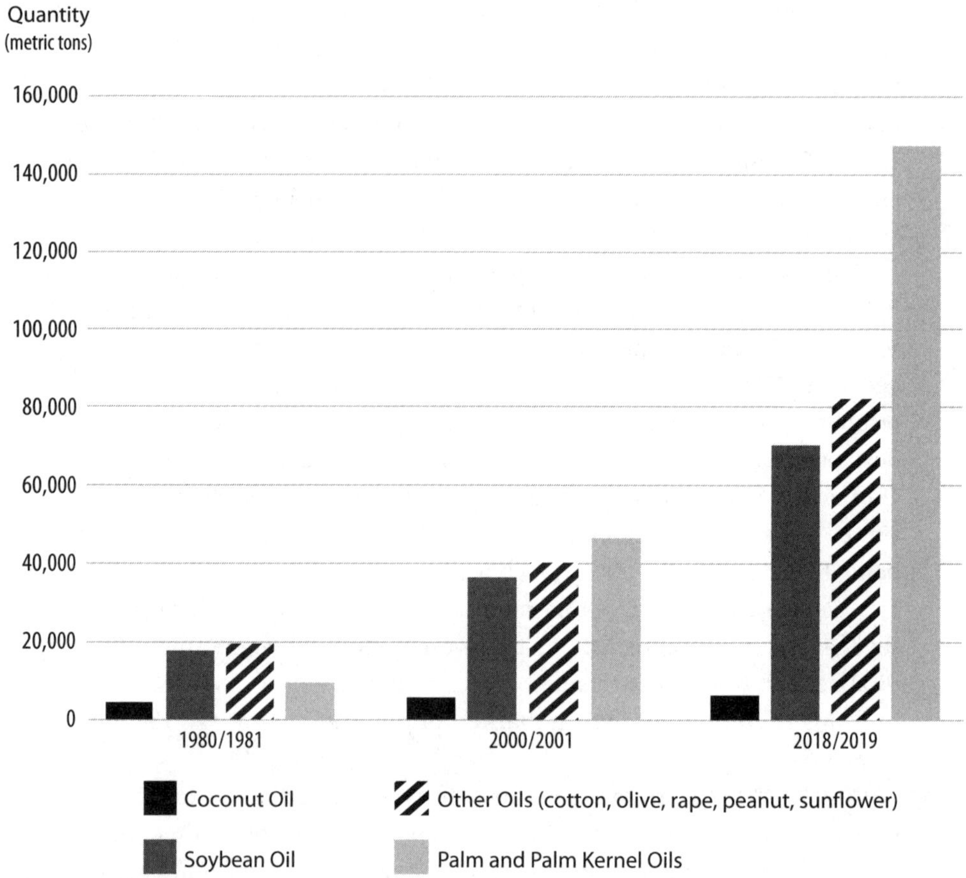

GRAPH 9.1 World vegetable oil supply by type, 1980–2018. Data from FAOSTAT.

Northern countries.[25] Like margarine, the name was something of a compromise, marking vanaspati as a new product rather than counterfeit ghee. Unilever's subsidiary in India started manufacturing vanaspati domestically in the mid-1930s, introducing the popular Dalda brand. Following the Northern marketing formula, it was sold in sealed tins with labels sporting palm trees and boasts about hygenic, all-vegetable ingredients. Cookbooks, street demonstrations, and print advertising helped sell Dalda across the subcontinent. In one promotional film, a poor wedding host surprises guests who think they've been feasting on ghee-laden dishes with a tin of Dalda.[26]

As India gained independence in 1947, critics renewed the attack on vanaspati, questioning the patriotism of a food made from imported fats. Mohandas Gandhi argued that it hurt small farmers and the dairy industry;

he also promoted a controversial study that found vanaspati devoid of any vitamins.[27] The Cow Protection League took up the cause, equating vanaspati with a war on cows. Vanaspati crushed demand for butter and ghee, they argued, and might hide fats from slaughtered animals. (In the 1980s, rumors that an importer had used beef tallow to make vanaspati caused national outrage and a 30 percent drop in sales.[28]) Despite the criticism, vanaspati triumphed—backed by a powerful domestic oilseed and hydrogenation industry. As the *Times of India* explained, "there is no country whose people can afford to draw their fats mainly from dairy products."[29]

The "Green Revolution" that did so much for wheat yields in India bypassed oilseeds, however. By the 1980s, Indian families spent up to 7 percent of their incomes on fats, even with per capita consumption well below regional averages.[30] India had to buy more and more imported fat. When World Trade Organization rules forced India to end a state monopoly and lower import tariffs in the 1990s, imports of palm oil surged, eventually making India the largest importer of it.[31] With support from the central government, Indian states and businesses recently launched oil palm cultivation programs, hoping to end the country's reliance on imports. Critics warn that it risks repeating the worst mistakes and injustices of plantation expansion in Southeast Asia.[32]

Africans also became consumers of palm oil in its industrial forms after 1950, participating in an economic system that sold soap and margarine in brightly decorated packages alongside promises of modernity and a better life. Advertising rarely hinted that the raw materials were harvested by African hands before their transformation in European-owned factories.[33] Africans continued to consume huge quantities of artisanal palm oil outside this industrial system, though experts can only guess at the scale: a 1981 report suggested that "wild" groves supplied perhaps 80 percent of Africa's consumption, more than a million tons annually.[34] The *dura* fruit common in western Africa didn't make as much as oil per hectare as *tenera*, but by most accounts, its oil tastes better and has a more appealing color. Consumers rate *dura* oil as less "fatty" than *tenera* oil. The extraction methods used by *dura* producers (and possibly differences in the plant itself) result in an oil with more olein than factory-made *tenera* oil, making a more fluid product with a different mouthfeel.[35]

A 2001 study in Ivory Coast found that nearly a third of households didn't even consider refined palm oil to *be* palm oil, equating it with peanut oil. Households consumed an average of 72 liters of refined oil annually, and only 17 liters of "red" palm oil. Most consumers insisted that unrefined red oil was the best, but refined oil cost half as much, thanks to a flood of imported palm oil from Southeast Asia.[36] Higher consumption of plantation oil seems inevitable as Africans eat more fat and more processed food. Still, traders do brisk business selling small bottles filled with artisanal oil in markets across Africa's oil palm belt, even exporting oil to serve expatriate communities in Europe and North America.[37] But even this niche market isn't safe from the plantation sector, which sells unbleached *tenera* oil with slogans like "The brand that brings you home."[38]

"Injected into the American Diet": The Oil War

The United States ended its import tariff on palm oil in 1959, but U.S. policy indirectly held back palm oil sales elsewhere. Under P.L. 480 (1954), better known as the "Food for Peace" program, the U.S. government gave free or steeply discounted food commodities to Cold War allies. Hydrogenated cottonseed and soybean oil poured into countries like Pakistan, dumping surplus oil from the American market and holding down prices for competitors like palm oil. By the 1970s, American farmers held more than 70 percent of the world market for oilseeds.[39]

American soybean farmers were alarmed when the United States suddenly became a major importer of palm and coconut oils after 1972. It was an excellent illustration of the long-range economic and ecological linkages of the edible fat industry. Animal fat production was declining, the result of leaner breeds developed in response to changing American tastes in meat and shrinking demand for tallow and lard. Meanwhile, high prices for soybeans in 1972–73 encouraged American and Brazilian farmers to plant lots of soybeans. The subsequent production glut drove down soy prices and convinced many farmers to plant maize instead the following season. The cut in soybean production was compounded by wet weather and an early frost in 1974. American soybean output dropped by 20 percent, sending the price of soybean oil spiking. For the first time in the soybean's postwar history, its oil was more expensive than palm oil.[40]

To fight palm oil, soybean farmers took aim at development aid. The much-trumpeted Jengka Triangle project in Malaysia (see chapter 10) brought publicity to the oil palm industry, and American farmers questioned why "they" (really, the World Bank and other international institutions) were giving competitors cheap loans. In 1976, a congressman proposed barring U.S. aid for any palm oil project, sparking a series of hearings. One senator said he was "raising Cain . . . about palm oil, see if we can't stop the flood of palm oil, because if I didn't, they would run me out of the State of Minnesota."[41] The proposed ban narrowly failed in the U.S. Congress. Another bill to reimpose tariffs on palm oil for any use other than cold-rolling and tinplating also failed. Yet another failed bill would have subjected palm oil to costly sanitary inspections and labeling rules.[42]

As pressure from the soybean industry mounted, the U.S. Department of Agriculture (USDA) sent experts to Malaysia to examine what was fast becoming a palm oil boom. Their forecast was sobering: Malaysia was poised for massive expansion, with only 3 percent of future output needed for domestic consumption. Noting that oil palms were a long-term investment, experts warned that plantations and smallholders would "absorb an abnormally large amount of pressure on their production margins" before switching crops.[43]

U.S. politicians proved reluctant to pass an outright ban on financing for palm oil, though they did agree on a resolution calling for the World Bank to focus on fighting malnutrition in developing countries instead of promoting palm oil exports to rich countries.[44] In the end, Congress passed a bill expanding American investment in development lending, with an official position opposing palm oil, citrus, and sugar projects that might compete with U.S. farmers.[45] World Bank officials politely ignored U.S. demands, convinced that the farm lobby would lose interest. They also pointed to the USDA's own prediction of a global edible oil shortfall in the 1980s, stressing a need for more palm oil, not less.[46]

During the 1976 hearings, a Texas congressman theatrically introduced a second element of the soybean farmers' attack on palm oil: cardiovascular health. "This morning," he said, "I thought about this hearing as I drank a glass of hot milk and had some raisin toast and then some white toast and when I began to spread on the margarine, I felt a little apprehensive. What did this margarine contain? Did it contain some palm oil?" He went on to say: "Why drink low-fat milk and eat unsaturated foods and then have this—as I would call it—dangerous type of oil injected into the American diet[?]"[47] The congressman connected palm oil with a long-brewing controversy over the role of fat consumption in atherosclerosis.

Initially, doctors focused on cholesterol as the culprit behind a postwar epidemic of atherosclerosis, a condition in which plaques obstruct flood flow in arteries. By the late 1950s, many began to suspect the real cause was saturated fat.[48] One scientist noted that there was sufficient evidence to link fat consumption with cardiovascular disease, but he added that the "real controversy lies in what kind of fat." "This," he said, "is where the pressures of industrial and agricultural interests start."[49] Egg, dairy, and meat lobbyists fought a brutal media campaign with vegetable oil producers over the saturated fat issue, each claiming to have science on their side. Marketers didn't wait for the science to settle, abusing new research findings to hawk products and disparage rivals. Several companies sold their all-vegetable, mostly unsaturated margarines by prescription in pharmacies and hospitals as a cure for high cholesterol levels (one brand was called "Emdee").[50] Advertisers ignored government warnings that "Laymen are not qualified" to use food labels like "unsaturated" or "cholesterol free" to treat disease.[51]

The lack of clear guidance from scientists and regulators, coupled with weak labeling rules, fueled confusion. Headline-grabbing reports touted the next "killer fat" or "miracle food" that would revolutionize diets.[52] One researcher (funded by the meat industry) observed: "Those citizens who suppose that scientists are cool, shrewd, quiet recluses making infallible decisions based on all the evidence must be perplexed by the present disheveled state of the Diet-Heart problem. Confusion and disagreement are rampant."[53] In 1976, the U.S. Food and Drug Administration (FDA) did force manufacturers to identify which fats actually went into "vegetable oil" or "shortening," a major goal of the soybean lobby. But the FDA allowed manufacturers to list all *possible* ingredients: they could use hydrogenated soybean oil one month, and switch to palm oil or tallow the next without changing the label. These "and/or" labeling rules kept ingredients obscure for all but the most committed shoppers, and let manufacturers enjoy the flexibility that interchangeable fats offered.

The disappearance of tallow and lard from industrial foods meant that the "tropical oils"—coconut, palm, and palm kernel oils—felt the brunt of attacks on saturated fat. At that point the majority of U.S. palm oil imports went into shortenings, replacing tallow. But refiners in Malaysia were starting to sell palm olein, threatening the liquid oils—soy, cotton, peanut, and rapeseed—that had been a reliable market for American as well as European farmers.[54] Experts predicted that the 1990s would see a glut of oil because of fast-growing world demand for meat, which in turn led to more oilseed production for animal feed, which in turn left behind oils as secondary products.[55]

American soybean farmers complained that palm oil importers misused "the good name of vegetable oils," taking advantage of consumers' frequent association of unsaturated fats with plant oils, and saturated fats with animal fats. Palm oil has significantly more saturated fat than soybean oil, though less than coconut and palm kernel oils.[56] The battle simmered on

FIGURE 9.1 A man dressed as a tropical planter sits next to a drum labeled "palm oil," part of an American Soybean Association campaign against "tropical oil" imports in the late 1980s. The advertisement states: "Meet the man who's trying to put you out of business." *Soybean Digest*, March 1986.

until 1986, when the American Soybean Association (ASA) launched a media blitz aimed at evicting "tropical oils" from the American market once and for all (figure 9.1). The ASA rallied farmers behind the campaign, claiming that imports were "taking $1,465 out of the pockets of every U.S. soybean farmer."[57] One side of the campaign targeted manufacturers, pressuring them to drop "unhealthy" saturated fats. The other side targeted consumers directly, warning them that companies were "hiding" unhealthy fats in popular foods. A paper distributed at a meeting of soybean farmers read: "We want to use the tropical fats issue to our advantage, to milk it for all it's worth."[58]

For the next three years, lobbyists, doctors, and representatives from food companies and palm oil producers assailed each other in the press and in congressional hearings over the "tropical oils" issue. Indonesia's ambassador charged that fat labeling proposals aimed at shaming companies into dropping palm oil were a "disguised, but pure form of protectionism." Malaysia's ambassador chimed in with a warning that disappointed oil palm smallholders might embrace communism, a claim echoed by Filipino representatives for the coconut industry.[59] A two-pronged counterattack spearheaded by Malaysian organizations challenged the health claims leveled against palm oil, while painting the palm oil industry as a free-market upstart fighting a coddled, state-subsidized soybean industry.[60] Food companies were exasperated: palm and coconut oils did useful things in foods. They kept cereal flakes crunchy in milk; they extended the shelf life of Twinkies. The "tropical oils" made up only 3.5 percent of America's fat intake in the 1980s.[61]

The ASA "was well aware of the emotive power of health issues," however.[62] I have a vivid memory of my mother skipping my favorite cookies in the supermarket, noting they had dangerous "tropical oils." My family wasn't alone in dropping palm oil. Though small to begin with, per capita consumption of palm oil in the United States fell by 41 percent after the ASA campaign. Companies voluntarily removed "tropical oils" from their products by late 1989—and boasted about it on new labels. The ASA campaign had no lasting impact on soybean prices, but the blow to palm and coconut exporters was "devastating."[63] Prices for crude palm oil didn't collapse overnight, but by 2001, they were half what they had been in the late 1980s. The U.S. "oil war" wasn't solely responsible for palm oil's woes, of course: the arrival of a popular new rapeseed oil from Canada ("Canola") and booming soybean production in America, Argentina, and Brazil kept the world market awash with fats.

But the fight over fat wasn't over. One lobbying group, the Center for Science in the Public Interest, celebrated the replacement of tropical oils with partially hydrogenated blends of soybean and other oils.[64] Partial hydrogenation converts some fatty acids into saturated forms but turns others into trans fatty acids. Normally found only in bacteria and the guts of ruminants, trans fatty acids remain unsaturated but behave more like saturated fats in foods. They let partially hydrogenated soybean oil mimick the physical characteristics of palm oil, with significantly less saturated fat to report on a label. As public health advocate David Schliefer points out, scientists questioned the safety of trans fats as early as the 1950s. Experts employed by the food industry intimidated researchers and tried to quash studies probing the health risks of trans fat consumption.[65] Yet by the 1990s, new studies showed a startling link between trans fats and cardiovascular disease. By 1994, the Center for Science in the Public Interest demanded regulation of trans fats, reversing course.

One food company employee told Schliefer, "We spent a million bucks and . . . proved the science was right, and then we had to do something."[66] That meant reformulating recipes for everything from margarine and ice cream to boxed cake mix. The trans fat problem wasn't just an issue in Northern countries, either. Thanks to the U.S. government's largesse with cottonseed and soybean oil, people across the global South ate trans fats in all sorts of products. Even palm oil was partially hydrogenated, especially in vanaspati, to get the right texture and melting point. Some vanaspati samples tested by researchers were more than 50 percent trans fats, a shockingly high figure.[67]

Palm oil emerged as the most economical replacement for trans fats in most food products. The Palm Oil Research Institute of Malaysia (PORIM, now the Malaysian Palm Oil Board, MPOB) funded studies challenging trans fats and defending palm oil in the 1980s and 1990s, laying the groundwork for the switch. When Dr. K. C. Hayes and colleagues in the United States published a 1991 paper showing that a blend of monounsaturated, polyunsaturated, and saturated fats could lower "bad" cholesterol (LDL) and raise "good" cholesterol (HDL) levels in monkeys, PORIM invited the scientists to Malaysia. Collaborating with Malaysian colleagues in human trials, they patented a cholesterol-lowering margarine recipe made with what was now branded "palm fruit oil" and other fats. Hayes recalled the typical response from the food industry at the time: "It's a tropical oil for God's sake, it would never sell in America." But under the "Smart Balance" brand, it did sell—and its market share surged as more and more news about trans fats appeared in the media.[68]

The innate properties of palm oil—rather than its price alone—mattered a great deal at this moment. With a balanced amount of saturated and unsaturated fatty acids, RBD palm oil could replace partially hydrogenated fat in bakery recipes and deep-fryers. Unilever ordered its factories to get rid of trans fats in 1994, reformulating more than 600 blends of fats for consumer products. After months of planning, the company replaced all partially hydrogenated fats with palm or palm kernel oil on a single day in 1995.[69] Unilever moved ahead of regulators, but other firms, like the small Baltimore bakery responsible for a local delicacy known as the "Berger Cookie," despaired as the FDA's 2018 date for a trans fat phaseout loomed. The owner tried to reformulate the recipe, but found the results "nasty . . . The texture's not there. It's an entirely different product." As the ban took effect, the bakers were surprised to find that their supplier already replaced trans fats in the cookie's fudge topping with "palm shortening" without any fanfare. The bakery and its customers hadn't even noticed.[70]

The soybean lobby's triumph over "tropical oil" was short-lived. By the 2000s, regulators around the world banned trans fats outright. With consumers unwilling to return to animal fats, palm oil was the only cost-effective solution.[71] Marketers in Malaysia also tapped into growing anxiety in the global North over genetically modified foods. Palm oil and palm olein replaced genetically modified soybean and rapeseed oil for brands seeking a "natural" or "GMO-free" label.[72] MPOB continues to tout the health benefits of palm oil around the world, pushing back against lingering ideas that palm oil is unhealthy with advertising and clinical studies. Skeptics in the World Health Organization, who still call for restricted saturated fat consumption, likened its work to the tobacco industry's defense of smoking.[73] But healthy or not, palm oil is here to stay in the food system.

Building New Markets: Oleochemicals and Agrofuels

Malaysian companies and institutions played a key role in expanding global markets for palm oil in the 1980s and 1990s. PORIM encouraged manufacturers everywhere to switch to palm oil, pointing to the benefits that manufacturers could claim on labels—all vegetable, cholesterol-free, "natural," GMO-free, and more.[74] "Halal" palm oil made without any trace of animal fat—not just in food, but in food-safe machine lubricants—appealed to companies selling to Muslim customers.[75]

The Malaysian government also used palm oil in diplomacy, offering palm oil on credit to developing nations.[76] In 1994, the country bought MiG-29 jets from Russia, paying for 20 percent of the price in palm oil. It bought more jets with palm oil in 2003 and negotiated with Russia and India in 2019 to buy yet more aircraft in exchange for promises to buy palm oil.[77] Malaysia also offered palm oil as a form of finance for overseas development projects. One unsuccessful proposal called for financing Istanbul's third Bosporus bridge with palm oil: Turkish partners would sell the oil in Turkey to raise capital, repaying Malaysia later. Packages like this were intended to ease oversupply pressures in Southeast Asia and to push back against competing commodities like sunflower and rapeseed oil.[78] Using palm oil in politics cut both ways, though. Malaysia could create new demand for palm oil with its diplomatic partners, but with many other oils available on the market, those partners weren't beholden to Malaysian suppliers. India's government dramatically cut import permits for palm oil in 2019–20 in retaliation for political spats with the Malaysian government.

PORIM and MPOB-backed research also created new markets in oleochemicals, especially biodiesel, conjuring up new demand for the vast supply of palm oil Malaysia was exporting. European and American soapmakers launched the oleochemical industry in the late nineteenth century, when they perfected methods for separating fatty acids and glycerin from fats. Glycerin and products derived from it wound up in resins and gums, explosives, and tobacco (as a sweetener). Other uses included sausage casing, cellulose film, gaskets, glues, candy, and paper.[79] Chemists found even more uses for fatty acids. Soap and candles consumed most of the supply well into the mid-twentieth century, but a new class of chemicals derived from fatty acids—surface active agents, or surfactants—had applications in "almost the entire gamut of industrial chemistry."[80] Surfactants change the way molecules interact. They help water lift stains from fabrics or dishes. They make your bathwater bubbly. In mining, surfactants separate light and heavy minerals; large mines in central Africa still use palm oil to extract copper and cobalt.[81] Textile and leather manufacturers use surfactants to spread dyes and oils over and into materials. Other surfactants work as emulsifiers, keeping products like salad dressing and lipstick from separating.

Most people had no idea how important the palm kernel was in the sudsy powder they washed dishes with in the 1920s or dumped into new electric washing machines in the 1930s. The early washing flakes and powders still relied on soaps (a simple surfactant), but they used the special properties of the lauric acid found in kernel oil. Lauric soaps dissolve in cold and mineral-rich water where other soaps fail, creating an impressive foam that consumers inaccurately associated with cleaning power. Palm oil could be replaced by all sorts of fats in soap, but palm kernels and coconuts are the only major sources of lauric acid.[82]

By the 1930s, chemists had developed ways to transform fatty acids into fatty alcohols, which were then made into new detergents like sodium lauryl sulfate. This and other synthetic detergents spread from laundry soap to dishwashing powder and liquid shampoo, all pressed on consumers with a relentless barrage of advertising. The revolution in cleanliness wasn't all good news for the oil palm industry, however. A pound of kernel oil transformed into detergent had the cleaning power of 3.5 pounds of palm oil turned into ordinary soap.[83] Even worse for the oil palm industry, chemists soon synthetized detergents from fossil fuels. As the price of petroleum fell

FIGURE 9.2 Oil derricks sprout palm fronds and fruit bunches in this 1943 Shell advertisement promising to "grow our own palm trees" in response to wartime palm oil shortages. The ad copy asserts that chemists were writing a "charter of freedom" from nature, turning petroleum into new substances without "the shortcomings of natural palm oil." Author's collection.

after 1945, the prospects for natural fats looked grim.[84] Unilever—despite owning oil palm plantations—jumped into fossil fuel chemistry as a matter of survival in the 1950s, battling the blockbuster success of Procter & Gamble's "Tide" laundry powder.[85]

As a Shell advertisement from 1943 asserted, chemists were writing a new "Charter of Freedom"—freedom from nature (figure 9.2). "When we could get palm oil and rubber from trees, and silk from the silk worm, it was natural to take what Nature provided. We won't do that any longer," Shell declared. Shell's own petroleum substitute for palm oil in tinplating vanished after wartime shortages eased, but a government-backed tallow substitute finally knocked palm oil out of the U.S. tinplate industry in the late 1950s. A few thousand tons remained in use annually in cold-rolling steel, however.[86]

Synthetic detergents were marvels of modern chemistry, but they created environmental problems. Petroleum-based detergents didn't biodegrade as quickly as plant-based ones. Cities found their sewers, treatment plants, and rivers overflowing with foam in the 1950s and 1960s. The surfactant properties that made detergents so effective created visible evidence of how consumer choices were impacting the natural world. Phosphates added to boost the cleaning power of detergents caused more problems, fueling algae blooms. Manufacturers blamed municipalities for not developing tools to treat detergent-laden wastewater, but public opinion in Europe and North America ultimately forced manufacturers to change their formulas. Most countries banned branched alkylbenzene sulphonate, the most common fossil fuel–derived detergent, in the 1960s.[87]

The backlash against petro-detergents breathed new life into fat-based oleochemicals, helped along by rising petroleum prices in the 1970s. By 2000, palm and coconut oils had reclaimed more than 60 percent of the market for fatty alcohols used in detergents and other products.[88] Beyond price and regulatory pressures, marketers saw advantages in touting the "natural" or "plant-based" content of products made with palm oils. By the 1990s, palm oil–based lipstick and "environmentally friendly" detergents made from palm kernels were joined by "green" plastics, agrofuels, and a host of more obscure products made from both of the oil palm's fats. One of the most ironic "green" uses for palm oil–based surfactants is in petroleum drilling and fracking.[89]

Tallow is still common in oleochemicals: unless a soap specifically claims to be plant based, it may well contain tallow. But the Northern factories processing most tallow no longer dominate the oleochemical industry. Asian producers using palm oil and kernels now have far more capacity, with Malaysia playing a leading role.[90] Malaysia built its first oleochemical plant in 1979, part of a broader strategy aimed at adding value to plantation products. Export taxes on unrefined palm oil meant that Malaysian refiners and oleochemical plants had as much raw material as they wanted, and tax holidays rewarded "pioneer" industries for setting up new plants.[91] By the 1990s and 2000s, European and American firms had sold off most of their oleochemical plants, focusing on consumer goods rather than intermediate products.[92] MPOB also sponsored key research that allowed palm oil to join kernel oil in detergents, fabric softeners, inks, and other products.[93]

MPOB invested early in what is now the fastest-growing segment of the oleochemical industry: agrofuels. (Agrofuels are a subset of biofuels; the latter includes wood and other nonagricultural fuel sources.) Palm oil isn't turned into fuel because it is cheaper than petroleum—not by a long shot. Instead, state mandates for biofuel blending or substitution created an artificial market. These policies, in turn, were driven by a mixture of political imperatives: pressure from environmental activists to reduce CO_2 emissions, as well as pressure from palm oil producers to boost consumption and prices.[94]

Palm oil and other oil palm products have a long history as fuels, of course. Africans burned palm and kernel oils in lamps, and they built fires with oily fruit fiber. Kernel charcoal is still an important fuel source for cooking in Africa's oil palm belt. And since the dawn of the oil palm plantation, oil mills have been firing their boilers with kernel shells and bunch stems, turning plantation waste into energy. POEM used cakes of fiber and shell to power locomotives in Gold Coast in the 1920s, and HCB used palm oil as fuel oil in riverboat boilers. The Belgian colonizers offered a 30,000 franc prize for a motor that could run on palm oil, hoping to ease pressure on forests caused by wood-cutting to fuel boilers.[95]

Palm oil works as fuel oil, but when it and all triglycerides combust, they leave behind soot and sticky residues that damage motors. They also lose their fluidity in cold weather. A Belgian scientist solved both problems by converting palm oil into a more volatile chemical, palm oil methyl ester (POME), in 1937. Colonial officials reportedly ran a bus for 20,000 kilometers on this fuel. It wasn't cost effective, though, and the idea languished.[96] Even hard palm oil fit only for soap was worth more than petroleum in the 1920s and 1930s.

Brazil was the first country to take a major step toward replacing petroleum, expanding a 1941 ethanol-blending policy for gasoline with a new diesel program in the 1970s as petroleum prices shot up. Yet Brazilian officials ultimately relied on the country's burgeoning soybean and cattle industries for fatty acids rather than on oil palm.[97] As one researcher pointed out in 1980, a ton of palm oil could buy two tons of petroleum. Brazil would have needed to plant more than five million hectares with oil palm to meet all of its diesel needs, covering more land than all the oil palm plantations in the entire world at the time.[98] And that was just Brazil. Replacing petroleum with palm oil across the world was staggeringly impractical.

Up to the 1990s, debates over biofuels centered on the cost and availability of petroleum—how far away the world was from "peak oil." That changed with growing concern over CO_2 emissions and climate change. Many countries introduced subsidies to promote biofuels, reckoning that they were "carbon neutral" or even carbon-sequestering. Colombia adopted a series of aggressive biodiesel goals in the 2000s, bringing together global interest in "green" fuels with the political elite's desire to support the country's fast-growing oil palm plantation industry. Indonesia, Malaysia, and Thailand also embraced biodiesel, ramping up POME production with an eye on export markets. For producing countries, the biodiesel market highlighted the oil palm's potential as a "flex crop." The term was coined to describe crops that could be shifted from one market to a significantly different one based on prices; the effect is to turn palm oil into something approximating a "universal commodity."[99] Palm oil had a long career "flexing" between food and soap, but oleochemicals and POME expanded the range of options dramatically.

The European Union created the biggest opening for POME in 2009, when it passed new biofuel mandates. Rapeseed, animal fats, and recycled cooking oils provided most of the raw materials for EU biodiesel, but palm oil became the fastest-growing input. Because large refiners in Southeast Asia and South America could "flex" their oil toward food or fuel, they were able to quickly take advantage of the new market. In 2010, only about 8 percent of the European Union's palm oil imports wound up as fuel. In four years, the figure shot up to 45 percent.[100]

Politicians who once condemned agrofuels as a radical environmentalist boondoggle now see them as a way of preserving the status quo. Agrofuels let societies keep fossil-fuel vehicles and infrastructure, while also propping up agribusiness.[101] But "flexing" food crops into fuel has consequences. A series of "biofuel food riots" in 2007 illustrated the dangers of letting rich countries burn food crops for energy, though ethanol rather than POME deserved the blame in this case.[102] So far, the main impact of POME has been in relieving overproduction, a problem created by the fast-growing oil palm plantation industry itself. It may well impact food prices in the future.

Proponents of POME argue that oil palm plantations capture nearly as much CO_2 as natural forests and that POME is actually a carbon-negative fuel, capturing more CO_2 in the plantation than burning POME releases. But many scientists disagree. One researcher who formerly advocated for agrofuels now argues (in a study funded by the petroleum industry, we should note) that they fall far short of expectations when all of the carbon emissions involved in agriculture and processing are counted.[103] The figures for palm oil are especially concerning. One estimate found that new plantations on peat soils would need to make POME for more than 400 years to replace the carbon released from disturbed peat.[104] As chapter 11 emphasizes, everything hinges on where and how oil palm is grown.

Sustainability as Commodity

Whether it takes the form of a fuel or food, palm oil enters most of our lives unnoticed. It moves in bulk tankers, sold by the ton, a uniform commodity that even manufacturers can't trace back to a single plantation or smallholder plot. As consumers, we have no idea where the palm oil in a bottle of shampoo or stick of margarine came from, who grew it, and under what conditions. And with no sensory connection to palm oil—we don't see it, taste it, smell it, feel it in the manufactured products we use—we have little to remind us about the oily stuff.[105]

Environmental and human rights activists tried to change this state of affairs beginning in the 1980s. Campaigns against deforestation in the Amazon and Indonesia's transmigration program (chapter 10) brought environmental issues to the forefront. Tropical forests became the "lungs of the world" in Northern rhetoric. Charismatic species like the orangutan served as mascots for the whole of tropical biodiversity. Humans also became symbols: forest peoples sometimes appeared as little more than "noble savages" threatened by corporate greed, rather than people with interests and goals of their own. Environmental and human rights organizations kept up a steady drumbeat of criticism against plantation development in the tropics through the 1980s and 1990s.[106]

But what really caught the world's attention was smoke. Haze from burning forests and peatlands blanketed Southeast Asia in 1997–98, drawing global attention to the scale of plantation expansion in the region.[107] As anthropologist Anna Tsing has argued, the fires were a "crisis of visibility." The choking haze forced people to confront political and economic forces that had long been driving deforestation in the region. Palm oil wasn't invisible anymore.[108]

Linking forest and biodiversity loss with climate change, activist groups—mostly based in the North—argued that tropical deforestation threatened the entire planet. Activists latched on to the idea that palm oil was "hidden" in everyday products: one Australian newspaper called it the "most evil ingredient hidden in foods."[109] A website warning about "deadly palm oil in your shopping trolley" insisted that "Consumers have the right to know if the oil used in a product is contributing to the destruction of rainforests and the slaughter of wildlife."[110] If photographs of sad orangutans didn't move readers, they couldn't help but notice warnings that palm oil contained large amounts of saturated fat, reviving the health claims of the "oil wars." Plantations and food conglomerates weren't just killing the forests; they were killing *you*. Lists of palm oil's "hidden names" still circulate widely on social media, presenting the many chemical lives of palm oil—palmitate, palm glycerides, stearate, palmityl alcohol, and many more—as a conspiracy to shovel an unwanted product into our bodies.[111]

The fast growth of social media gave activists a powerful tool to raise awareness. One notorious video produced by Greenpeace in 2010 targeted Nestlé and Sinar Mas, a major plantation company. An office worker tears open a Kit-Kat wrapper and removes an orangutan's finger. Colleagues look on in horror as he bites down, sending blood spurting. Riffing on Nestlé's marketing slogan for the Kit-Kat, the video asks viewers: "Give the orangutan a break. . . ." The UK supermarket chain Iceland won a viral advertising coup of its own in 2018, when regulators barred it from airing an anti palm oil advertisement on television, resulting in millions of online views. The cartoon features an orangutan bounding around a girl's room; through a flashback, the ape shares a harrowing vision of machines destroying the forest to make way for a plantation. The girl launches a palm oil boycott to protect what's left of the orangutan's home. The ad was reminiscent of another widely circulated video purporting to show an orangutan "fighting" an excavator in a fresh forest clearing.[112]

As Peter Dauvergne points out, "shock campaigns" like these create outrage, not organizations. They succeed in making visible the invisible, of exposing our personal complicity in commodity webs that can cause great harm to distant people and places. But moving from guilt to action has historically not been a simple process. Abolitionists boycotting "slave sugar" in the 1820s didn't end slavery any more than activists boycotting Nike in the 1990s ended sweatshops.[113] When companies acknowledge consumer pressure, they usually respond with slogans and certification programs rather than systemic changes in the way they make, buy, and sell things.[114] In short, they mitigate the worst excesses of the system without reforming the underlying power disparities that make exploitation of people and natural resources possible.

Representatives of the oil palm industry in Southeast Asia claim that environmental concerns merely disguise Northern protectionism: "European vegetable oils . . . cannot compete [with] the palm oil."[115] But Northern consumers represent only a modest share of global demand for palm oil.

Boycotts—like the European Union's vote to ban POME in 2020—can move prices, but they can't restructure the political economy of the plantation system.[116] As one director of an environmentally-minded palm oil company put it, "the rest of the world doesn't give a shit" about environmental issues.[117]

As chapter 11 shows, people living alongside the oil palm industry care deeply about environmental issues and their connections with land and labor rights. Boycotting palm oil does nothing to help them. It's clear that palm and palm kernel oil are going to be major features of the twenty-first-century economy. Chemistry continues to open new markets for palm oil, and the liberalization of trade—coupled with rising incomes in the global South—means that more people than ever will come to rely on palm products in their lives. Oil palms will feed, wash, and perhaps fuel the world. Unlike in the last century, they won't do it invisibly. Consumers in the North and South have unprecedented access to information about the commodity chains linking them to plantations and smallholder farmers. What remains to be seen is what they'll do with that knowledge.

10 The Oil Palm's New Frontiers

"Slaves of Our Own Native Land"

In the 1970s, the World Bank and other development organizations pitched oil palm as a perfect crop for developing the tropics: it produced much needed food and promised to be an endless source of cash as the global market for palm oil grew. But oil palm didn't inevitably bring wealth and prosperity. People in the Philippines sent an urgent letter to the Commonwealth (formerly Colonial) Development Corporation in 1982, pleading with the agency to halt its funding of a new plantation. The authors claimed they had been tricked into signing away their land with "sugar-coated promises," accusing the plantation management of trying "to make us slaves of our own native land." The plantation's security forces answered their protests with beatings, rapes, and murders.[1]

The plantation in question was NGPI, a CDC-financed partnership between Guthrie and the Philippine government in Mindanao.[2] NGPI's managers had hired a paramilitary group called the Lost Command for security. Led by Col. Carlos Lademora, the Lost Command was supposed to be fighting the communist New People's Army, but his men spent their time terrorizing and extorting civilians.[3] Earlier, protests by farmers had scared off other plantation companies seeking land in the region. NGPI used Lademora's men to acquire land titles. "The arms they carry along bring out the message very clearly to the farmer," wrote a local priest.[4] Having sold their land at gunpoint, farmers found jobs on the new plantation "a precarious thing."[5] Chief Datu Lumintap, head of a displaced indigenous community, asked: "Where do we go? Surely the British will not welcome us in England, wherever that is."[6]

As the story began to attract media attention in Britain, a CDC spokesman dismissed the complaints as "the sort of grievances you get in most places." The CDC had "an excellent record in projects of this kind" and could "make a genuine contribution" in Mindanao.[7] But NGPI's manager, Bruce Clew, candidly told British officials "that the acquisition of land from otherwise reluctant peasant farmers, squatters and minority tribesmen would

FIGURE 10.1
A figure with a Union Jack superimposed attacks another figure with a bayonet in this poster (ca. 1984) urging readers to "Stop this plantation!" Text below the image warns that huge tracts of land in the Philippines were being seized for cash-crop export production financed by the CDC. "That's why the Third World starves," the poster declares, adding: "Your taxes are paying for it." OD 71/106, TNA. Reproduced courtesy Campaign for Human Rights in the Philippines UK.

have been impossible without Lademora's help." When asked why he had hired such vicious thugs, Clew responded: "They exist."[8] Clew saw the approximately 2,000 families displaced by the plantation as squatters with no legal claim.[9] In the eyes of one official, he was "an old-fashioned planter" who showed "considerable contempt for the inefficiencies of the small Filipino farmer." "I cannot say even now that I like some of his methods," the official remarked.[10]

British humanitarian activists received the same harrowing letters describing evictions, murder, torture, and exposure to toxic chemicals.[11] Together with environmentalist allies, they argued that the CDC—a government-backed investment body—was funding the exploitation of Filipinos and their forests to feed Northern consumers (figure 10.1). They were wrong about the ultimate purpose of the oil: it was all slated for domestic consumption. But their campaign added palm oil to a growing list of commodities seen in Britain and throughout Northern countries as threats to both the environment and human rights.[12]

The CDC dismissed the Lost Command as an "extraneous matter," but to soothe anger over the plantation's land grab, they offered what had become the CDC's hallmark: a Nucleus Estate-Smallholder (NES) project. Having pioneered the concept, the CDC boasted it knew "how to co-operate with rural people and help them out of poverty."[13] Flying British MPs over the area, CDC manager Derek Nesbit contrasted the orderly plantation with the scars of forest fallow cultivation. The visitors asked why the CDC hadn't stopped this "slash and burn" agriculture. Nesbit replied: "they just might if influential organisations who knew nothing about the subjects stopped interfering with their plans."[14] In a later report, the CDC asserted that planting oil palm repaired "previously degraded land," claiming that "ecologically, the effect of a perennial tree crop like oil palm is similar to that of reafforestation."[15]

When local activists pushed the CDC to abandon the plantation and develop the whole area for smallholders, the CDC retorted that without the plantation, there would be no smallholder oil palm industry. Indeed, they insisted that "more enlightened policies" toward smallholders relied on the success of a plantation.[16] Father Ton Zwart, a leader in the local opposition, noted that the NES scheme would chain smallholders to the company and a single commodity that they couldn't eat, process, or sell themselves. "CDC may well be the best corporation of its kind," he stated, "but it is the kind which is the problem."[17] Under "a general atmosphere of fear," the priest and nearly 700 others signed letters opposing the NES plan.[18]

In private, CDC and Philippine officials agreed that Lademora and the Lost Command were "a great embarrassment."[19] They pressed NGPI to hire a new security force, and NPGI also agreed to pay people more for their land. Clew forced plantation workers to choose between the money and their jobs, however. Lademora's men lingered around the plantation, too. Workers reported intimidation, unfair wages, and unsafe working conditions, particularly for those spraying chemicals.[20] One man was murdered and several others tortured in 1983, likely at the hands of the Lost Command. One British official accused Clew of being in "basic sympathy with Col. Lademora."[21]

When someone cut down fifty-two hectares of oil palms and plastered towns with antigovernment posters, the Lost Command answered with violent reprisals.[22] As many as twelve farmers were shot and killed in an area slated for plantation development.[23] The CDC insisted that the victims were New People's Army (NPA) guerillas, and that its model of "enlightened development" was exactly what Mindanao needed to fix a "progressively deteriorating situation."[24] Neither the Lost Command nor NGPI's new guards could keep the NPA away, however: NPA fighters seized weapons and burned down several structures in a daring December 1984 raid on the plantation.[25] Clew—who survived a brush with "people's justice" in an impromptu trial—claimed that the raid was "a catharsis in the NPA's attitude to the plantation." The CDC suspected he had simply paid the NPA off, as he had done with Lademora's men.[26] By this point, the CDC decided to withdraw from Mindanao: it refused to pull existing loans for NGPI, but it canceled the NES extension, pending "a change in the political scene."[27] Government troops swept in following the December raid. Their presence brought no peace: paramilitaries killed at least five men in 1985, including Rogelio Noble, a labor organizer and father of seven who was shot in front of his family for planning a strike.[28] The CDC washed its hands of the bloody affair, acquitting itself and Guthrie—which soon sold the plantation—of any wrongdoing in a 1987 report.[29]

Not all parts of the oil palm frontier were this contested or bloody. Oil palm cropland grew to cover a vast area in the postcolonial era: from 3.6 million hectares in 1961 to 10.3 million in 2000, topping 20 million hectares by 2018.[30] This new oil palm belt was geographically and politically diverse, but the NGPI episode highlighted key elements of the oil palm's new history across the tropics. Organizations and ideas inherited from the colonial era shaped the future for plantations and smallholders alike. Former colonial officials often staffed international development agencies, and they deployed old ideas like NES projects as panaceas for development problems. Britain's CDC held stakes in at least thirteen oil palm projects by 2000, and the scientific and managerial influence of France's IRHO spread just as far.[31] Colonial-era companies like Guthrie, SOCFIN, Unilever, and Harrisons & Crosfield could be found nearly everywhere oil palms sprouted, though they were joined by upstart competitors.

The Mindanao case also showed the importance of local conditions in shaping outcomes. Although the politics of the Cold War overshadowed new oil palm projects up to 1990, local needs—notably domestic demand for fats—took precedence over global imperatives in project planning. National governments, rather than international lenders or multinational firms, made key decisions about how, where, and at what cost oil palms would grow. And while smallholders and plantations pressed deeper into tropical forests and peatlands in many parts of the globe, in other places, they successfully repurposed old banana plantations, cattle pastures, and rubber plots to make room for oil palm. The histories of rural spaces, the strength of local and national institutions, and the presence or absence of state (or private) violence were key factors in determining whether new oil palm landscapes looked like NGPI in Mindanao, or were more positive stories, like the smallholder oil palm booms in Costa Rica or Thailand.

Part I: Old and New Frontiers in Asia

Plantations and Smallholders in Malaysia

Experts and industry insiders often point to Malaysia's postcolonial experience with oil palm as a model for economic development. Competent technocrats oversaw agricultural and then industrial development, funded by palm oil exports. Private-sector plantations led a shift to *tenera* palms that—combined with new agronomic practices—quadrupled palm oil yields between 1950 and 1990.[32] Planters and government agencies collaborated to study and introduce the African oil palm weevil in 1981, ending nearly a century of reliance on hand-pollination and the mediocre performance of indigenous insects (most importantly, *Thrips hawaiiensis*).[33] Instead of nationalizing foreign-owned plantations, Malaysia's government oversaw their gradual sale to *bumiputera* ("sons of the land") owners.[34]

Malaysia's political elites simultaneously backed the smallholder-settler model for oil palm, hoping "that Malay small-scale agriculture could replace foreign plantations as a dynamo for export growth."[35] The NES experiment at Kulai had proven the viability of the model, and officials were eager to replicate it across Malaysia. Farmers were eager, too, applying to join new settlements and writing for help switching from rubber or coconuts to oil palm.[36] Agencies like FELDA sought out "the best land for the best people," putting what they viewed as "waste" land to good use.[37] FELDA dropped the nucleus estate from its projects early on, however. It took space away from potential settlers, and settlers proved more reliable in delivering fruit than the NES model assumed. They had nowhere else to sell their fruit, and they faced eviction if they failed to meet standards.[38]

By 1961, FELDA realized that unleashing machete-wielding settlers on the jungle was slow and expensive. From then on, contractors cleared trees, planted crops, and built houses for settlers. Instead of earning land with their sweat, settlers had the cost of development added to their debts, to be repaid out of future earnings.[39] The size of settler allotments shrank to four hectares. Separate *dusun* plots for fruit trees vanished, covered in more oil palms. Even rubber settlers had to put in oil palms on their *dusun* land, wherever a nearby factory could handle the fruit.[40] Most significantly,

FELDA converted land titles into shares, lumping land into eighty-hectare units worked by teams.

Critics warned that FELDA treated smallholders "as if they were the indentured labour on a commercial estate."[41] One settler who built an unauthorized addition on his FELDA-issued home had to tear it down to maintain uniformity.[42] Settlers themselves complained about managerial discipline, and they vociferously opposed the switch from landholding to shareholding. Some said: "[Prime Minister] Tun Razak promised us land, not a piece of paper."[43] But discontent with the new system didn't stop FELDA from planning more settlements into the 1970s.

The most emblematic project was the gargantuan Jengka Triangle scheme, which broke ground in Pahang in 1965. The World Bank–funded project was projected to host 100,000 settlers by 1985, and both Bank and U.S. officials saw it as a key step in demonstrating the success of capitalist development in a corner of the world where the Cold War was heating up.[44] While officials depicted the region as an uninhabited stretch of swamp and forest, at least 110 indigenous families used the area, and seventy-five people were displaced early in the project. They weren't offered—and probably didn't want—a place in the settlements.[45]

The forest was no match for contractors armed with chainsaws, bulldozers, and herbicides. "The jungle is falling like grass in Jengka these days," wrote one journalist. Workers "[pushed] back the forest almost magically—picking up fruit and orchids as they go, and sending wild animals scuttling." Tun Razak declared, "The forest must make way for development. The people's hunger for land must be satisfied."[46] He wasn't shy about linking economic development with politics, promising voters in Kelantan their own "massive land development scheme on the scale of the Jengka Triangle" if they voted for his party in 1969.[47]

According to the Bank, the environmental impact of Jengka Triangle was "less severe" than expected. FELDA controlled soil erosion with cover crops and contouring. New effluent treatment systems kept palm oil mill waste out of waterways. Still, it was clear that cutting down 100,000 hectares of forest "had a considerable effect in terms of reduction of wildlife population." At least twenty-four elephants, forty forest oxen, and five Sumatran rhinos lost their homes. The elephants didn't take the invasion passively: one herd returned to feast on 79,000 young palms before it was relocated at great expense.[48]

By the 1980s, the trees had matured, and settler incomes were reaching three times the rural poverty level. The plantations and oil mills were financially sound.[49] But critics noted that FELDA had "helped the favored few" rather than the neediest families. Moreover, the project was fantastically expensive. At $15,000 U.S. dollars per family, it was the most expensive rain-fed agricultural project the Bank had ever financed.[50] Bank officials weren't optimistic, noting that settlers' children were unlikely to live under FELDA discipline.[51] A later study concluded that the Jengka Triangle would "export thousands of youths to the major urban centres."[52] Women shared wealth only in the context of marriage: men had to be married to get into settlements, and titles remained in men's names. Women gathered loose fruit that fell from harvested bunches but had to negotiate with their husbands to see any income from their efforts.[53] As settlers got richer, they started new businesses and hired migrant workers to tend the palms, a typical outcome in Malaysia's smallholder projects.[54]

World Bank and Malaysian officials often pointed to FELDA's work in West Malaysia as an exemplar for the world. Yet the model was never successfully replicated elsewhere, even in East Malaysia.[55] FELDA wasn't sanguine about exporting its model to Sabah or Sarawak: its experience in West Malaysia involved "settling a group of people whose ethnic traits and attitudes toward work are known and well understood." FELDA worried that "We have not understood the prospective settlers [in East Malaysia] well enough."[56]

The oil palm era began in East Malaysia during the last days of British rule, with a CDC project at Mostyn in Sabah. As the project matured, it was supposed to add smallholders before selling off the plantation "to the Sabah public."[57] Harrisons & Crosfield partnered with CDC on the project, though it only put up £50,000 next to the CDC's £1.5 million. Unilever soon opened its own plantations in Sabah.[58] Sabah farmers weren't enthusiastic about oil palm, however. Smallholders around Mostyn chased a cocoa boom instead; one man recalled that "Every coffee shop . . . spoke [of] nothing else except the planting of cocoa and its lucrative returns."[59]

Sarawak's first big oil palm project was another CDC-backed NES scheme, launched in 1968.[60] British officials agreed to finance it explicitly because of the benefits it promised to smallholders, yet the Sarawak government soon abandoned the settler component.[61] By the time CDC sold its stake in 1990, the project was simply a giant plantation in the hands of private owners, Sarawak Oil Palms Bhd. FELDA's own foray into Sarawak in the 1980s through a subsidiary company didn't even make a pretense of including smallholders, and it met strong resistance from Dayak landowners.[62] FELDA's "Sahabat" complex, launched in 1980 on a thinly populated peninsula in eastern Sabah, did include smallholders, and it was backed by $70 million from the World Bank.[63]

But the 100,000-hectare project was mainly an exercise in converting forest to plantation; only five of the thirteen planned villages were ever settled.[64] Most of the plantation was worked by migrant laborers.[65] The smallholder-focused "development era" that began with Kulai proved short-lived.

Part of the problem with the NES model in East Malaysia was resettlement: communities were not as pressed for land as their compatriots in West Malaysia, and West Malaysians showed little interest in migrating east.[66] Sabah and Sarawak smallholders did take up oil palm in the 1970s and 1980s, though, especially when the prospects for cocoa soured with falling world prices and the arrival of the damaging cocoa pod borer. Many participated in "in situ" development projects, where state agencies worked with private firms to incorporate entire villages into new plantations, turning farmers into either smallholders or shareholders in the venture. These projects had the advantage of working within customary land tenure systems (or at least official interpretations of them), but respect for customary land often proved to be a cover for selling off huge tracts of forest for the benefit of private firms.[67] Oil palm often followed on the heels of logging, and elites in Sabah and Sarawak amassed great fortunes through logging concessions.[68]

While the Cold War inspired the United States, World Bank, and other actors to support the Malaysian palm oil boom, it success hinged on factors particular to Malaysia. Politicians giving out smallholder plots through NES projects were more interested in winning votes than in fighting the distant threat of a communist insurgency, and smallholders were often enthusiastic participants in projects that turned them into small-scale oil palm capitalists. Along with big plantation companies, state-backed smallholders pushed the oil palm frontier in Malaysia far beyond what private firms had accomplished up to Malaysia's independence in 1957.

By the mid-1980s, Malaysia was the world's leading exporter of palm oil, set on a trajectory that doubled Malaysia's palm oil output between 1980 and 1990 and nearly doubled it again by 2000. Yet this incredible growth was soon eclipsed by an even bigger boom in Malaysia's neighbor to the south, Indonesia.

Indonesia: Transmigration and the Forest Frontier

Indonesia followed Malaysia down the road of state-led smallholder development. While Indonesian politicians insist on the novelty of their smallholder model, it was in fact another iteration of the colonial-era NES concept.[69] In terms of size, Indonesia's oil palm project has been a smashing success: beginning at 100,000 hectares in 1970, Indonesia's oil palm coverage grew to 1 million hectares in the early 1990s and to 10 million hectares by 2015. There may well be more oil palms in Indonesian monocultures today than in all of Africa's remaining "wild" palm groves.[70]

Perhaps 40 percent of Indonesia's oil palms are held by smallholders in one form or another, and much of the rest grew out of "nucleus" plantations.[71] The NES choice, researchers Rob Cramb and John McCarthy argue, wasn't born out of idealism or faith in peasant agriculture. Rather, it reflected a convergence of factors: "the interests of a developmental state with limited resources, plantation companies needing access to land, smallholders wanting to improve their livelihoods, but lacking finance and technology, and donor agencies promoting broad-based economic growth."[72] Those donors played an important role. The World Bank invested more than half a billion dollars in Indonesia's oil palm industry, helping resettle tens of thousands of people to grow oil palm, in addition to other projects focused on rubber, cocoa, and rice. That's more than the Bank spent on *all* of its oil palm projects in Africa, Latin America, and the rest of Asia combined.[73]

Indonesia's two key resources, according to a Bank report, were "Javanese labor and unexploited land in the other islands."[74] After independence, the government began a "transmigration" program settling Javanese farmers in Sumatra, Kalimantan (Borneo), and other islands.[75] Suharto put Indonesia on the "right" side of the Cold War in the aftermath of a 1965 coup, and the Bank rushed in with money for ambitious new projects, including oil palm plantations. Existing plantations seized from colonial companies in Sumatra were the first targets, but Bank officials urged Indonesian officials to follow the Malaysian model and use NES oil palm settlements for new transmigration schemes in Sumatra and Kalimantan.[76] From the Indonesian point of view, smallholder participation justified expropriation of land by state companies. It also ensured that Bank financing continued to flow. By the late 1980s, though, Indonesia increasingly entrusted new settlements to private companies, creating "large-scale conglomerates controlled by Suharto's political, military and corporate allies."[77]

The Suharto regime (1966–98) promoted oil palm with two goals. First, it wanted to keep food—especially cooking oil—cheap for Indonesian consumers.[78] Second, it aimed to transform "waste" lands into productive assets. Anticipating criticism from environmental activists, Indonesian officials—and some Bank officials—argued that plantations weren't actually bad for the environment. They claimed that "slash and burn cultivation practiced by indigenous people was far more destructive" than forest clearance for oil palm monoculture, echoing colonial sentiments.[79] Planners expected Javanese migrants to bring permanent agriculture and the tools of the Green Revolution to outlying islands.[80] And as one scientist noted, "the settlers themselves are no lovers of forest. Forest is alien to most Javanese and is perceived as a source of spirits, ghosts, and pests. As a result, they are quite happy to see it felled."[81]

Transmigration and plantation expansion created conflicts everywhere they went.[82] The Dutch never produced a countrywide register of land in the colonial era, and the Indonesian government found that "empty" land usually had occupants. Sometimes companies and officials struck deals with locals to sell land or to include them as smallholders. Some farmers found that their farms and trees were added to NES projects "without their consent and without compensation."[83] Sometimes developers—backed by the army—just took the land. The state-backed business sector that emerged in the 1970s–80s became, as anthropologist Anna Tsing describes it, "a predator . . . born from the mix of nepotism, international finance, and military muscle, and feeding on cheap resources ripped illegally from rural communities."[84] With a total price tag of more than $7 billion, transmigration projects offered endless opportunities for graft. Foreign lenders marveled at how money "vanished into thin air."[85]

Suharto's regime stifled most opposition to transmigration and plantation development, but it could not stop all protests. In one incident in Sumatra, the army killed more than 100 transmigrants in clashes over land: the migrants wanted fertile hillsides, rather than the valley they had been assigned. Only a few days earlier, settlers in another project chased off plantation workers with swords and spears.[86]

Forest communities often felt the worst effects of oil palm expansion. Like their counterparts in Malaysia, Indonesia's forest peoples usually lacked formal title to land.[87] In the 1980s, some officials—including the Minister of Transmigration—took explicitly chauvinistic stances, promising that "The different ethnic groups will in the long run disappear because of integration," resulting in "one kind of man."[88] Indonesia officially recognized customary law (*adat*), but *adat* law was no guarantee that people could actually protect their land against powerful officials, businessmen, or elites in their own community.[89]

One Bank-funded oil palm project in Sumatra targeted lands of the Orang Rimba people. Though they "had asked to be left alone" in 1984, the government seized their forests for an oil palm NES project.[90] When Orang Rimba representatives protested, they were told "to go complain to Suharto."[91] Despite having newly adopted rules on the protection of indigenous peoples, the Bank ignored Orang Rimba complaints.[92] The Bank suggested resettling them in a nearby forest reserve (now Bukit Duabelas National Park), but even this was under threat by the 1990s as loggers and hunters preyed on its resources. One study predicted that the Orang Rimba would soon "resemble their counterparts in Peninsular Malaysia . . . surrounded on all sides by oil palm" without room to forage and practice forest fallow agriculture.[93]

Indonesia's plans for oil palm, rubber, and rice projects in Irian Jaya (the western half of Papua) in the 1980s inspired some of the fiercest resistance. Building on criticism of Bank-funded settlements in the Brazilian Amazon, international NGOs publicized the links between plantation and NES development with environmental degradation and the displacement—often backed by violence—of forest peoples.[94] While Papua's forests lacked charismatic megafauna, forests in Sumatra and Kalimantan were home to species like the orangutan, Sumatran tiger, and Sumatran rhino. The voracious rate of forest clearance in Indonesia—perhaps 1 million hectares annually since the 1960s—threatened the survival of these iconic species along with hundreds of human communities.[95] As NGOs realized, defending indigenous rights was a way of defending ecosystems, and vice versa.

Yet international activists tended to remain stuck in what one critic called an "orang-utan phase."[96] Asia's great ape became a symbol of tropical biodiversity under threat, but the emphasis on orangutans often obscured the stakes for people—as forest dwellers, settlers, workers, and consumers of palm oil. The orangutan-centric strategy provoked outrage in the North, but little tangible action. International pressure and domestic reform did little to protect either the environment or indigenous peoples from the oil palm industry's growth, which accelerated at an unprecedented pace into the 1990s and 2000s.

Papua New Guinea

The young oil palm industry in Papua New Guinea (PNG) attracted much less attention from environmentalists; today it's often promoted as a case study in best practices for both environmental and labor issues. PNG was an Australian colony until 1975. The World Bank saw Malaysia's NES projects as a perfect model for this rural and underdeveloped country, so long as oil palm projects were "established under the commercial management of experienced private enterprise."[97] Harrisons & Crosfield (H&C) signed a partnership with Australia for the first NES project at Cape Hoskins on the island of New Britain in 1967.

Getting people to Hoskins was a daunting affair, involving the resettlement of thousands of people with little or no experience growing cash crops. Oil palm was an alien plant for everyone.[98] A sociologist studying the early settlers noted their high expectations for the crop and warned that they would be disappointed.[99] According to an H&C employee, there had been no serious thinking about how settlement would play out, and the settlers were a highly diverse set of migrants from Papua rather than the relatively homogenous groups selected for settlement in Malaysia or Indonesia. Among other early blunders, H&C's staff accidentally set the Dagi River ablaze several times with leaking mill effluent and chemicals.[100]

Rumors and conspiracy theories swirled among the settlers, and the first years were full of brawls. Clan justice required men to avenge attacks on their comrades, inspiring cycles of retaliation.[101] In 1971, Highlanders from Papua got into a fight with a Tolai supervisor, Peter Tavip, and killed him with an axe. When a detachment of Australian police arrived, 300 men stepped forward to confess to the crime, foiling the investigation. No one was surprised when a Highlander turned up dead, likely the victim of a revenge killing.[102] Other killings followed, and police and settlers traded arrows and gunfire at one point.[103] Settlers also feuded with local Nakanai families, who had sold land for the project. A militant cargo cult formed among the Nakanai, and in 1993, one Nakanai group stormed palm smallholdings and evicted settlers.[104]

Subsequent NES projects funded by the World Bank, CDC, and other lenders in Papua also led to conflicts between settlers and locals.[105] Cultural tensions sparked many fights, but the deeper issue was land. Under PNG law, land held in customary tenure is inalienable and can only be leased, in exchange for gifts. In swidden agriculture, the stakes of land gifts were low: most plots were used for a few years at most. Oil palms were effectively permanent, sparking demands for the return of land or higher rents.

None of these problems meant that Hoskins, Popondetta, and other NES sites were failures. New H&C management at Hoskins salvaged the operation, helped by high prices for palm oil in the 1970s.[106] Across PNG, the oil palm projects gave settlers cash incomes and generated export revenue for PNG's government. Settlers weren't trapped in an endless cycle of ethnic hatred and violence, either. They came to joke about one anothers' "ethnocentrism," and Hoskins settlers pushed the management to adopt a 1977 ordinance granting powers to expel troublemakers. Today, children in ethnically mixed community schools identified as members of the "*wel pam* (oil palm) tribe."[107]

Settlers also experimented with oil palm agronomy, testing different methods rather than trusting the company's advice.[108] Families adapted to the individualistic nature of the NES system, segregating oil palm work and earnings from the customary sharing of food and labor.[109] Company managers eventually abandoned settlements and shifted to a model akin to the "in situ" smallholder projects in Sabah and Sarawak.[110] New Britain Palm Oil Ltd., the Hoskins plantation company, developed a particularly innovative way of addressing gender inequities in its "Mama Lus Frut" system. The company only paid male household heads for fruit, but men didn't want to spend time picking up fruit fallen from bunches. Their wives had no legal claim to fruit earnings, but they could withhold their labor. "Why should we help the men buy beer?" asked one woman.[111] Armed with a company-issued "Mama card," women began to sell loose fruit directly to the mill. The money went into their own bank accounts. Men started paying women with extra fruit in exchange for weeding and other work: fruit was easier for men to part with than cash, which men faced great social pressures to spend. Although the PNG experience with oil palm was far from perfect, it showed that the NES model could be adapted to work in cultural and political contexts that were radically different from those for the pilot at Kulai.

Part II: Latin America

New oil palm industries took root in Central and South America for many of the same reasons that Malaysia, Indonesia, and PNG embraced the tree. As in Southeast Asia, the Cold War was an important political backdrop for the oil palm story in Latin America. The cast of international institutions was similar, though they had no colonial baggage in the region. Yet national and local dynamics were especially important in determining where and how oil palm took off as a cash crop.

Colombia, the region's biggest palm oil producer today, launched its industry with domestic capital backed up by foreign expertise. In contrast, one giant multinational firm, United Fruit Company (UFC, now Chiquita Brands), was responsible for the introduction and expansion of oil palm in Central America, particularly in Honduras and Costa Rica. Across the region, new oil palm projects created the same issues that they did in Asia: land disputes, violence, and environmental degradation. Unlike Asian cases, however, the Latin American oil palm industry took root in a region with a long history of effective labor and peasant organization, limiting the ability of firms to impose plantations on communities—and raising the stakes of conflict in ways that sometimes resulted in horrific violence.

Central America: Costa Rica and Honduras

United Fruit Company and its rival, Standard Fruit, dominated the economies, politics, and landscapes of Central America for much of the twentieth century. Bananas made the companies rich, though the same couldn't be said for the places and people that grew those bananas. As one UFC employee put it, "the chasm between what companies like United Fruit put into an indigenous system and what they take out" was "the stuff that revolutions are made of."[112] And as workers mobilized and fought for higher wages, sometimes forging alliances with peasant farmers eager to take back land from plantations and landlords, revolutions came.

Banana companies owned more than a million hectares by the 1930s, in part because bananas were profitable, but also because nature was working against them.[113] Panama disease, caused by *Fusarium* fungi, forced the companies to practice a form of shifting cultivation across the region's lowland forests. When *Fusarium* ruined a tract, the companies would pick up everything—down to the railroad tracks—and move on.[114] By the 1950s, Panama disease had practically wiped out the desirable *Gros Michel* banana, and another plague, Black Sigatoka (caused by *Mycosphaerella fijiensis*), was on the march. To fight Sigatoka, companies sprayed fields with "Bordeaux mixture."[115] The chemical itself (copper sulphate and slaked lime) was not highly toxic. The problem was the sheer amount used in what turned out to be a futile effort to stave off Sigatoka. Workers sprayed so much of it that they reportedly turned blue. Repeated treatments sterilized land with copper, making it useless for bananas or food crops.[116] The combined threat of fungal disease and copper contamination meant that bananas were "a profitable but temporary crop."[117]

One of the few crops that could tolerate elevated copper levels was the oil palm. UFC had been experimenting with oil palms since the 1920s, when it collected seeds from Africa and Sumatra. Diseases and rats damaged early test plots, and UFC showed little interest in the crop. That changed during the Second World War, when UFC responded to U.S. government calls for new sources of palm oil.[118] By 1945, UFC was planting oil palms alongside cocoa and abaca, searching for the next big plantation crop for Central America. While *Fusarium* can attack oil palms, the company didn't see it as a serious threat. Because Sigatoka didn't infect palms at all, abandoned banana land could be profitably repurposed. UFC could even reuse the cableways and trams that moved banana bunches for palm fruit.[119]

By the 1950s, UFC's diversification campaign was in full swing, with the wholehearted support of the U.S. government. UFC's oil palm project promised "to revolutionize the vegetable oil situation" in Central America, with palm oil selling for half the cost of lard.[120] The company had more than 17,000 acres of oil palms (almost all Deli *dura*) across the region by 1958, but it grew less convinced that oil palm was the crop of the future. UFC had twice as much land under cocoa and five times more under sugar, along with its 145,000 acres of bananas.[121] As its plantations matured, the company argued that oil palm "shows lower returns both to the local economy and to the company than banana operations," defending its decision to stick with bananas as long as possible.[122]

UFC approached Unilever about an oil palm partnership in 1960, but Unilever turned down the offer. Afterward, UFC's new plantings slowed down significantly.[123] The company realized that it needed new *tenera* palms and expeller mills to compete with Asian competitors in the export market, but it wasn't willing to take the risk alone. On top of this, UFC faced growing labor problems. A 1954 strike in Honduras—and the unwillingness of the government to crush it—marked the beginning of the end for the UFC plantation system in Central America.[124]

In both Costa Rica and Honduras, UFC did find eager customers for its palm oil. Numar, a Costa Rican margarine firm launched by an American entrepreneur, bought all the oil UFC could make. Numar popularized margarine in the region, where most people had never heard of it or "believed it was inferior butter." Numar spent $5,000 a month on advertising and cooking demonstrations to convince people to buy the stuff. It worked so well that UFC bought Numar in 1965.[125] Feeding the domestic market in Costa Rica and neighboring Honduras was profitable, thanks to stiff tariffs. But as historian Patricia Clare Rhoades has argued, these tariffs made UFC "a prisoner of its own protected trade." In Costa Rica, the government included margarine in a "basket" of consumer goods, setting prices and restricting UFC's ability to export palm oil.[126] UFC and its successor companies made money, but they had few incentives to invest in a big expansion until market liberalization took hold in the 1980s.

Farmers, rather than plantations, were responsible for Central America's next phase of oil palm production. Hungry for land that banana companies were hoarding, farmers demanded agrarian reforms in the 1960s and 1970s.

Worried about a repeat of the Cuban revolution, governments tried to find ways to give farmers land without killing off the lucrative plantation sector. With the approval and financial support of Uncle Sam, experts brought forward the ready-made NES model. The idea of coordinating smallholder cash-crop production with a larger industrial venture was not new to Latin American: banana companies often bought fruit from smallholders. An even more similar precedent was the *colono* system practiced in parts of Caribbean, in which sugar smallholders sold their highly perishable crop to crushing mills hosted by larger plantations.[127]

Rather than hand over the big plantations in the 1950s and 1960s, the Honduran government encouraged farmers to colonize forests and "waste" lands. One such area was the Aguán valley, which UFC abandoned in the 1940s after Panama disease struck. With U.S. funding, the government relocated thousands of families and built a series of palm oil mills to serve smallholder cooperatives. Settlers agreed to accept low prices for their fruit in exchange for cooperative ownership of oil mills, though it took a 1980 strike before farmers finally secured control.[128]

Honduran settlers found that although they owned their palms and the oil mills, companies run by a handful of powerful families monopolized downstream oil refining and manufacturing. This held prices, and farmers' profits, down.[129] And the Aguán was far from a paradise. Settlers carved out homes in harsh conditions, struggling through a damaging 1974 hurricane and large debt repayments. Few of the cooperatives were commercial successes, plagued by debt as well as "labyrinthine webs of embezzlement and fraudulent accounting."[130]

Costa Rica saw a similar shift toward smallholders in the 1970s as peasant activists led invasions of vacant fruit company land. By 1975, twenty cooperatives with more than 1,100 members had been formed to produce palm oil, along with 350 independent farmers. Like small-scale farmers everywhere, they preferred to balance oil palm with food crops. Farmers let cattle graze under their palms and only put half of their land under oil palms, despite the exhortations of Palma Tica, UFC's successor and the monopoly buyer of palm fruit in Costa Rica.[131] Even though more people now owned land, their economic conditions didn't always improve. Some smallholders found themselves trapped in "disguised peonage" after accepting seedlings and loans from Palma Tica. Earnings in the oil palm sector were typically lower than in the heyday of the banana boom, and fewer people found work. In Costa Rica's Pacífico region, nearly 15,000 people worked on UFC's banana estates in the 1950s; only 6,000 worked in oil palm by 2003.[132]

South America: The Colombian Experience

Like the Central American states, South American oil palm ventures were motivated by food security. The UN's Food and Agriculture Organization urged countries across the region to grow oil palm in the 1950s, but most governments moved slowly to promote the crop.[133] In Colombia, however, different strands came together to create the conditions for an oil palm boom. Despite amenable environmental conditions, *E. guineensis* didn't take root in Colombia under Spanish rule or in the post-independence period. Afro-Colombians reproduced African systems of "agroforest" cultivation in the lowlands of western Colombia—one anthropologist said he thought he "was entering a rain forest" on his first visit to their farms—but oil palms weren't included.[134] Instead, Afro-Colombians exploited native palms, including the "vegetable ivory palm" (*Phytelephas* species), harvesting nuts for the world's button-making factories and eking out a living in a nation that denied them formal ownership of land.[135]

The oil palm arrived in Colombia several times in the early twentieth century. Catholic missions planted a few as ornamentals, and state research stations hosted experimental plots. In the interwar period, Florentino Claes, a Belgian scientist, provided new seeds and urged the government to take up oil palm. But the government was committed to another oil crop, the castor bean. Claes told a government minister: "the castor bean only makes shit," referring to its effect on the digestive tract. But his pitch for oil palm failed.[136] One oil palm plantation dating to the 1930s was cut down in 1955, and UFC opted to abandon its wartime oil palm experiments in favor of bananas.

A Colombian entrepreneur, Moris Gutt, was a key figure in getting the oil palm industry off the ground after 1945. Gutt's firm, GRASCO, produced soap, margarine, and chemicals. He was unhappy about his dependency on Philippine coconuts and Peruvian fish oil. Colombia imported an average of 50,000 tons of edible oil every year in the 1950s.[137] Gutt approached the government for help, and Colombia turned to the FAO. The FAO in turn sent Maurice Ferrand (a veteran of Belgian Congo's oil palm sector) to tour the country in 1958–59. Ferrand decided Colombia's Pacific coast was "of first order quality for oil palm."[138] He argued that it was "of great social and economic interest" to include marginalized Afro-Colombian farmers in oil palm development, suggesting an NES project.[139]

Meanwhile, Gutt had been connected with the IRHO, which dispatched experts of its own to identify a plantation site in the center of the country.[140] Gutt's "Indupalma" at San Alberto was a straightforward plantation with no smallholder involvement. Local farmers complained about being tricked or coerced into selling their land for the project, and Gutt employed a former military commander as manager, who reportedly visited landowners "flanked by two gun-slingers."[141] Indupalma soon set about transforming forest and scrub into a scientific plantation. IRHO experts supplied *tenera* seeds, tackled nutrient deficiencies in the soil with fertilizers, and applied chemicals to eradicate unwanted plants and insects. *Imperata*, the invasive grass so loathed by Southeast Asian planters, proved no match for doses of 2,4,5-T (one of the active ingredients in Agent Orange). When experts saw beetles preying on young palms, they sprayed Heptachlor (now banned in most places). When caterpillars crawled into fruit bunches, the answer was aerial spraying with Carbaryl powder. Armed with science and technology, planters thought they were winning a war on nature.[142]

But natural forces proved to be allies of Colombia's nascent oil palm plantation sector. By the late 1960s, cheap Peruvian fish oil threatened to drown the Colombian market just as the country's palms matured. Relief came in the form of *El Niño*. The 1972–73 weather event nearly wiped out Peru's fishing industry. Fish oil disappeared from the market, leaving buyers hungry for palm oil. The Colombian government urged even more plantation development and offered subsidies for estates larger than 500 hectares.[143] Palm oil's moment in the sun came too late for Afro-Colombian smallholders working on Ferrand's experiment, however. Most had abandoned oil palm by 1970, finding that they could not afford to wait for the palms to mature. The smallholder share of Colombian palm oil dwindled to insignificance in the 1980s.[144]

In addition to battling the environment, Indupalma fought to control the nearly 2,000 workers it needed to operate the plantation and oil mill. As workers unionized and demanded better wages, the company turned to subcontractors working without benefits. In the midst of bitter disputes, the company accused union leaders of murdering the plantation's chief of staff in 1971. It was an early incident in what became a string of killings. One of the most dramatic episodes was the 1977 kidnapping of Indupalma's manager in Bogotá by the M-19 movement, in solidarity with the union. The company ultimately gave in to union demands and stopped subcontracting.[145] In the meantime, the company allegedly turned to paramilitaries for help, labeling union activists as guerilla sympathizers. Nearly 100 Indupalma workers and six union presidents suffered violent deaths in the 1980s and 1990s.[146]

The dynamics that inspired much of the killing in Colombia—conflicts over land and labor—weren't unique to that country. Guatemala and Honduras also saw spates of deadly violence as oil palm plantations grew at the expense of smallholders. In contrast, smallholders in Costa Rica and Ecuador generally had been able to keep their land, selling fruit to mills or participating in cooperatives. Nothing in the ecology or economics of the oil palm created violence; instead the arrival of the oil palm industry exposed the tensions and injustices already present in rural communities. Whether oil palm functioned as a tool for development or for oppression hinged on local and national histories, not on global political actors and economic forces.

Part III: African "Failures" in Context

History also shadowed the oil palm's career in postcolonial Africa. In many accounts, this is a story of failure, caused by political intrigue or corruption. The continent's once-mighty exporters collapsed while dynamic competitors in Southeast Asia seized the world market. The 1980s were a "lost decade"; by the end of the century, palm oil exports from Malaysia and Indonesia were worth more than "the total value of *all* agricultural exports from *all* of sub-Saharan Africa."[147]

Yet writing off the African oil palm experience as an episode in "entrepreneurial failure" or "stagnation" misses the story.[148] Violent conflict wrecked the oil palm industry in Africa's two biggest producers, Nigeria and Congo. Big increases in populations everywhere meant that oil formerly destined for export never left the continent. Just like projects in Latin America and Asia, internationally funded projects emphasized domestic food security, reflecting a "basic needs strategy" that dominated Western aid policy in the developing world by the 1970s. Well-fed people, the theory went, were less likely to upset the international order.[149] Many oil palm projects for Africa designed by foreign experts were failures—some disastrously so—but there were success stories. Big plantations played a role, but smallholders proved to be dynamic entrepreneurs once conditions for oil palm production grew more favorable.[150]

Nigeria and Congo

Nigeria's share of world palm oil exports was still more than 30 percent at independence in 1960, and it was 25 percent by the time civil war broke out in 1967. Exports fell to practically nothing in the 1970s, and the country began importing huge quantities of vegetable oil to meet domestic demand.[151] Congo, which held another quarter of the export market in 1960, met a similar fate. As chapter 8 emphasizes, the primary reason for this collapse was political rather than economic or ecological. Nigeria's civil war ravaged the oil palm belt. Hundreds of thousands died, and many more were driven off their land. The few oil palm plantations that existed were destroyed along with their oil mills, and people cut down palms to put in badly needed food crops.[152]

Congo's bitter conflict with the breakaway state of Katanga had no direct impact on the oil palm industry, but the uprising in Kwilu in 1963–68 (which cost somewhere between 60,000 and 100,000 lives) was very much linked to the colonial legacies of oil palm. Fruit cutters formed a militant corps of organized radicals, aiming to topple the remnants of the colonial system and the economic inequality it entailed. The revolt, led by a former HCB trainee, spread across a broad area as migrant workers brought the movement home from labor camps.[153] The failed Kwilu rebellion and others like it didn't lead to an immediate collapse in Congo's oil palm exports, however. That came after 1973, when General Mobutu nationalized foreign plantations. Unilever and other European firms lost their estates, and even though they got some of them back by 1977, the damage was done. Years of underinvestment and decline ensued.[154]

Congolese staff struggled to maintain what had been INEAC's research station at Yangambi, but they never got the funding to resume large-scale work on oil palms.[155] Officials approached the World Bank for help in the late 1970s, but their experts wrote off what was left of the elderly "grove" palms once coveted by Lever. Even doubling prices for fruit couldn't convince men to climb such tall trees. The Bank urged the government to abandon palm groves and the smallholders who worked them (and made perhaps a third of the country's palm oil). The best bet for producing oil—and paying off loans—was "rehabilitation of existing commercial oil palm estates."[156] Foreign lenders paid half the cost of a nearly-$50 million project to replant plantations owned by Unilever and other firms. Unilever sold off other properties, dating to Lever's 1911 treaty, instead of replanting.[157] Yet the new *tenera* palms didn't produce enough to feed the country, and by the 1990s, Congo was firmly among the ranks of exporters-turned-importers of palm oil.[158]

Nigeria's experience with oil palm rehabilitation was somewhat better, though it still wasn't enough to stave off imports. By 1980, Nigeria imported more than 50,000 tons of palm oil, a figure that quadrupled before a 1986 import ban took effect. Imports surged again in the 1990s as firms imported crude (rather than refined) palm oil, dodging the ban.[159] As in Congo, the key issue was where to spend money: plantations or smallholders? As early as 1965, the CDC was separating investments in nucleus estates and smallholder schemes in Africa, preferring to finance the former. One official called plantations "the more important part of the [NES] venture."[160] The World Bank wound up paying for smallholder components but took a pessimistic view of their economic value in Nigeria.[161]

The Bank and other foreign lenders did ultimately agree to finance a huge "grove rehabilitation" project in Nigeria, but the civil war ruined those plans. When foreign lenders returned to Nigeria in the mid-1970s, their expectations for oil palm were limited, focused on solving "domestic shortages and Nigeria's impending dependence upon foreign supplies."[162] Injections of money kept the Nigerian Institute for Oil Palm Research (NIFOR) running, but efforts to self-finance its research through sales of seedlings and oil from its experimental plots didn't come close to meeting the institute's needs.[163] Meanwhile, NIFOR's expatriate staff had decamped to Malaysia, where public and private research institutes reaped the benefits of decades of experience in Nigeria, funded by taxes on African farmers through the marketing boards.[164]

The Bank supported four smallholder oil palm projects in southern Nigeria in the late 1970s.[165] SOCFIN planned one project in Rivers States and another in Imo State, calling for farmers to use hybrid tree varieties, chemical fertilizers and pesticides, cover crops, and other trappings of the modern plantation.[166] There was no room for intercropping or "wild" palm groves.[167] As the projects unfolded, SOCFIN complained about "poor smallholder discipline," urging the government to remove all old palms and ban food crop interplanting.[168]

After ten years, the Bank compared the results of the Rivers and Imo State projects. Although the nucleus plantations did well, smallholders in both projects fell 30 percent short of fruit-delivery targets. SOCFIN's experts insisted that this proved the need for a plantation to keep oil mills running. They also stressed the benefits of professional (i.e., expatriate) management, blaming Nigerian staff for inefficiency and political intrigue.[169] Smallholders had no voice in project design, but they succeeded in convincing managers and NIFOR experts to change some policies. The Imo project managers "accepted, albeit reluctantly" farmers' demands to grow food crops among the palms.[170]

Farmers saw clear benefits in growing a diverse range of crops to hedge against disease, drought, and market fluctuations.[171] Importantly, Imo farmers agreed to sacrifice their old palms so long as they could grow food. They saw the value of *tenera* palms, so much so that farmers continued to plant them even after the cheap fertilizer provided by the Bank disappeared in 1985. Farmers outside the project area also planted *tenera* seedlings. Bank officials praised Imo farmers for their "enthusiasm and business-like attitude" toward oil palm. Many were in fact veterans of late colonial tree-replanting projects, with plenty of knowledge about *tenera* palms and "scientific agriculture." Families still made oil in the household when local prices were attractive, but they sold to the mill and even harvested palm groves outside their own plots when the mill offered better prices.

In contrast, the Rivers State project, Risonpalm, banned intercropping. Land for a nucleus estate was seized from eleven communities, who complained about the loss of fallow farmland and resource-rich forests.[172] Risonpalm staff found that smallholders were reluctant to fell old palms and plant new ones; they had ample access to "wild" groves in addition to palms on their farms. Risonpalm's managers quickly adapted to this reality. The mill machinery was designed for *tenera* fruit, but local farmers began harvesting so many grove palms that the mill began accepting mixed *dura* and *tenera* fruit.[173] Risonpalm bought thirty trucks to collect fruit from what was becoming a bustling "wild" palm district, instead of the two that the budget had called for. Risonpalm also allowed farmers to buy up to one ton of palm oil for every twenty tons of fruit they delivered, recognizing the importance of women's oil-selling income in local marketplaces.[174]

Neither farmer initiative nor adaptable management could protect the oil palm sector from the other kind of oil Nigeria is famous for. Petroleum might have made Nigeria rich. Instead it fed a kleptocratic elite and polluted huge areas of the Niger Delta.[175] With an overvalued currency, Nigeria's exports became uncompetitive. Domestic wages skyrocketed in the 1980s, and young men flocked to towns, stripping plantations and smallholders alike of the labor needed to harvest oil palms.[176] The nucleus plantations and the smallholders they were supposed to serve suffered major neglect in the 1980s and early 1990s, as politicians squabbled over what to do with these debt-ridden projects.

Ivory Coast: An Oil Palm Success Story

The World Bank and other international lenders failed to resuscitate the palm oil export economy in Nigeria and Congo, contributing to a growing view in the West that Africa was a "hopeless continent." The Ivory Coast followed a different path. Today at least 40,000 farmers operating small and medium-sized oil palm plots work alongside plantations. Ivorian smallholders produce about 70 percent of the country's industrially processed palm fruit, using 72 percent of the land planted with oil palms. Together with the plantations, Ivorian smallholders have achieved what no other African country has managed in recent years: domestic self-sufficiency in palm oil along with a big surplus for export.[177]

Ivory Coast embarked on NES-type projects shortly after independence, guided by experts from the IRHO and funded by European and international lenders. The initial focus was on domestic food needs, but the government also hoped to diversify the country's export sector, which was dominated by cocoa and coffee. Oil palms could grow on soils unsuited for coffee or cocoa, opening up new land for cash crop production.[178] The government's *Plan palmier* envisioned "Industrial Plantations" working alongside smallholders organized into "Village Plantations." The latter got loans to plant and maintain *tenera* palms, with a six-year grace period (the palms bore fruit in year 4 or 5). Smallholders and plantations both benefited from state land policy, as the government opened up colonial-era forest reserves. It also "streamlined property rights," offering secure land titles to "whoever puts it to productive use." Traditional chiefs worked with government officials to enforce new property rules.[179]

Ivory Coast initially partnered with two colonial-era companies, SOCFIN and Blohorn, to run state-owned plantation and oil-milling firms. It bought out the private firms in 1976 but invited them back a few years later.[180] The *Plan palmier* called for the replacement of colonial-era hydraulic oil presses with expellers, which entailed the replacement of *dura* palms with *tenera*.[181] With help from the IRHO, experts ensured that plantations and smallholders planted high-yielding trees selected for the regional environment. The World Bank, CDC, European Union, and the *Caisse Française de Développement* all provided capital to finance the projects, which today cover more than 250,000 hectares. The original plan called for smallholders to take over the plantations, but this was scrapped in favor of running them with wage labor.[182] Good roads built for the project ensured that outlying farmers could sell their produce to the mills, located up to twenty kilometers away in the "industrial plantation" zones.[183]

Relations between smallholders and the Société pour le Développement du Palmier à Huile (SODEPALM), the state plantation company, weren't always smooth. Initially managers banned intercropping, but they soon gave up on enforcement. Farmers mixed oil palms and food crops as they had done for centuries.[184] SODEPALM also struggled to convince farmers to use chemical fertilizer. In historic cultivation patterns, oil palms *are* fertilizer, building up the soil with organic matter. Farmers were skeptical about the need for chemicals, the cost of which came out of their fruit sales. Yet they were obligated to buy fertilizer from SODEPALM by their fruit-selling contracts; many simply resold it.[185] Still, state firms kept prices fairly constant, riding out the highs and lows of the global market until the government liberalized pricing in the 1990s. When smallholders thought prices were too low, they often sold fruit in local markets, though their *tenera* fruit proved less popular than *dura*.[186]

Ivoirian farmers also saw palm wine as an important source of income. Foreign experts often wrote about wine-tapping as a "wasteful" practice—harkening back to colonial debates over palm wine—but farmers weren't acting irrationally.[187] Palm wine was the logical end of a crop cycle, one that began with cassava and other food crops, shifted to oil palm as the trees matured, and ended with wine-tapping and land-clearing to start the cycle again.[188] A 1983 study found that a hectare of palms felled for wine in Ivory Coast was worth four times what a year's worth of fruit earned, without taking into account the cost of harvest labor. In places like southeastern Nigeria where more benign forms of tapping are practiced, palm wine is seen as a permanent "money spinner," paying tree-owners more than oil and creating jobs in tapping, transport, sales, and distillation.[189] Palm wine is still an important part of the oil palm story in Africa, but it's one that foreign agencies and experts ignored or actively opposed.

Uneven Development across the Oil Palm Belt

The NES model didn't deliver miraculous results in other countries. In the early 1980s, SODEPALM moved across the border into Liberia in a $37 million project financed by a cast of international lenders. A Malaysian plantation firm provided managers, rounding out a truly global development project. It was a fiasco from the start.[190] Global palm oil prices slumped, crushing hopes for exports. The Bank's experts recognized that making oil by hand and selling it locally proved far more lucrative for smallholders than selling fruit to the mill. Still, Bank officials griped that Liberians were corrupt and lazy, practicing "poor husbandry." Liberia's government never got a chance to respond. As the Bank's report tersely states, "The Borrower could not provide [a response] due to the ongoing civil war and the collapse of the Government."[191]

Civil war in Sierra Leone (1991–2002) also wrecked oil palm projects, on top of the tens of thousands of lives lost. The World Bank and CDC tried NES projects in the country beginning in 1972, hoping to recycle colonial-era Pioneer oil mill equipment.[192] CDC experts designing the project relied on lingering colonial stereotypes, and some were in fact former colonial officials who failed the grasp the gendered economics of harvesting, making, and selling palm oil in Africa.[193] As in the Liberian case, farmers found that they earned more selling fruit locally or by tapping "valueless" family labor to make oil at home.[194] Men told researchers that they no longer knew how—or cared to—climb oil palms.[195] When the project closed in the early 1980s, the Bank suggested that Sierra Leone drop smallholders and focus on plantations, though the start of the civil war halted new work.[196]

To the east, Ghana escaped the horrors of civil war. Yet its policies toward oil palm were, in the words of one critic, "comparatively schizophrenic and extremely modest" when contrasted with the "evangelical zeal" for oil palm shown in Ivory Coast.[197] The first major oil palm project was the 1975 Ghana Oil Palm Development Company (GOPDC). Relying on Ivoirian and IRHO expertise, GOPDC hoped to revive a defunct Nkrumah-era "state farm."[198] Unilever agreed to a similar project in southwestern Ghana in the same year (Benso Oil Palm Plantation).[199] Early settlers fought with GOPDC over food crops, and the company abandoned resettlement in favor of contracting with "outgrowers," independent smallholders not formally tied to the NES project.[200] Together, plantation workers and outgrowers launched a modest recovery in Ghanaian palm oil production, though by 1990 their output still couldn't meet domestic demand, to say nothing of an export surplus.

Cameroon came closest to matching Ivory Coast's success with oil palm. A mixture of private plantations and government-owned estates dating back to the German occupation fed the domestic market and generated respectable export earnings into the 1970s. At the urging of the World Bank, plantations took on outgrowers, though with less enthusiasm and on a smaller scale than in Ivory Coast.[201] Economic and environmental woes battered the industry in the 1980s, however. High wages kept workers job-hopping, preventing the development of the skilled workforce needed to prune and harvest aging oil palms.[202] A drought in 1983 slashed palm oil yields, and by the mid-1980s, plantations and smallholders were locked in a price war with each other and with imported palm oil from Malaysia. Smallholders won the fight, and the plantations were forced "to dump their produce onto the world market, suffering major financial losses."[203] By 1990, Cameroon had the distinction of remaining self-sufficient in palm oil but had little left over for export. Unilever, the biggest private plantation owner, sold off its Cameroon holdings as part of a strategic shift away from plantation ownership.[204]

Setting the Stage: Postcolonial Development and the Structures of Growth

National and local politics, economics, and ecologies were what made or broke oil palm projects in case after case. The conflicts among farmers, workers, and plantation owners in places as far apart as Mindanao and Colombia were superficially similar, but they reflected histories that long predated the arrival of *E. guineensis*. What links the stories of peasant activists invading banana-company land in Costa Rica with those of Javanese settlers in Kalimantan or Nigerian smallholders negotiating with state-owned companies was, of course, the crop they all raised—and the forces that pushed that crop above others. It could have been cocoa, coffee, rubber, cotton, or a host of other cash crops. But oil palm had special appeal for policymakers at national and international institutions, because it seemed so versatile. It grew in many tropical soil types, met pressing domestic food needs, and offered the dream of cash-earning exports in the future. It was the perfect crop for planners looking for "permanent" crops that could anchor farmers—and local and foreign capital—in an industry with seemingly unlimited demand.

As young palms spread in neat rows across three continents, they were always overshadowed by the state. Corporations did press the oil palm frontier outward—think of Guthrie in Mindanao; Unilever and a clutch of others in Sabah and Sarawak; Indupalma in central Colombia—but the role of the state was important everywhere. At the most basic level, states continued to open land for development, classifying it in ways that denied forest peoples, "squatting" farmers, and others a legal stake. NES projects provided political cover to plantation companies: for the CDC, World Bank, and other lenders, smallholder participation justified loans that paid for infrastructure, mills, and trees that could well have been financed by private capital.

All told, smallholders who stuck with oil palm did fairly well through the 1980s. But the "neoliberal turn" of the late 1980s and 1990s led to stark reversals in many places. Farmers lost subsidized loans and fertilizer, and they faced new threats from imported palm oil as their governments implemented austerity programs and trade liberalization. More startling, state-owned plantations wound up in private hands as governments sold off assets at fire-sale prices. Many of Indonesia's state plantation and NES ventures wound up in the hands of domestic and foreign investors. Malaysia's FELDA transformed into a global holding company, trading in smallholder development for mega-agribusiness. And in Africa, old companies like SOCFIN—joined by new competitors headquartered in Singapore, Jakarta, and Kuala Lumpur—got what they had never been able to achieve in the colonial era: legal title to big tracts of land for plantation oil palms.

11 Globalization and the Oil Palm

It was a familiar scene: across several oil palm plantations in Lever's Congo concessions, villagers shared harrowing tales of abuses and stolen land. But the year was 2019, not 1919. A century after they were taken, these lands were in the hands of Feronia, a Canada-based agribusiness. With money from the CDC and other European development lenders, Feronia bought what was left of Lever's Congo empire to "save it from extinction."[1] The CDC insisted that Feronia restored "severely neglected trees;" rescued 8,000 jobs held by "barefooted" workers; and repaired houses, clinics, and other infrastructure. Yet workers and community members told NGOs about intimidation, inadequate wages, exposure to toxic chemicals, mill effluent that poisoned waterways, and several murders associated with Feronia's plantations. Police harassed people carrying palm fruit, accusing them of stealing from the plantation's trees.[2]

The CDC and the other lenders insisted they were "committed to tackling" the charges, disputing some of the allegations and acknowledging the need for improvement in other areas. Yet Feronia and its creditors never acknowledged a fundamental point raised by Congolese activists: that the plantations were stolen property, built on land taken by colonizers without consent in 1911. The CDC's old favorite—an NES project that would put land back in the hands of smallholders—wasn't even mentioned in replies to Congolese and international NGOs. Instead of investing in reforms, Feronia sold out to another CDC-backed firm in 2020, which vowed to press on with plantation development.

Feronia's revival of colonial-era plantations was one of many episodes in an oil palm boom that is now in its fourth decade. It would take another book to do justice to this recent history.[3] Places that had few oil palms in 1980—Colombia, Thailand, Ecuador, Honduras, Guatemala, and more—have joined the club of big producers. Instead of presenting a comprehensive review of the new oil palm frontier, this concluding chapter reflects on how the oil palm's past influenced its future. Ideas and organizations dating back to the colonial era continue to shape the industry, but the era of "globalization" in the 1990s brought significant changes as well.

Palm oil rode on earlier waves of globalization, back to the first casks shipped from Africa half a millennium ago. Long-distance trade linked Africa's oil palms with industries and ecologies across the globe in the nineteenth century. The colonial plantation complex sent capital, people, and plants shuttling around the globe, along with palm oil and kernels. The recent era of globalization is in some ways a return to form for the oil palm industry, after a postcolonial interlude dominated by development-minded states. In this new political and economic context, old plantation firms became multinational agribusinesses; the roles once filled by "coolie" labor were taken up by undocumented migrants. Capital poured into tropical regions as raw materials flowed out.

One hallmark of contemporary globalization has been neoliberal economic reform, stripping away state interventions in the world economy. These reforms transformed the relationships among states, international organizations, and smallholders. At institutions like the World Bank and CDC, funding for smallholder projects evaporated in favor of plantations. Through "structural adjustment programs" (SAPs), the Bank and IMF forced many countries to sell off state-owned plantations. Privatization put huge swaths of land in unaccountable hands, land that in many cases had been taken from communities in the name of development and food security.

Another feature of the neoliberal era has been land reform. The phrase means different things around the tropics, but the trend is toward secure individual title for land. Where earlier land reforms had broken up large estates, as in Latin America, neoliberal policies created conditions that allowed the revival of plantations. In the rest of the tropics, land reform aimed to supplant or formalize traditional tenure systems, opening up communally owned land for sale and development. Sometimes communities benefited from official recognition of their ownership; others found themselves victims of "land grabs" by powerful firms.

The ecological impact of the oil palm industry—its "footprint" on our planet—began to attract a great deal of attention in the late 1990s. Plantation monoculture transformed forests and peatlands, releasing huge amounts of carbon into the atmosphere. Ranks of *tenera* palms advanced at unprecedented speed, threatening biodiversity. That lack of biodiversity, especially among plantation palms, came back to haunt the industry when the first devastating disease outbreaks hits in the 2000s. But unlike earlier decades, when environmental degradation went unnoticed and unchallenged, the oil palm industry faces a growing number of critics among global and local NGOs. These groups have used new technologies to document and disrupt plantation expansion, cultivating alternatives to a plantation-centered future for the oil palm.

Privatization and the Return of the Plantation

In 1992, FELDA broke with more than thirty years of practice and planted oil palms without recruiting any settlers to harvest them. Resettlement costs had shot up to more than $19,000 per family, and fewer Malaysians were willing to uproot their lives for a new start among oil palms. FELDA continued to exist as an agency, but its plantation arm joined a collection of other investments in oleochemicals, hotels, and far-flung ventures like Canadian rapeseed plantations, all held by the publicly traded Felda Global Ventures (now just FGV).

FELDA's abandonment of smallholder-focused development and the return of the plantation system was part of a global trend.[4] Indonesia's state agencies made a similar pivot toward the private sector in the 1990s, forming private-public partnerships or using state power to award new concessions to politically connected firms.[5] Initially, Indonesian policy insisted that new plantations set aside up to four hectares for smallholders for every one hectare of plantation land.[6] The post-Suharto government abandoned this policy in the name of economic liberalization and political decentralization, giving lower-level politicians more power to award concessions.[7] Land-sharing ratios fell from 1:4 in favor of smallholders to 2:1 and then 4:1 in favor of the plantation.[8] In 2020, Indonesia abandoned the requirement to set aside land for smallholders altogether.

The plantation's comeback relied on access to both land and labor. Government land policies (see below) provided the former; haphazard enforcement of immigration rules contributed to the latter. Equally important was the fact that industrial and agricultural development across Southeast Asia drove more and more people into a "reserve army" of the landless, laboring on plantations far from home in conditions reminiscent of the colonial period.[9] Permanent plantation employment with good housing and social amenities was a rarity by this point, with labor contractors managing workers for most plantations. Workers have little power to unionize or change jobs: contractors might hold a worker's documents hostage or threaten to report undocumented workers to authorities.[10] The debts and coercion that kept colonial-era workers bound to plantations reappeared, too: an exposé published in 2020 alleged that leading firms turned a blind eye to debt slavery, child labor, and violent punishments on plantations.[11] Past efforts to develop labor-saving harvest machinery have not resulted in much beyond improved cutting tools. New experiments with drones that apply fertilizer or even cut fruit bunches with lasers are symptomatic of an industry grappling with its utter reliance on cheap labor.[12]

The state also retreated from oil palm development in Africa, where SAPs forced governments to sell public assets and cut smallholder subsidies. Multinational companies rushed in to exploit what the Bank and others described in language that could have been from the 1890s: an underdeveloped continent, ripe for investment.[13] In the name of state-led development, communities in many parts of the oil palm belt agreed (or were forced) to give up land in the 1970s and 1980s. Now these state plantations were sold to unaccountable firms owned by foreign investors or domestic elites.

Nigerian President Obasanjo wound up with several oil palm properties, including the remains of the CDC's failed "BCC" scheme at Kwa Falls. SIAT, a Belgian firm, bought Risonpalm as well as GOPDC in Ghana. Across the continent, nucleus plantations became plantations, plain and simple. In many cases, privatization involved "widespread displacement of smallholder production systems" as plantations kicked "squatters" off "reserve" land within concessions.[14] At Okomu Oil Palm Estate in Nigeria (on land expropriated in 1979 for a state plantation and sold to SOCFIN in 1990), tensions over plantation jobs and land took a violent turn in the 2000s as "youths" deposed the chief and engaged in deadly clashes with the plantation security force.[15] The smallholders that these former nucleus estates were supposed to serve lost out as SAPs cut fertilizer and seedling subsidies. At the same time, market liberalization policies opened the continent up to a veritable tidal wave of cheap palm oil from Southeast Asia.[16]

Latin American countries also endured SAPs. While there weren't state-owned oil palm plantations to privatize, reforms cut subsidies for smallholders while opening up the oil palm industry for new investment. Military and police units served as "escorts" for politically connected elites, helping secure titles to land and suppressing labor organizations.[17] But not all reforms were bad for small farmers. In Costa Rica, the era of neoliberal reform opened up the oil palm industry to new entrants. Deregulation and trade liberalization broke the monopoly of Palma Tica (United Fruit's successor) on oil milling, and new mills—including cooperatively-owned facilities—encouraged more and more farmers to plant oil palm.[18] Deregulation similarly opened doors for smallholders in Ivory Coast.

International development aid, taxpayer money, and the sweat of smallholders helped build plantation complexes across the tropics. Yet overall, smallholders saw few benefits from privatization and reform in the 1990s. Farmers in Papua New Guinea got no direct payments when Harrisons & Crosfield sold off its stake in New Britain Palm Oil Ltd. for $120 million in 1996, for example. Nor did they get a penny when the new owners sold their shares to Sime Darby in 2015 for $1.7 billion.[19]

Land Reform

The privatization of state plantations gave local and multinational firms new footholds across the oil palm belt. But in many places, land reform had an even greater impact. In Sarawak and Sabah, plantations grew at breakneck speed in the 1990s and 2000s, as land reform opened up forest reserves and tracts held under traditional landholding systems. Some indigenous groups fought against loggers and plantation developers with road blockades and protests, but others bought their own chainsaws and planted oil palm seedlings. Dayak elites in Sarawak worked closely with firms and the state to develop their lands; when Malaysia announced a nationwide halt to new forest clearance in 2018, the Dayak Oil Palm Producers' Association protested that customary land "must be systematically developed." As the association's president pointed out, land is the community's only asset.[20]

Indonesian authorities also reformed land law and policy, encouraging state and private companies to work with landowners rather than importing settlers.[21] A recent survey in Kalimantan found that "frontier" communities who turned over land held under traditional tenure usually wound up worse off than before oil palm arrived, though plantations in more heavily populated areas did lead to higher incomes.[22] In Sulawesi, farmers who signed over their land got rents worth less than 20 percent of what they might have made working their own smallholdings. Some had to take jobs as plantation laborers anyway.[23]

Land reform moved in different directions in other countries. Only a few years after the NGPI debacle in the Philippines, the government enacted a sweeping land reform that returned the plantation to local farmers—though only on paper, as they had to accept collective titling and lease the land back to the company. Indigenous Manobo landowners were left out entirely, and critics argue the measure was more about gutting a militant labor movement than building farmer-owned cooperatives. This ineffective land reform left palm monoculture in place, and workers continued to protest working conditions, wages, and profit-sharing into 2020. Paramilitaries still prowl the area, "disappearing" activists. Land reform shifted the costs of maintenance and replanting from the company to workers, and the farmer cooperatives now face hard decisions as their 35-year-old palms reach towering heights.[24]

Land reform was especially important in the Latin American oil palm story. In 1992, the Honduran government—responding to pressure from U.S. and international lenders—rewrote its agrarian laws, striking down limits on plantations and opening up the sale of land in settlement zones. The strength of farmer cooperatives waned as elites snatched up palm lands from indebted settlers. By the mid-1990s, the Aguán region "suffered the highest levels of land re-concentration in the country."[25] Distressed and coerced land sales surged in the wake of 1998's Hurricane Mitch, which toppled palms and flooded fields. Women—who had been excluded from the management of oil palm cooperatives—found themselves without food gardens as men sold land for cash.[26]

In the 2000s, peasant activists fought back with land occupations and legal challenges. Despite fluctuating palm oil prices and violence against organizers, the cooperatives remained committed to oil palm. For some, "the palm tree has become a sign of development."[27] But the main beneficiaries of land reform in Honduras were plantation companies. With high yields, their estates have turned Honduras into central America's leading exporter of palm oil. A similar story is playing out in Guatemala, where the end of the country's long-running civil war in 1996 was immediately followed by land reform. Framed as a way of giving title and capital to poor farmers, it instead allowed wealthy elites to buy up land, even in the face of ongoing title disputes.[28]

In Colombia, Cold War–era land reforms were "ineffectual attempts . . . at holding back the ogre" of communism.[29] These reforms continued into the neoliberal era, as the "War on Drugs" escalated. Colombia's oil palm trade group (FEDEPALMA) promoted "productive alliances" between smallholders and plantations as a way of ending the civil war and drug-related violence. FEDEPALMA's president (later, Colombia's minister for agriculture) drew a dubious analogy with Malaysia: "Remember that Malaysia, before the African palm, was a country infested by guerillas and affected by subversion, today it's a country of peace. We'll imitate the Malaysians."[30]

Working through new land laws, Indupalma turned over its entire plantation to workers in the 1990s, frustrated by labor costs and the long running dispute with the union. Company billboards declared: "Whoever sows palms, harvests wealth and peace."[31] The land reform didn't solve all conflicts, but homicide rates fell significantly as hostilities among labor activists, the company, and paramilitaries waned.[32] The move also saved Indupalma money, letting farmers deal with harvest costs, fertilizer, and replanting. Farmers could only sell their fruit to Indupalma's oil mill, though they did win the right to raise food crops amidst their palms, weakening the company's power over them.[33]

Despite official support for smallholding in Colombia, plantations continued to grow, sometimes with the aid of paramilitary violence. One farmer was told to sell his land, "or we'll buy it from your widow."[34] Land reform took an especially pernicious turn for Afro-Colombians in the late 1990s. A new law granted their communities legal title to more than 4.5 million hectares of land in western Colombia, yet just as "these rights were being granted on paper . . . they were disappearing on the ground."[35] Colombian investors—including drug cartels seeking new ways of laundering money—relied on "threats, targeted killings, massacres, and individual and collective forced displacement" to drive out residents. Companies then boasted of replacing "weeds with palm" on these stolen lands.[36]

The African experience with land reform is still unfolding. Early reforms, like those imposed by Nigeria's military regime in the 1970s, sought to formalize landholding and concentrate power in the hands of the state. New titling systems were cumbersome and subject to lengthy court battles; people continued to rely on chiefly authority for land transfers. Under pressure from the World Bank and IMF, more countries moved to formalize traditional land-tenure systems in 1990s, with varying results. In Ivory Coast, a 1998 reform recognized land rights solely for Ivorian citizens, overriding customary pathways to landholding for the country's large population of migrant workers and their descendants. The move helped spark a coup (1999) and civil war (2002–4).[37] The World Bank continues to press for land reform across Africa, arguing that titles and maps will provide more security than customary tenure.

Customary tenure is far from perfect: it is subject to the whims of powerful leaders, and to hierarchies of gender, class, and ethnicity. But it acknowledges a basic right to land that disappears when land becomes just another commodity in the market. In too many cases, land reform simply provided legal cover for land grabbing. While the 1.8 million–hectare "Kalimantan Border Oil Palm Mega-Project" is currently defunct, bulldozers are already moving on gargantuan projects like the 280,000-hectare Tanah Merah project in Irian Jaya, and Atama's 470,000-hectare project in the Republic of Congo—projects that rely on the colonial myth that tropical forests are empty, unpeopled spaces.[38]

The Ecological Footprint

The new plantation boom's impacts on smallholders, workers, and indigenous groups in the tropics were intensely local, but the ecological impacts of oil palm development had global significance. Scientists speak of a "mass extinction" crisis, as plantations and logging transform tropical habitats.[39] Huge amounts of carbon locked away in plants and soils are released as carbon dioxide and methane when tropical forests are disturbed. As one report put it: "If tropical deforestation were a country, it would rank third in [CO_2 equivalent] emissions."[40]

Nothing made the environmental costs of oil palm more visible than fire and smoke. Three percent of Borneo's forests burned in 1997–98, with many fires originating in plantation or smallholder land-clearing.[41] Fires during the 2015–16 El Niño season on Borneo and Sumatra were even worse, burning a million more hectares than during the 1997–98 season.[42] Haze from these fires blanketed cities like Singapore, Kuala Lumpur, and Jakarta, bringing the costs of forest-clearing home to corporate offices, government ministries, and millions of urban consumers far from the oil palm frontier. Oil palm is not the only culprit: other industries, especially pulpwood, share the blame for deforestation in Southeast Asia. In some cases, oil palm concessions are merely a ruse to cut timber. Several PNG logging companies obtained oil palm concessions and hired workers to tend palm nurseries before leaving "baby oil palms abandoned like weeds" when the timber ran out.[43] Industries like logging and pulpwood deserve scrutiny and regulation, but that doesn't mean oil palm is innocent or insignificant in deforestation.

Importantly, many fires—nearly 90 percent of recent cases documented on Sumatran oil palm concessions—burned peatlands. When peat burns, the soil itself combusts, smoldering for weeks or months. In the short term, smoke inconveniences people and damages their lungs. In the long run, it releases staggering amounts of carbon into the atmosphere, with major consequences for the global climate. When they don't burn, drained peatlands still release large amounts of carbon as organic material breaks down in a newly oxygenated environment. Oil palm is one of a few commercial crops that can tolerate deep peat soils. As Helena Varkkey points out, firms moved into peat because no new mineral soils were left, but also because peatland was usually unencumbered by traditional forms of land tenure.[44] The swidden cultivation practiced by many of Southeast Asia's forest peoples is limited to better-drained land. Across the tropics, extensive strategies—putting more land under oil palms—remains the largest driver of growth in the industry for planters and smallholders alike.[45]

But not all palm oil is the direct product of deforestation. In Malaysia, tens of thousands of hectares of rubber have been converted to oil palm since the 1950s, aided by state replanting programs. In Ivory Coast and Ghana, oil palm often grows on land previously cleared for cocoa; millions of hectares could be converted from cocoa to oil palm without any new deforestation (though at great cost to the world's chocolate lovers).[46] So many forests in Latin America were cleared for logging, cattle ranching, and bananas over the past four centuries that oil palm usually didn't replace primary forests. In Colombia, half of all oil palms grow on former pasture; less than 20 percent replaced "natural" grasslands or forest.[47] In southern Mexico, communities see leafy palm plantations as an aesthetically and economically desirable improvement over open pasture.[48]

Like their counterparts in Africa, Afro-Brazilians in Bahia often grow oil palms in mixed plots, fostering biodiversity.[49] Afro-Colombian farmers pursue similar strategies, reproducing "the heterogeneity and complexity found in primary or secondary forests" in farm plots that increasingly include oil palms.[50] Around the globe, smallholders are often wary of using costly fertilizers and pesticides, and they are willing to accept lower yields in exchange for the food security provided by intercropping other plants with oil palm.[51] For Bahian farmers—and many African farmers—biodiversity includes oil palms themselves. *Dura* palm varieties are valued for their taste as well as their apparent resistance to disease.[52]

It's unfair to romanticize smallholders as environmental guardians, however. The Honduran settlers of the Aguán, to take one example, had no qualms about clearing forest to plant oil palms and other cash crops.[53] Oil palm plots tend to host less biodiversity and accumulate less biomass than comparable coffee or cocoa farms, and contracts with oil mills require many smallholders to grow palms in a monoculture. As one Colombian farmer put it, "Palms live alone."[54]

The ecological footprint of oil palm extends beyond the plantation or smallholder plot, too. Mill effluent, the watery runoff remaining after palm oil is extracted, is a serious concern in many areas. Making a ton of palm oil leaves up to 100 tons of effluent behind. When oil mills dump effluent into rivers or treatment ponds prone to flooding, they contaminate waterways and create fish kills that stretch for miles downstream.[55] This is one area where large plantation companies led the way, developing better treatment methods. New Britain Palm Oil Ltd., which once set a river on fire with its effluent, was an early leader in generating electricity from biogas made in effluent treatment ponds.

Big companies have also reined in the use of dangerous pesticides like paraquat. They promote best practices that reduce fertilizer use and control runoff. They invest in developing new, higher-yielding plant varieties that produce more oil per hectare, easing pressure on forests. But the oil palm's voracious appetite for nutrients in palm monocultures poses a long-term problem. Petrochemical feedstocks, energy-hungry ammonia plants, and African and South American phosphate mines rarely figure in calculations of the oil palm's ecological footprint, but the industry would struggle without them. Fertilizer can amount to a quarter or even half of the total cost of growing and harvesting palm fruit.[56]

The environment is not a static thing that humans impose their will on, however. Sometimes it bites back. Diseases have dogged the plantation industry since its start in Sumatra a century ago. No disease seemed to pose an existential threat, however. That changed with a massive outbreak of "bud rot" (*pudrición del cogollo*, PC) in Colombia between 2004 and 2015. PC was probably to blame for outbreaks that struck plantations across the tropics from 1920 into the 1990s. "Lethal yellowing," possibly caused by the same pathogen, hit new plantations in Brazil's Pará state in the 1970s.[57]

The recent Colombian outbreak was the largest by far, turning 70,000 hectares of leafy palms into eerie galleries of headless trunks, open to the sky. The episode cost producers some $250 million.[58] Colombian experts blamed an organism called *Phytophthora palmivora*, though researchers have associated the disease with fungi, bacteria, and nutrient deficiencies in the past. In the Colombian case, heavy rainfall created ideal conditions for *P. palmivora* to attack oil palm monocultures, killing most infected trees. The most common *tenera* type proved highly susceptible to PC. No major outbreaks have been reported since, but the Colombian PC outbreak is a warning of the dangers of plantation monoculture. A deadlier form of *P. palmivora* or another pathogen could ravage the world's plantation industry in short order. Climate change may help long-established threats like *Ganoderma* thrive, too.[59]

Optimists can point to scientific accomplishments that make the oil palm's future brighter, however. Botanists have spent decades bioprospecting for resistant traits in wild *E. guineensis* and *E. oleifera* populations, and new breeds are tougher and higher yielding than the early *tenera* varieties. In palms with "virescens" genes, fruit turns from green to orange when ripe, taking the guesswork out of harvesting. Drought-resistant traits helped

Southeast Asian planters endure the fierce 2015–16 El Niño cycle with only moderate losses.[60] Cloned oil palms promise even higher yields than conventionally-bred varieties, though cloning creates new problems with undesirable epigenetic mutations.[61] In any case, these innovations are for plantations, reaching smallholders slowly, if at all. Many of the high-performing hybrids require hand-pollination, a labor-intensive regression to the plantation industry's early days.

Still, millions of smallholders and plantation workers rely on palm oil for their livelihoods. Environmental campaigns that urge consumers to boycott palm oil won't help them. They also reflect an environmental double standard: habitat destruction in Europe and North America in past centuries is a sunk cost. Meanwhile, people in the tropics are denied the same opportunity to build their fortunes with natural resources. Palm monoculture is undeniably an efficient way of making oil and money. Oil palms cover about 5 percent of oil cropland but make more than 30 percent of the world's edible oil supply. If palm oil were to simply disappear, the soybean oil needed to replace it would take ten times more land than oil palm currently occupies. Spare a thought for the palm kernel, too: it provides vital lauric oils to industry, supplementing a slow-growing and lower-yielding coconut industry.

But ultimately, one-to-one comparisons between palm oil and other oil crops are misleading. Oil palms can't grow in the same places as rapeseed or sunflowers, and vice versa. For soybean farmers, oil is a secondary product. Demand for animal feed drives the soybean industry, not cooking oil.[62] And of course, land is not a uniform thing. Ten hectares of rapeseed on the Canadian prairie presents a very different case—in terms of biodiversity as well as carbon emissions and social impacts—from one hectare of oil palm on a Borneo peat swamp. Oil palm can be a good thing for people and the environment on degraded grasslands, cattle pastures, or old rubber plantations.[63] It's another matter when those palms replace primary forest, especially on peat.

Global Organizations and the Oil Palm Boom

While the age of neoliberal globalization brought a wave of plantation development to Africa, Asia, and the Americas, it also empowered local activists and global NGOs to push back. Northern NGOs often took a one-sided, environment-focused approach to plantation development through the 1990s and early 2000s. (Recall the bloody orangutan finger at the end of chapter 9—all about conservation, with no consideration of farmers' or workers' livelihoods.) But some NGOs changed course in the 2000s.[64] One strategy sought to widen the group of "stakeholders" involved in planning and implementing top-down policies. The most important of these projects was the Roundtable for Sustainable Palm Oil (RSPO), which began in a series of meetings between NGOs, industrial consumers, and plantation companies from 2001 to 2004.[65] The RSPO developed a definition for "sustainable" palm oil and launched a program to certify producers. Industry giants have taken the RSPO seriously. Unilever, Colgate-Palmolive, Procter and Gamble, and Nestlé are among a long list of consumer firms that have committed to buying "sustainable" palm oil. Nestlé, for example, used certified palm oil for 79 percent of its 2019 consumption.[66]

Critics argue that consumer firms—nearly all of whom sold off their plantation portfolios over the past half-century—are shifting the costs of sustainability downstream. Suppliers have to meet new standards or lose business.[67] For end-consumers, certified palm oil becomes a new commodity, advertised with palm tree logos on cookies and shampoo bottles. Critic Oliver Pye calls it "a magic solution to a problem that is structural, political, and social."[68] Although 19 percent (14.5 million tons) of the world's internationally traded palm oil supply carried an RSPO certification by 2019, most buyers were unwilling to pay a higher price for it. From 2013 to 2018, only half of RSPO-certified oil sold at premium prices.[69] This is a weak incentive to participate in certification programs, though the threat of boycotts and sanctions has undoubtedly convinced some firms to sign on. The European Union has set a 100 percent sustainable goal for palm oil, for instance. By 2018, forty-five of the biggest plantation and oil milling companies were RSPO members, though only twenty-five had met most of the basic benchmarks for sustainability.[70]

The structure of the palm oil industry makes certification difficult, however. Oil mills buy fruit from multiple plantations and smallholders, and refiners bulk palm oil into huge tanks. How can a consumer be sure they are buying sustainable palm oil? Nestlé could trace only 62 percent of its 2019 purchases back to specific producers—still an impressive feat, given that it bought palm oil from eighty-eight suppliers operating more than 1,624 oil mills. Some experts see great potential in blockchain technologies, which would create secure identities for every lot of oil. Consumers could scan a code on a package and follow the oil all the way down the commodity chain.[71] This vision for techno-governance relies on the assumption that people—end consumers or industrial buyers—will actually pay more for sustainable palm oil once they know about it. But unlike a product like coffee where variety and *terroir* serve as selling points, this industrial palm oil is a featureless commodity. It's hard to build an imaginary relationship with producers through a thing that can't be seen, tasted, or smelled.

Tracing and certification regimes also risk leaving smallholders behind. It's much simpler to inspect a 5,000-hectare plantation than it is to visit 1,000 farmers on their own 5-hectare plots. Farmers have reasons for avoiding certification programs, too: registration is the road to taxation.[72] Others lack the titles necessary to prove land ownership. Building safe storage facilities for chemicals and buying safety equipment presents yet another hurdle. And many farmers are still clearing trees to plant new palms. The RSPO bans forest clearance and has no place for forest fallow, even where it might be ecologically sustainable.[73]

Some of the RSPO's fiercest critics argue that it is, in effect, certifying deforestation. Those who cleared land prior to joining have no obligation to replace the ecosystems their oil palms displaced. Several studies allege that certified plantations are no better than the rest in protecting biodiversity or suppressing fires.[74] But the alternative to top-down, voluntary regulation of big agribusiness is often presented as an unregulated free-for-all. An RSPO official bluntly said: "The CEO of a company does not wake up thinking 'how can I abuse workers?' . . . The smaller growers do wake up thinking how to screw over a worker. These are the faceless suppliers with no name. . . . Those are the ones I am worried about. The ones not at the table."[75]

The table is not as sturdy as corporate membership might suggest, though. When activists challenged IndoFood through the RSPO complaint mechanism, the company quit RSPO altogether. Meanwhile, when consumer-goods firms like Unilever and Nestlé drop "bad" suppliers—often with great fanfare—the plantations they formerly bought from continue to exist and grow. Such firms lose any power they had to influence plantation operations, while the uncertified oil "leaks" into the market through myriad refiners and traders.[76] Indonesia and Malaysia have further weakened the RSPO by launching their own certification programs, objecting to an NGO setting the terms of sustainability. They argue that national agencies can better balance environmental protection with the interests of palm oil producers.[77]

There is another side to the NGO story, however: a ground-up approach that connected local activists and NGOs with global organizations. When

Nigerian women protested the arrival of Pioneer oil mills in 1950, news trickled slowly through official channels and media; a concerned British MP pressed the Colonial Office for details and received only a cursory report.[78] The matter generated little interest in the British press. When Risonpalm prepared to move into a forest reserve containing one of Nigeria's last intact rainforests in the 1990s, however, Nigerian and European NGOs organized. Together they convinced the European Union to withdraw funding for the project.[79] While projects backed by private capital are harder to influence, many oil palm projects—especially in Africa—rely on funding from groups like the CDC, which are subject to pressure campaigns from activists ready to provide evidence of land grabbing, pollution, and labor abuses with smartphone cameras and satellite imagery.

In Sierra Leone, one community used local courts to reverse a major plantation concession. Singapore-based Siva Group signed deals with chiefs and other landowners, but several groups challenged the documents and the legitimacy of the signers. Landowners asked Namati, an international NGO dedicated to guiding plaintiffs through legal systems, for help. The World Rainforest Movement organized a workshop for women in 2017, resulting in an alliance of local activists led by women. Together they planned, fought, and won their case against the plantation, getting back 41,500 hectares.[80]

The ability of activists to connect with global NGOs and media was especially important in a case like Herakles Farms in Cameroon. This U.S.-based company hoped to clear 73,000 hectares of "degraded" forest for oil palm. Cameroon's government overruled local opposition to the lease, awarding Herakles a ninety-nine–year concession. Local activists collected stories of land grabbing, violence, and deforestation, passing them to international NGOs to publicize in major media outlets. Dogged by investigations, Herakles suspended new operations in 2015 and accepted higher rents and a much-reduced concession.[81]

In Southeast Asia, law and policy are supposed to be protecting farmers from predatory agribusinesses. Activists with groups like Indonesia's Sawit Watch have documented and challenged hundreds of cases of illegal forest destruction, land grabbing, and human rights abuses. But legal victories have been few and far between: allegations of bribery and alliances between planters and the military, police, or criminal gangs are common.[82] The RSPO offers another pathway for activists to challenge plantation firms, but the RSPO's database of publicly available cases show that many were punted off to Indonesian courts.[83]

Activists in Latin America have also used the RSPO to challenge oil palm plantations, but several factors limit its utility in policing the industry there. First, most plantations are locally owned. Domestic elites with ties to government, the military, or paramilitaries face no serious pressure from outside shareholders or government regulators, as multinational corporations do. Second, the expansion of the oil palm sector in Latin America came with more overt violence than in much of Africa or Southeast Asia. Dozens of activists have been murdered in and around oil palm developments stretching from Guatemala to Peru. Yet despite the dire circumstances, local NGOs rooted in a long tradition of peasant (*campesino*) organization have led an effective resistance to encroaching plantations.[84] They have occupied illegal plantations, fought court battles, and appealed to national and international bodies for help.

The island of Papayal in northwestern Colombia offers an especially compelling case highlighting the power—and limits—of NGO activism. *Campesino* activists seized the "Las Pavas" cattle ranch in the 1990s after it was abandoned by its drug-trafficking owner. Paramilitaries drove the farmers off in 2003, but the latter returned and applied for legal title under land reform laws. Meanwhile, the official owners sold Las Pavas to an oil palm company that brought in paramilitary forces (and later, riot police) to drive out the "squatters."[85]

Campesino activists reoccupied the land again after 2009, while also appealing to national courts and international media. By this point, the consumer-goods brand Body Shop was the major buyer of the plantation's organic-certified palm oil. After northern NGOs organized protests, Body Shop dropped its Colombian suppliers. But this was the limit of consumer activism: the plantation was still there, still selling oil. Real action came from *campesinos* challenging the plantation on the ground and in courts. They have won legal victories, although to date, the plantation has ignored court rulings. The community remains in a tense standoff, accusing plantation staff of destroying farmers' crops and homes. Several community leaders travel with state-provided bodyguards.

Blockchain traceability or a scannable code on a package of soap will not help these farmers: their survival relies on a national government taking effective action. As anthropologist Tania Li warned in a recent study of Indonesian plantations, "bad" companies can't be certified into "good" ones in places where business and political elites have the power to take land, abuse workers, and flout regulations. Entire governance systems need reform.[86]

Sustainable Futures

In 1986, during the early stages of Indonesia's oil palm boom, World Bank officials debated the best way to respond to NGOs opposing new plantations. One Bank official declared in a memo: "forests are important, Indonesia's tribal people are important, and the 40 million poor people in Java are important too. All development entails trade-offs, and in this case particularly, there are no unambiguous rights and wrongs. There are only opportunities for trying to maximize benefits while minimizing adverse results."[87]

Did the benefits outweigh the costs? People like the Orang Rimba of Sumatra didn't think so. Hemmed in by oil palm on all sides, Orang Rimba families are forced to forage for palm fruit and hunt pigs on plantations where both hunter and prey are unwelcome; others have given up on the forest in favor of a sedentary future in towns. Their way of life has been eradicated by oil palm.[88]

But this isn't the whole story of oil palm. More than 2 million families in Indonesia and Malaysia earn their livings as oil palm smallholders, independently or attached to plantation projects. Not all are well off, with many struggling with debts for land, fertilizer, and seedlings.[89] But a substantial number earn a good living: they have chosen oil palm over cocoa, rubber, and other crops that are equally accessible to smallholders. Many smallholders want more oil palm, not less. A 2010 survey in Nigeria found that a large majority of oil palm smallholders were "comfortable," with only 12 percent indicating that they struggled to support their families. At least 1.8 million Nigerians grow, make, or trade palm oil (the government claims as many as 4 million).[90] While plantation companies and international development agencies often cite job creation as a major benefit of plantations, a country like Nigeria would need at least *9 million* hectares of new plantations to match the employment provided by the smallholder sector.

Smallholders in Africa and Bahia have the added benefit of a vibrant domestic market for their palm oil. Almost none of this oil is exported or counted in official statistics, but it is very much a part of the oil palm story. So is palm wine: the common plantation practice of poisoning old trees with glyphosate before knocking them over wastes an important resource in Africa. Making and selling oil locally is especially important for women. Men tend to cut women out of family income when they sell fruit to an industrial mill.[91] New machines, like a hand-operated expeller press that doesn't require fruit crushing, have helped make small-scale production more profitable. "Don't send your mother and family to work in the dirty, backbreaking country pit [for fruit crushing]", urges one Liberian ad for the machine.[92]

Small-scale palm oil isn't going to satisfy global demand for food, oleochemicals, and biofuel. Rather, it's a reminder that there are alternatives to plantation methods—just not cheaper ones. When the eccentric firm of Dr. Bronner's Magic Soaps decided to pursue an organic source for the few hundred tons of palm oil it needed annually, the company didn't turn to RSPO suppliers. Instead, it set up a subsidiary in Ghana, repeating early-twentieth-century experiments with standalone mills. There is one major difference: the company pays whatever it takes to get palm fruit ("crazy prices" at times, according to the general manager). The mill is partially unmechanized to maximize employment, especially for women. The company pays for social services like a clinic. It does influence how farmers work their land—maintaining a USDA organic certification is key for the company—but it doesn't own land or trees.[93] Paying high prices for palm oil is a simple way to promote sustainable smallholder practices. RSPO advocates have lambasted big companies for their unwillingness to pay more for "green" palm oil, too. But paying more for palm oil runs against the reason palm oil is in so many products: its cheapness.

Thailand, a country with a fledgling oil palm industry in 1980, offers a more realistic pathway for equitable development on the scale currently required by the world market. Today Thailand is the third biggest exporter of palm oil, and smallholders make about three-quarters of it. The plantations that do exist are small, most under 1,000 hectares. Many of the palms replaced rubber trees. Thanks to the proliferation of independent oil mills competing for fruit, farmers have a good deal of bargaining power.[94] The Thai state looms large over the oil palm industry, however. Legislation gives farmers secure title to land while limiting the size of plantations. The government promotes biodiesel on a huge scale, consuming a bigger share of Thai palm oil than food industries do. In the midst of the COVID-19 pandemic, the government ordered power stations to burn crude palm oil in place of fuel oil to prop up sagging prices. Thailand's success with oil palm is a reminder that state-led development has worked for oil palm smallholders in the past, and it can work again—when it's done well.

Palm Oil Planet?

Disappearing up a chimney to send electrons coursing down wires: palm oil could hardly be consumed more invisibly than in a power station. It's an unlikely fate for a commodity that began its global career in the open, a cherished product rubbed on dry hands by Europeans and smeared on the enslaved before they were sold like cattle. Oil palms were many things for Africans, but in the nineteenth century, they became the first link in a commodity chain connecting African labor and ecologies with European—and later global—industries and consumers. Palm oil became an industrial raw material in the nineteenth century because it was cheap and versatile. Chemical technologies rendered it even more versatile and invisible, relegating the stuff to little-understood and often unmentioned roles in soap, candles, tinplate, and adulterated butter.

The plantation finally brought palm oil into the industrial food system. As one consultant put it, the oil palm became little more than a chemical "reactor" in a plantation: "You feed a number of things into it and it gives out oil."[95] Grown far away and transformed into entirely new products, palm oil quietly seeped into one sector after another, gaining market share as consumers and companies turned against animal or petroleum-based products for cultural, economic, and environmental reasons. Palm oil didn't always stay invisible, as the fights over saturated fat illustrate, but it remained the cheap ingredient of choice for many companies.

Palm oil regained its visibility in the twenty-first century because of how it was produced. Building on the linked campaigns in support of environmental protections and human rights of the 1980s, campaigners—mostly, but not all, in the global North—identified palm oil as an especially destructive commodity. In their rhetoric, the oil palm industry ravaged the tropics and slaughtered orangutans, all to put "hidden" ingredients in our soaps and candy bars. The fact that palm oil is perceived as being *in* things, rather than as a thing in its own right, produced a hostile and ultimately unproductive narrative in consuming countries. Cocoa, beef, coffee, and soy pose similar threats tropical ecosystems. All have ugly histories with regard to labor and land rights. Yet few groups call for outright bans on these things. Palm oil is an easy target precisely because it is a substitute, a replacement commodity. Boycotters don't ruminate on what palm oil replaced and why, obscuring the choices that led manufacturers to switch in the first place.

It's clear that palm oil consumption and the oil palm industry will only get bigger in the coming decades. The pertinent questions are: How big? What shape will it take? And what impact will it have on the planet? The first question hinges on agrofuel policies, population growth, and consumer choices, along with the growth in other oil industries. One authority guesses that the world might need 240 million tons of palm oil by 2050, about twice the amount produced today.[96]

The last two questions depend on farmers, governments, and corporations. For a moment in the 1960s, palm oil really delivered on the promise of development. Smallholders across three continents planted oil palms, backed by international lenders and (in Africa and Asia) newly independent governments eager to transform agrarian life. The swing back toward the plantation in the 1990s reversed the trend. But plantations are not inherently more efficient than small farmers in raising oil palms. Instead, they are an effective way of extracting value from land and labor for investors, more lucrative than building machines to process smallholders' fruit. While plantations produce much more oil today, smallholders deserve to be taken seriously. They are not merely an "epiphenomenon of plantation and state scheme development," as some authors argue.[97] Smallholders are the leading producers of cocoa, coffee, and other crops. Cumulatively, smallholders have cut down more forest than plantations in their eagerness to plant cash crops. They could cultivate much more oil palm. Their calculus is not just about the price of palm oil but about its value relative to rubber, cocoa, coffee, and other products—including artisanal palm oil and palm wine in Africa—that circulate in completely different commodity webs than industrial palm oil.

The world would probably be better off if the massive oil palm expansion predicted for Latin America and Africa looked more like the Thai or Ivoirian experience than the Indonesian example.[98] But farmers across the oil palm frontier are not passive and powerless, waiting for governments and corporations to decide their fate. They are making their own history, standing up to plantation land-grabs even in the face of official repression and assassinations.[99] Consumer boycotts of palm oil will do nothing to help them, or save orangutans, or protect carbon-sequestering peatlands. But neither will blind consumption, merely using palm oil without questioning the structures that put farmers and our environment in such precarious positions in the first place.

Acknowledgments

Every global history is a community project, even if this book appears under a single name. Over the past ten years, I've become indebted to many researchers, archivists, librarians, and others across four continents. I particularly thank the staff at Ghana's Public Records and Archives Administration Department; Malaysia's Arkib Negara; the UK National Archives; the Royal Botanic Gardens, Kew; the London Metropolitan Archives; the British Library; the SOAS Library and Archives; Unilever Art, Archives and Records Management; the Hagley Museum and Library; the US National Agricultural Library; the Cornell University archive and libraries, and the many other libraries and institutions I visited or received material from.

For permission to reproduce archival material, I thank Elementis Plc. and the Unilever Collections at Unilever Art, Archives and Records Management, Port Sunlight. UARM staff reviewed this book as a condition of accessing and reproducing archival material, but they did not comment on the text or request changes. I also thank the Board of Trustees of the Royal Botanic Gardens, Kew, and the staff of the Library and Archives and Economic Botany Collection for permission to reproduce text and photos of materials in their collections. Figure 4.1 was reproduced by kind permission of the London Borough of Lambeth, Archives Department.

An earlier version of chapter 7 appeared as "Shallow Roots: The Early Oil Palm Industry in Southeast Asia, 1848–1940" in the *Journal of Southeast Asian Studies* 51, no. 4 (2020). Other sections of the book are based on work published in the *Commodities of Empire Working Paper Series*, *African Economic History*, the *Journal of World History*, and *Environmental History*. I'm grateful to the editors and anonymous reviewers at these journals for their comments.

Thanks are also due to the UNC Press team. Brandon Proia helped me work through ideas long before I knew what the book was going to look like. I've also benefited enormously from comments, criticism, and new work shared by colleagues at meetings of the American Society for Environmental History, the African Studies Association, the Agricultural History Society, and the Commodities of Empire Workshop series. David Biggs provided a warm introduction to Southeast Asian Studies at the SE Asian Natures workshop in 2018.

A number of grants made the research behind this book possible. The project began with a Hagley Museum & Library exploratory grant in 2012, and grew during my time as the 2016 Cornell College of Human Ecology Dean's Fellow in the History of Home Economics. The American Philosophical Society/British Academy Joint Fellowship supported research in the United Kingdom in 2018, supplemented by a Research Excellence Fund grant from Michigan Tech. The manuscript was written thanks to one big grant from the University of Connecticut Humanities Institute.

Michigan Tech granted sabbatical leave, and UCHI hosted me as an external faculty fellow for 2018–19. Thanks to Michael Lynch, Alexis Boylan, Jo-Ann Waide, and Nasya Al-Saidy for all their work supporting me during that year. I could not have written the book without that time and the community of scholars. Thanks also to Betty and Ken Metzler for making our time in Connecticut much brighter.

For advice and support, I especially thank Carol MacLennan, Nancy Langston, Derek Byerlee, and Corey Ross. So many colleagues shared conservations, criticism, and their own research with me. They include Stephen Buckley, Philip Cass, Paul Chassé, Jim Clifford, Max Counter, Rob Cramb, John Garrigus, Spring Greeney, Andrew Hardy, Anya King, Kevin Lanagan, Josh MacFayden, Anthony Medrano, Nate Millet, Matt Minarchek, Ruel Pagunson, Erin Pischke, James Putzel, Geoffrey Pakiam, Beatrice Quarshie Smith, Peter Thompson, the Rev. Neville Threlfall, Helena Varkkey, Shakila Yacob, and many others whose contributions I forgot to properly document. Sorry! Hermann von Hesse and Laura Vidal Chiesa provided research assistance. James Fenske and Enrique Martino generously shared digitized archival material. Nancy Shoemaker shared crucial archival notes and taught me how to make soap with palm oil. I am especially grateful to Case Watkins for sharing his work and providing map 1.1. His new book, *Palm Oil Diaspora: Afro-Brazilian Landscapes and Economies on Bahia's Dendê Coast* (Cambridge University Press, 2021), tells the full story of the oil palm in Brazil, which I only sketch out here. Finally, thanks to Clare, for living with this project and all that it's taken over the past decade.

Notes

Introduction

1. James Hall, "To Emigrants," 40; "The African Trade," 165; Tindall, *South Carolina Negroes*, 157.

2. The standard work on the oil palm is Corley and Tinker, *Oil Palm*, but see earlier editions by Hartley for significantly different historical treatments.

3. On commodity webs, see MacFadyen, *Flax Americana*, 16.

4. Defined as a family owning an amount of land that can be worked with no or few hired workers. See Clarence-Smith, *Cocoa*, 125.

5. Borras et al., "Rise of Flex Crops"; Alonso-Fradejas et al., "Inquiring into the Political Economy."

6. Mosbergen, "Palm Oil Is in Everything."

7. Lian and Wilcove, "Oil Palm: Disinformation Enables Deforestation," 68.

8. Teltscher, *Palace of Palms*, 261.

9. Sime Darby, "Palm Oil Facts and Figures," c. 2015, http://www.simedarby.com/upload/Palm_Oil_Facts_and_Figures.pdf.

10. Behrendt, Latham, and Northrup, *Diary of Antera Duke*, 267; for the 9 lb figure, see Sarbah, "Oil Palm." Maier, "Precolonial Palm Oil," states that a gallon weighs 7.7 lbs, but this is a U.S. gallon.

Chapter 1

1. Winterbottom, *An Account*, 58.

2. On the idea of a humanmade or "technological" sublime, see Nye, *American Technological Sublime*; for a vivid description of an Indonesian plantation, see Li, "After the Land Grab."

3. Wolf, *Injuries*.

4. Stumpf, "Biosynthesis of Fatty Acids," 39–41.

5. Zeven, "On the Origin of the Oil Palm (*Elaeis guineensis* Jacq.)"; Clark, "Prehistoric Populations," 80–81; Shaw, "Early Crops in Africa," 113; Wickens, "Preliminary Note."

6. Singh et al., "Oil Palm Genome Sequence Reveals Divergence of Interfertile Species in Old and New Worlds"; for a debunked theory of a Brazilian origin, see Cook, "Brazilian Origin."

7. Sarbah, "Oil Palm," 249.

8. Zeven, "On the Origin of the Oil Palm (*Elaeis guineensis* Jacq.)"; Zeven, *Semi-Wild Oil Palm*; see also Hartley, *Oil Palm*, 1st ed.; and Corley and Tinker, *Oil Palm*.

9. As Zeven noted, even swamps weren't guaranteed to be "natural." Humans changed waterways and watertables, meaning that some "swamp" trees might merely be relicts of a drier past. Zeven, *Semi-Wild Oil Palm*, 23; Zeven, "The Partial and Complete Domestication of the Oil Palm (*Elaeis guineensis*)"; for a premodern example of hydrological engineering, see Hubbard, *The Sobo*.

10. Hartley, *Oil Palm*, 2nd ed., 5.

11. Translated in Usoro, *Nigerian Oil Palm Industry*, 21–22.

12. Kone, "Le Palmier à Huile En Côte d'Ivoire."

13. Carney and Rosomoff, *Shadow of Slavery*, 6; quoting Frobenius, *Volksdichtungen aus Oberguinea*, 75.

14. Afigbo, *Ropes of Sand*, 41, 128.

15. Shaw, "Early Crops in Africa," 131.

16. Harris, "Traditional Systems," 351.

17. Oas, D'Andrea, and Watson, "10,000 Year History of Plant Use at Bosumpra Cave, Ghana"; D'Andrea, Logan, and Watson, "Oil Palm and Prehistoric Subsistence." Chimpanzees do eat palm fruit and can crack kernels, but they may have learned this skill from humans. Mercader et al., "Chimpanzee Sites."

18. Sowunmi, "The Significance of the Oil Palm (*Elaeis guineensis* Jacq.) in the Late Holocene Environments of West and West Central Africa."

19. Sowunmi, "The Significance of the Oil Palm (*Elaeis guineensis* Jacq.) in the Late Holocene Environments of West and West Central Africa"; Alabi, "Environment and Subsistence of the Early Inhabitants of Coastal Southwestern Nigeria."

20. For the typical view, see Harlan, de Wet, and Stemler, "Plant Domestication"; for a summary of the critique, see Ickowitz, "Shifting Cultivation"; Fairhead and Leach, *Misreading the African Landscape*; Fairhead and Leach, *Reframing Deforestation*.

21. Salzmann and Hoelzmann, "The Dahomey Gap"; Maley and Chepstow-Lusty, "*Elaeis guineensis* Fluctuations"; Maley, "Impact of Arid Phases"; Oslisly et al., "Climatic and Cultural Changes in the West Congo Basin Forests over the Past 5000 Years"; Oslisly and White, "Human Impact"; Flenley and Bush, *Tropical Rainforest Responses to Climatic Change*, 156–58.

22. Oslisly and White, "Human Impact," 357.

23. Bostoen, Grollemund, and Koni Muluwa, "Climate-Induced Vegetation Dynamics and the Bantu Expansion"; Bostoen et al., "Middle to Late Holocene Paleoclimatic Change and the Early Bantu Expansion in the Rain Forests of Western Central Africa."

24. Baeke, "Wuli Witchcraft," 38–39.

25. Zeven, *Semi-Wild Oil Palm*, 25.

26. Fairhead and Leach, *Misreading the African Landscape*; Watkins, "Landscapes and Resistance in the African Diaspora."

27. "Yes, these countries are now more or less wooded," they affirmed, "because, since always, we have facilitated the development of these forests by our works of cultivation; there is on the other hand a number of artificial forests created of all sorts by ourselves or by our ancestors." Vanderyst quoted in Spinage, *African Ecology*, 348; see also Vanderyst, "Origine Des Palmeraies"; Vanderyst, "Les Palmeraies."

For critical reappraisals of declensionist theories, see Fairhead and Leach, *Reframing Deforestation*, 17–18 and passim; Richards, *Indigenous Agricultural Revolution*; and on fire, see Munier, *Le palmier a huile*, 27.

28. Hellermann, "Reading Farm and Forest," 105–6.

29. Spinage, *African Ecology*, 352.

30. Zeven, *Semi-Wild Oil Palm*, 91–92; Jones, *Slaves to Palm Oil*, 48; most writers attribute the two-rope system to Ibibio or Igbo climbers, but it did appear elsewhere, possibly brought by Nigerians. See, for example, Bristowe, "Gold Coast Palms."

31. Nadel, *Black Byzantium*, 72.

32. Basden, *Among the Ibos*, 158–59.

33. MacCormack, "Control of Land," 36.

34. Basden, *Niger Ibos*, 403.

35. Chuku, *Igbo Women*, 51–52.

36. Agbasiere, *Women in Igbo Life and Thought*, 30.

37. Moloney, Druce, and Shelley, *Sketch*, 42–43.

38. See Igbo calendar in Uzozie, "Tradition and Change," 198; see also Martin, *Palm Oil and Protest*, 22–24; Bay, *Wives of the Leopard*, 196.

39. Moloney, Druce, and Shelley, *Sketch*, 42–43; Sarbah, "Oil Palm."

40. Bradbury, *Benin Kingdom*, 135.

41. This figure doesn't include time spent gathering water or firewood, and it was recorded in Dahomey's rather unique groves. Lynn, *Commerce and Economic Change*, 49; Rao, "Changing Position," 22; for the Dahomey case, see Reid, "Warrior Aristocrats," 348; for other estimates, see Maier, "Precolonial Palm Oil," 15; Forde, "The Rural Economies," 51; Northrup, "Compatibility of the Slave and Palm Oil Trades," 361; Oriji, "A Re-Assessment," 541.

42. Inikori, "Africa and the Globalization Process."

43. Junker, *Travels in Africa*, 10–11.

44. A. Jones, *German Sources*, 52.

45. "Lagos Palm Oil. (*Elæis guineensis*, Jacq.)," 206.

46. See discussion in Robins, "Oil Boom."

47. Jones, *German Sources*, 33; for contemporary uses in protecting skin, see Fentiman, "The Anthropology of Oil."

48. Jones, *German Sources*, 205.

49. Equiano, *Interesting Narrative*, 17.

50. Roberts, "Medical Exchange," 489.

51. Havik, "Hybridising Medicine," 192; Simon, "A Luso-African Formulary of the Late Eighteenth Century," 113.

52. Bosman, *New and Accurate Description*, 485–86.

53. Chuku, *Igbo Women*, 53.

54. Reynolds, *Trade and Economic Change*, 21.

55. Jones, *German Sources*, 232.

56. Jones, *German Sources*, 224.

57. Johnston, *George Grenfell and the Congo*, 742.

58. Newbury, *British Policy 1786–1874*, 446; Bradbury, *Benin Kingdom*, 135.

59. Fairhead and Leach, *Misreading the African Landscape*, 183.

60. For example, Sato, “Hunting of the Boyela,” 11.

61. Akiga and East, *Akiga’s Story*, 95.

62. Isichei, *Igbo Worlds*, 216–17; Talbot, *Shadow of the Bush*, 271.

63. Alpern, *Amazons of Black Sparta*, 70; Isichei, *Igbo Worlds*, 102.

64. L. Wilson, *Krobo People*, 42.

65. Torday, “Bushongo Mythology”; and see the more detailed, but uncited, version in Ergo, “Histoire de l’Elaéiculture,” 13–14.

66. Jones, *German Sources*, 52.

67. Purchas, *Purchas His Pilgrimes*, 976.

68. Gruca, van Andel, and Balslev, “Ritual Uses of Palms.”

69. Buckley, *Yoruba Medicine*, 98, 110; see also McKenzie, *Hail Orisha!*

70. Henderson, *King in Every Man*, 118.

71. Jones, *German Sources*, 120. The practice continued in the Yoruba diaspora as well; see Parés, *Formation of Candomblé*, 117.

72. Pratten, *Man-Leopard Murders*, 59.

73. Talbot, *Shadow of the Bush*, 130; Agbasiere, *Women in Igbo Life and Thought*.

74. Basden, *Among the Ibos*, 160; Hofstra, “Social Significance,” 114.

75. Sweet, *Recreating Africa*, 256n53.

76. Gruca, van Andel, and Balslev, “Ritual Uses of Palms.”

77. See Robins, “Imbibing the Lesson of Defiance.”

78. Friedel, “Sur des matières grasses,” 648–50.

79. Lucas, *Ancient Egyptian Materials and Industries*, 269, 271.

80. Banks and Hilditch, “Note on the Composition.”

81. Buckley and Evershed, “Organic Chemistry of Embalming Agents in Pharaonic and Graeco-Roman Mummies,” 838.

82. Kantor, “Plant Ornament,” 291.

83. Hartley, *Oil Palm*, 1st ed.; Hartley cited Raymond, “Oil Palm Industry”; who in turn cited Fickendey, *Die Ölpalme an der Ostküste von Sumatra*, who cited Friedel, “Sur des matières grasses.”

84. Berger and Martin, “Palm Oil,” 397.

85. *The Oil Palm*, “History and Origin.”

86. Fife, *Shocking Truth about Palm Oil*, 11.

87. Hughes, *Natural History of Barbados*, book iv, 112–113.

88. Winterbottom, *An Account*, 60.

89. Herodotus, *Histories*, book 2, chapter 86. For the erroneous translation see MacKendrick and Howe, *Classics in Translation*, vol. 1, 117. The Rawlinson (1860), Macaulay (1890), and Godley (1920) translations all use “palm wine.”

90. Diodorus Siculus, *Library of History*, book 1, chapter 91. Another Greek text, the Periplus of the Erythraean Sea, contains a reference to “palm oil” exported from East Africa, but this was likely coconut oil.

91. Copley et al., “Detection of Palm Fruit Lipids”; Ibrahim and Baker, “Medemia Argun.”

92. Translation and background information kindly provided by Anya King. Lewicki’s translation simply states that the fruit was “very fat.” Lewicki and Johnson, *West African Food in the Middle Ages*, 108; also see translation in Shams al-Dīn

Muḥammad ibn Abī Ṭālib Dimashqī, *Manuel de la cosmographie du moyen âge*, 341. On "Samaqanda" and the problem of geographical sources for West Africa, see Bühnen, "In Quest of Susu."

93. Some authors confused coconuts with oil palms and wrongly insisted that the latter were known in the Arab world. See Ghesquière, "L'Elaeis guineensis Jacq. est-il africain ou américain?," 341.

94. Girshick and Thornton, "Civil War in the Kingdom of Benin, 1689–1721," 359.

Chapter 2

1. Cà da Mosto, *Voyages of Cadamosto*, 42–44.

2. Pacheco Pereira, *Esmeraldo de Situ Orbis*, 128–29.

3. Garcia da Orta described "palmitos" that may have been palm kernels, but considered them a kind of coconut (*Colóquios*, 240). Clusius is often credited with identifying oil palm as "nux indica" (*Aromatum*, book 1, chapter 123) but his earliest description is a Latin translation of da Orta. Clusius updated his description of the oil palm several times but never got it quite right, with an accurate description appearing in a posthumous correction (*Curae posteriores*, 85). See Opsomer, "Les Premières Descriptions," who refutes claims made by Ghesquière in "L'Elaeis guineensis Jacq. est-il africain ou américain?"; see also Schultes, "Taxonomic, Nomenclatural and Ethnobotanic Notes on Elaeis," 175.

4. Purchas, *Purchas His Pilgrimes*, pt. 2, 936; on scientific "noise," see Schiebinger, *Plants and Empire*, 84.

5. Zeven, *Semi-Wild Oil Palm*, 6–7; on coconuts in Europe, see Kennedy, "Gripping It by the Husk."

6. M'Cormac, "On the Oil Called Palm Oil."

7. Sloane, *A Voyage to the Islands*, vol. 2, 114.

8. Blackwell, *Curious Herbal*, vol. 2, plate 363; Madge, "Elizabeth Blackwell—the Forgotten Herbalist?"

9. Jacquin, *Selectarum stirpium Americanarum historia*, 280–82.

10. On trading forts, see Osei-Tutu and Smith, *Shadows of Empire*.

11. See, for example, Law, *The English in West Africa, 1691–1699*, 169, 211, 511.

12. Law, *The English in West Africa, 1691–1699*, 171.

13. Bosman's original Dutch is less poetic: "not bad to eat." Bosman, *Nauwkeurige beschryving*, pt. 2, 65; Bosman, *A New and Accurate Description*, 285. While the passage is often attributed to Barbot, it was copied from Bosman.

14. Reputedly translated from a Dutch text ca. 1600, in Purchas, *Hakluytus Posthumus*, 274.

15. Atkins, *Voyage to Guinea*, 71.

16. Donelha, *Account of Sierra Leone*, 81.

17. Purchas, *Purchas His Pilgrimes*, pt. 1, 416.

18. Froger, *Relation of a Voyage*, 9.

19. Law, *The English in West Africa, 1691–1699*, 439–40.

20. Jones, *German Sources*, 224 and passim; also see Bosman, *A New and Accurate Description*, 286.

21. Law, *The English in West Africa, 1691–1699*, 6, 322.

22. Law, *The English in West Africa, 1691–1699*, 444.

23. Quoted in Winterbottom, *An Account*, 61.

24. Apeh and Opata, "Oil Palm Wine Economy"; Béhi et al., "Vin de Palme"; Akyeampong, *Drink, Power, and Cultural Change*.

25. See Klein, *Atlantic Slave Trade*, 7–10.

26. Manning, *Slavery and African Life*, 3.

27. Vogt, "Early São Tomé–Principe Slave Trade," 454, 466.

28. Reproduced in A. Jones, *Brandenburg Sources*, 195; see also Zaugg and Koslofsky, "Ship's Surgeon."

29. Printed in Churchill, *Collection of Voyages*, 234. Branding was not universally practiced. A Danish account from 1760 described captives being shaved and rubbed with palm oil before inspection by slavers, who "may spend four hours examining a single slave." Rømer, *Reliable Account*, 226.

30. Dalrymple-Smith and Frankema, "Slave Ship Provisioning."

31. Aubrey, *Sea-Surgeon*, 127–28, 130.

32. Quoted in Watkins, "Afro-Brazilian Landscape," 167.

33. Avity, *Estates, Empires, & Principallities*, 1100.

34. Carney and Rosomoff, *Shadow of Slavery*, 67; I am grateful to Case Watkins for clarifying the meaning of "coquinho," the term used in the source cited by Carney and Rosomoff (Alencastro, *O trato dos viventes*, 252). For another example, see Biron, *Curiositez de la nature*, 108–9.

35. Sloane, *A Voyage to the Islands*, vol. 1, 53.

36. Barham, *Hortus Americanus*, 130.

37. Watkins, "Afro-Brazilian Landscape," 144.

38. Carney and Rosomoff, *Shadow of Slavery*, 9.

39. Moreau de Saint-Méry, *Description*, vol. 1, 54. John Garrigus kindly provided this reference. O. F. Cook suggested that most captives weren't familiar with palm oil and did not pursue it in America, but this wasn't true. Cook, "Oil Palms," 16–18.

40. P. Thompson, "Henry Drax's Instructions," 586.

41. Reproduced in Churchill, *Collection of Voyages*, 249.

42. Ligon, *Barbados*, 69.

43. P. Thompson, "Henry Drax's Instructions," 601.

44. Sloane, *A Voyage to the Islands*, vol. 2, 114.

45. Barham, *Hortus Americanus*, 130.

46. On linguistic evidence, see Hall, *Slavery and African Ethnicities*.

47. See Inikori, "Africa and the Globalization Process," 82; and more generally in Inikori, *Africans and the Industrial Revolution*.

48. Tomich, "Une Petite Guinee"; Handler and Wallman, "Production Activities."

49. Berlin and Morgan, "Introduction," 9.

50. Long, *History of Jamaica*, 740.

51. Roughley, *Jamaica Planter's Guide*, 419–20; Prince, *The West Indies*, 5; Higman, *Jamaican Food*, 171.

52. Carney and Rosomoff, *Shadow of Slavery*; Watkins, "Afro-Brazilian Landscape," 154, 198.

53. Hughes, *Natural History of Barbados*, book iv, 111.

54. Heywood and Thornton link this incident to Angola, but the practice was common on the Gold Coast as well. *Central Africans, Atlantic Creoles*, 92, 292.

55. Wolsak, "Of Fishpots, Bonnets, and Wine," 55–57.

56. Ligon, *Barbados*, 77. Case Watkins suggests this palm was *Roystonea regia*.

57. Watkins, "Afro-Brazilian Landscape," 165.

58. Imray to Hooker, 30 M1855, Director's Correspondence, RBGK, http://plants.jstor.org/stable/10.5555/al.ap.visual.kmdc1336; and March to Hooker, 20 March 1856, http://plants.jstor.org/stable/10.5555/al.ap.visual.kmdc1402.

59. Patiño, "Informacion Preliminar," 10.

60. American botanist O. F. Cook was a leading proponent of the American-origins theory. See Cook, "Brazilian Origin."

61. Watkins, "Afro-Brazilian Landscape," 156–57.

62. Watkins, "Afro-Brazilian Landscape," 158.

63. Dampier, *Voyage to New Holland*, vol. 3, 71.

64. Carney and Rosomoff, *Shadow of Slavery*, 141; on the problems of identifying African ethnicity in the historical records of slave societies, see Hall, *Slavery and African Ethnicities*.

65. Watkins, "Afro-Brazilian Landscape," 199–200.

66. Murrell, *Afro-Caribbean Religions*.

67. Parés, *Formation of Candomblé*, 117.

68. Schwartz, *Slaves, Peasants, and Rebels*, chap. 4.

69. Carney and Rosomoff, *Shadow of Slavery*, 237n33.

70. Watkins, "Afro-Brazilian Landscape," 165–66.

71. Price, "Subsistence on the Plantation Periphery," 112.

72. Berlin and Morgan, "Introduction."

73. Georgia Writers' Project—Savannah Unit, *Drums and Shadows*, 52.

74. Goucher, *Congotay!*, 23.

75. Bleichmar, "Books, Bodies, and Fields."

76. A. Jones, *German Sources*, 16.

77. A. Jones, *German Sources*, 122.

78. Donelha, *Account of Sierra Leone*, 81.

79. Havik, "Hybridising Medicine," 197–99.

80. Quoted in Watkins, "Afro-Brazilian Landscape," 168.

81. W. Smith, *A New Voyage to Guinea*, 163. Most of this text was plagiarized from other authors.

82. Roberts, "Medical Exchange," 502, argues that African materials in the European pharmacy were "largely stand-ins for more commonly used reagants," but as I argue, the widespread use of palm oil suggests otherwise.

83. Biron, *Curiositez de la nature*, 108–9, uses the word "Aouara," which today describes an American palm. He describes the palm in question as "much more common in Africa than in America."

84. Atkins, *A Treatise on the Following Chirugical Subjects*, 196, 203, 208.

85. Chakrabarti, *Materials and Medicine*, 144.

86. Ligon, *Barbados*, 51.

87. Schiebinger, *Secret Cures of Slaves*, 42, 47.

88. Long, *History of Jamaica*, 740.

89. Handler, "Slave Medicine and Obeah," 59–60.

90. Voeks, "African Medicine and Magic in the Americas," 66.

91. Hogarth, *Medicalizing Blackness*, 87; Parrish, *American Curiosity*, 274.

92. Chakrabarti, *Materials and Medicine*, 153–55; Sheridan, *Doctors and Slaves*, 78–79.

93. Handler, "Slave Medicine and Obeah," 66–67.

94. Hogarth, *Medicalizing Blackness*, 88.

95. Stilliard, "Legitimate Trade," 8.

96. Kerr, *General History*, 334–35.

97. J. V., *Golgotha*, 19; Salmon, *Medicina Practica*, 1–2, 38, 108, 116–17; E. R., *Experienced Farrier*, 354.

98. Harvey, *The Family-Physician, and the House-Apothecary*, 134.

99. See, for example, *Daily Post* (London), 4 February 1729.

100. Behrendt, Latham, and Northrup, *Diary of Antera Duke*, 84n40.

101. W. Smith, *A New Voyage to Guinea*, 163.

102. Lewis, *Experimental History*, 472–73.

103. Pomet, *Compleat History of Druggs*, 136. This translation renders the French "amande" as palm kernel, when it in fact refers to the palm fruit.

104. Quincy, *Pharmacopoia*, vol. 1, 237; Lemery, *Traité Universel*, 603; Miller, *Botanicum Officinale*, 326–27.

105. "Extracts from the Records of Surrey County," 79.

106. Bancroft, *Natural History of Guiana*, 269.

107. Lewis, *The New Dispensatory*, 191–92; see also Lewis, *Experimental History*, 472–73.

108. Dalzel, *History of Dahomey*, xxvi.

109. *Collecção de Noticias*, vol. 2, 88; Barbot, *Description of the Coasts*, 106. Barbot's quote is likely an embellishment of the Portuguese source.

Chapter 3

1. Drescher, *Econocide*; Brown, *Moral Capital*; see also Palmer, "How Ideology Works."

2. House of Lords Deb., 16 May 1806, vol. 7, col. 233.

3. Behrendt, Latham, and Northrup, *Diary of Antera Duke*, 81.

4. Brown, "Origins of 'Legitimate Commerce'"; Dalrymple-Smith, "Commercial Transition," chap. 4.

5. Coleman, *Romantic Colonization*; Brown, *Moral Capital*, 315; Ciment, *Another America*, 82.

6. MacQueen, *Colonial Controversy*, 110–14.

7. Fairhead et al., *African-American Exploration*, 129.

8. "The African Trade," 165.

9. Fairhead and Leach, *Reframing Deforestation*, chap. 2; Fleur, *Fusion Foodways*.

10. See the evolution of arguments in Law, *Slave Trade*; Law, Schwarz, and Strickrodt, *Commercial Agriculture*.

11. Austen, *African Economic History*, 97; a view also found in Hopkins, *Economic History*.

12. See Lynn, *Commerce and Economic Change*, 46.

13. Falola, *Colonialism and Violence*, 20.

14. *Report from the Select Committee on the West Coast of Africa*, pt. 1, 44.

15. Quoted in Stilliard, "Legitimate Trade," 46–47.

16. Laird, *Narrative of an Expedition*, 106–7.

17. See Sherwood, *After Abolition*, chap. 1.

18. Collier to Croker, 12 March 1821, enclosed in *Further Papers Relating to the Slave Trade*, 4–5.

19. Britain demanded bonds certifying the purpose of casks, but this was a weak enforcement method. See *Correspondence with Foreign Powers*.

20. "Report of the Case of the Schooner 'Sete de Avril' . . ." 15 November 1839, in *Correspondence with the British Commissioners*, 123–24; voyages 3090 and 2637 in slavevoyages.org.

21. Canot, *Captain Canot*, 127.

22. Canot, *Captain Canot*, 99–100, 126.

23. *Report from the Select Committee on the West Coast of Africa*, pt. 1, 96.

24. Reid, "Warrior Aristocrats," 202.

25. Northrup, "Compatibility of the Slave and Palm Oil Trades," 360.

26. Quoted in Dike, *Trade and Politics*, 53.

27. Reid, "Warrior Aristocrats," 201.

28. Forbes, *Dahomey*, 25.

29. Forbes, *Dahomey*, 55.

30. Reid, "Warrior Aristocrats," 346.

31. *Report from the Select Committee on the West Coast of Africa*, pt. 1, 125.

32. Law, "Introduction," 10.

33. Reynolds, "Agricultural Adjustments on the Gold Coast after the End of the Slave Trade, 1807–1874," 317.

34. Manning, "Slave Trade," 207–10.

35. Mann, *Lagos*, 24; Falola, *Political Economy*, 115.

36. Latham, *Old Calabar*, 32–34; Hargreaves, "Political Economy," 34–36; on canoes and aquatic cultures generally, see Dawson, *Undercurrents of Power*.

37. Hargreaves, "Political Economy," 95–96, 104, 112–13; G. Jones, *Trading States*, 47; Northrup, "Nineteenth-Century Patterns," 6, 16.

38. *Report from the Select Committee on the West Coast of Africa*, pt. 1, 34.

39. Laird, *Narrative of an Expedition*, 341; Lynn, *Commerce and Economic Change*, 29; Nwaubani, "The Political Economy of Aboh, 1830–1857," 108; Köler, *Einige Notizen Über Bonny*, 138–40.

40. Stilliard, "Legitimate Trade," 116; Law, "Introduction," 11.

41. Dike, *Trade and Politics*, 100; Lynn, *Commerce and Economic Change*, 71.

42. Hargreaves, "Political Economy," 107; Lynn, *Commerce and Economic Change*, 166.

43. J. King, "Details of Explorations," 260.

44. Lynn, *Commerce and Economic Change*, 26.

45. Lynn, *Commerce and Economic Change*, 91.

46. J. Adams, *Sketches*, 100–101.

47. *Report from the Select Committee on the West Coast of Africa*, pt. 2, 115.

48. Carnes, *Journal*, 267.

49. See Strickrodt, *Afro-European Trade*.

50. Reid, "Warrior Aristocrats," 158.

51. For one shopping list, see J. Adams, *Sketches*, 100–101; Lynn, "Liverpool and Africa."

52. Bold, *Merchant's and Mariner's African Guide*, 82, 85.

53. *Report from the Select Committee on Africa (Western Coast)*, 196; Strickrodt, *Afro-European Trade*, 216; Sanders, "Palm Oil Production," 62; Lynn, *Commerce and Economic Change*, 56.

54. Cruickshank, *Eighteen Years on the Gold Coast of Africa*, 43–44.

55. Agiri, "Aspects of Socio-Economic Changes," 469.

56. Hogendorn and Johnson, *The Shell Money of the Slave Trade*, 73.

57. Lovejoy and Richardson, "'This Horrid Hole'"; see the example in Langdon, "Three Voyages."

58. William Oates, "First Voyage to Bonny," 10 October 1853, D_O-17, Merseyside Maritime Museum, accessed through *Slavery, Abolition and Social Justice* database.

59. Hargreaves, "Political Economy," 94.

60. Quoted in Simmons, "Ethnographic Sketch," 5.

61. Lovejoy and Richardson, "From Slaves to Palm Oil."

62. Dike, *Trade and Politics*, 161–62; Noah, "Political History"; Nair, *Politics and Society*, 15.

63. Harvey, "Chopping Oil in West Africa."

64. Dike, *Trade and Politics*, 109.

65. Dike, *Trade and Politics*; Johnston, "British West Africa," 108; Newell, "Dirty Whites."

66. Waddell, *Twenty-Nine Years*, 261; see also Lynn, *Commerce and Economic Change*, 89–90.

67. Dike, *Trade and Politics*, 110.

68. Dike, *Trade and Politics*, 111.

69. Stilliard, "Legitimate Trade," 127.

70. Lovejoy, *Transformations in Slavery*, chap. 8.

71. See Law, "'There's Nothing Grows in the West Indies but Will Grow Here.'"

72. Quoted in Sherwood, *After Abolition*, 121.

73. Sutton, "Commerical Agriculture," 464; Mann, *Lagos*, 133–35; Falola, *Political Economy*, 97–98.

74. Lynn, *Commerce and Economic Change*, 51–52.

75. Quoted in L. Wilson, *Krobo People*, 94.

76. Sutton, "Commerical Agriculture," 468.

77. Maier, "Precolonial Palm Oil," 12.

78. Manning, "Slave Trade," 221.

79. Obichere, "Women and Slavery in the Kingdom of Dahomey," 16.

80. Reid, "Warrior Aristocrats," 420–21; Manning, *Dahomey*, 13.

81. Quoted in Shields, "Women Slaves," 186.

82. Quoted in Law, "Trade and Gender in Yorubaland and Dahomey," 199.

83. Ohadike, *Anioma*, 186.

84. *Report from the Select Committee on the West Coast of Africa*, pt. 1, 90.

85. Northrup, "Nineteenth-Century Patterns," 9.

86. Lovejoy, *Transformations in Slavery*, 184.

87. *Report from the Select Committee on Africa (Western Coast)*, 91.

88. Shields, "Women Slaves," 187.

89. Lynn, *Commerce and Economic Change*, 54.

90. S. Martin, "Slaves, Igbo Women, and Palm Oil," 181.

91. Korieh, "The Nineteenth Century Commercial Transition in West Africa," 607; Ukegbu, "Nigerian Oil Palm Industry," 49.

92. O'Hear, "Enslavement of Yoruba," 64–65; Johnson and Johnson, *The History of the Yorubas*, 324–25; McIntosh, *Yoruba Women*, 133.

93. *Report from the Select Committee on Africa (Western Coast)*, 358.

94. "Lagos Palm Oil. (*Elaeis guineensis*, Jacq.)," 207; see discussion in Mann, *Lagos*, 133–35.

95. Shields, "Women Slaves," 194; Robin Law agrees that male slaves made palm oil in Dahomey, but cites evidence for only female labor in Yorubaland. Law, "Trade and Gender in Yorubaland and Dahomey," 205–7.

96. Shields, "Palm Oil and Power," 97–98.

97. Nelson, *Congo Basin*, 36.

98. *Report from the Select Committee on the West Coast of Africa*, pt. 1, 50.

99. Quoted in Shields, "Women Slaves," 194.

100. Mann, "Owners, Slaves," 149.

101. Delaney, *Official Report*, 36.

102. Hargreaves, "Political Economy," 97.

103. Shields, "Palm Oil and Power," 74; McIntosh, *Yoruba Women*, 134.

104. Crow, *Captain Hugh Crow*, 197.

105. G. Jones, *Slaves to Palm Oil*, 47–48.

106. Compare descriptions in Addison, "Palm Nut Tree"; Ukegbu, "Nigerian Oil Palm Industry," 52–53; S. Martin, "Slaves, Igbo Women, and Palm Oil," 182; S. Martin, "Gender and Innovation," 419.

107. Dop and Robinson, *Travel Sketches from Liberia*, 509.

108. Cowan, *Liberia*, 101.

109. Lynn, *Commerce and Economic Change*, 49; Zeven notes figures as high as 630 workdays. *Semi-Wild Oil Palm*, 94.

110. Maier, "Precolonial Palm Oil," 15.

111. Ukegbu, "Nigerian Oil Palm Industry," 45, 58–60; S. Martin, "Gender and Innovation"; Maier, "Precolonial Palm Oil."

112. Farquhar, *Oil Palm*, 30; Lynn, *Commerce and Economic Change*, 49–50.

113. Lynn, *Commerce and Economic Change*, 56–57.

114. Chuku, *Igbo Women*, 52–53.

115. Quoted in Maier, "Precolonial Palm Oil," 14.

116. Farquhar, *Oil Palm*, 30.

117. Law, "Wheeled Transport in Pre-Colonial West Africa," 255.

118. Obichere, "Women and Slavery in the Kingdom of Dahomey," 8.

119. Holman, *Travels*, 402.

120. Knapp, *Chemical Technology*, 430; see also Muspratt, *Chemistry*, 618.

121. Baikie, *Narrative*, 301.

122. Köler, *Einige Notizen Über Bonny*, 151; Cole, *Life in the Niger*, 16–17.

123. "Lagos Palm Oil. (*Elaeis guineensis*, Jacq.)," 205–6.

124. 1877 Gold Coast report, reproduced in Moloney, Druce, and Shelley, *Sketch*, 47.

125. See, for example, Maier, "Precolonial Palm Oil," 6, 22.

126. S. Martin, *Palm Oil and Protest*, 33–34.

127. Brannt, *Animal and Vegetable Fats*, 336.

128. Whitford, *Trading Life in Western and Central Africa*, 94.

129. Morel, *Affairs of West Africa*, 79.

130. Mockler-Ferryman, *British Nigeria*; but this is based on Moloney, Druce, and Shelley, *Sketch*.

131. Moloney, Druce, and Shelley, *Sketch*, 57.

132. Schnapper, *La politique et le commerce*, 121; Chevalier, *Documents sur le palmier à huile*, 15.

133. Demeur, "To John Demeur"; *Patents for Inventions*, 69.

134. Fyfe, "Charles Heddle"; Fyfe, *Sierra Leone*, 258; Brooks, "Peanuts and Colonialism," 48; Johnston credits an unnamed Liberian with the first kernel exports in 1850 but provides no corroborating information. *Liberia*, 405.

135. Mitchell, "Trade Routes."

136. Fyfe, "Charles Heddle," 235.

137. Heddle to Baikie, 8 May 1857, in *Reports made for the Year 1856 to the Secretary of State*, 179–81.

138. Fyfe, "Charles Heddle," 241–42.

139. Cowan, *Liberia*, 102.

140. N. Richardson, *Liberia's Past and Present*, 249–50.

141. Delaney misidentified Herring as "Herron." Delaney, *Official Report*, 32.

142. N. Richardson, *Liberia's Past and Present*, 249–50; see also Herring's obituary in *Sierra Leone Weekly News*, 7 September 1895, 12.

143. Cowan, *Liberia*, 102.

144. Moloney, Druce, and Shelley, *Sketch*, 65.

145. S. Martin, *Palm Oil and Protest*, 47.

146. Manning, *Dahomey*, 99.

147. Lynn, *Commerce and Economic Change*, 124.

148. Grace, *Domestic Slavery in West Africa*, 12–13.

149. Manning, "Slave Trade," 210; Cochard et al., "Geographic and Genetic Structure of African Oil Palm Diversity Suggests New Approaches to Breeding."

150. *Report from the Select Committee on the West Coast of Africa*, pt. 1, 34.

151. *Report from the Select Committee on the West Coast of Africa*, pt. 1, 225.

152. *Report from the Select Committee on the West Coast of Africa*, pt 1, 340.

153. *Report from the Select Committee on the West Coast of Africa*, pt. 1, 249.

154. Lynn, *Commerce and Economic Change*, 55.

155. Fairhead and Leach, *Reframing Deforestation.*

156. Burton, *Abeokuta and the Camaroons Mountains*, 263–64.

157. See Spinage, *African Ecology*; Mccann, "Climate and Causation."

158. L. Wilson, "Bloodless Conquest"; Reid, "Warrior Aristocrats," 211.

159. *Report from the Select Committee on the West Coast of Africa*, pt. 1, 169.

160. *Report from the Select Committee on Africa (Western Coast)*, 141.

161. Manning, *Dahomey*, 98; for photographs, see Sarbah, "Oil Palm"; Adam, *Le palmier à huile.*

162. Quoted in L. Wilson, *Krobo People*, 78.

163. Quoted in L. Wilson, *Krobo People*, 81.

164. Field, "Agricultural System"; L. Wilson, *Krobo People*, 72–73.

165. Getz, *Slavery and Reform*, 59–62.

166. Hamilton, *Excelsior*, 339.

167. Sarbah, *Fanti Customary Laws*, 69–70, 153–56.

168. Lloyd, "Problems of Tenancy," 96–97; Johnson and Johnson, *The History of the Yorubas*, 95–96, 329; Fenske, "Land Abundance and Economic Institutions."

169. Delaney, *Official Report*, 32.

170. Shields, "Palm Oil and Power," 164–65.

171. Mann, "Owners, Slaves," 149.

172. Forbes, *Dahomey*, 27.

173. Bay, *Wives of the Leopard*, 1.

174. Ross, "Career of Domingo Martinez," 80; Bay, *Wives of the Leopard*, 171.

175. Juhé-Beaulaton, "La palmeraie du Sud Bénin"; Soumonni, "Dahomean Economic Policy."

176. Stilliard, "Legitimate Trade," 56; Lynn, *Commerce and Economic Change*, 43.

177. Guézo, "Impact of British Abolition," 147.

178. Soumonni, "Compatibility of the Slave and Palm Oil Trades," 83.

179. Alpern, "Dahomey's Royal Road," 17–18; Dissou, *Économie de la culture*, 45; Reid, "Warrior Aristocrats," 164–65.

180. Manning, *Dahomey*, 80; but see Wartena's critical assessment in Wartena, "Making a Living," chap. 6.

181. Burton, *Mission to Gelele*, 128; Skertchly, *Dahomey as It Is*, 16

182. Quoted in Law, "Trade and Gender in Yorubaland and Dahomey," 198–99.

183. Burton, *Mission to Gelele*, 237; Dissou, *Économie de la culture*, 47; Wartena, "Making a Living," 267.

184. Hallett, *Niger Journal*, 296.

185. *Report from the Select Committee on the West Coast of Africa*, pt. 1, 334.

186. Partridge, *Cross River Natives*, 143.

187. Ukegbu, "Nigerian Oil Palm Industry," 31–32.

188. Thomas, *Anthropological Report*, 96; for the Yoruba custom, see Ukegbu, "Nigerian Palm Oil Industry," 37.

189. S. Martin, "Slaves, Igbo Women, and Palm Oil," 186.

190. On the erosion of communal rights, see Oriji, "A Re-Assessment," 539–47; S. Martin, *Palm Oil and Protest*, 82; for strengthening, see Green, *Ibo Village Affairs*, 133; Bradbury, *Benin Kingdom*, 23.

191. Chubb, *Ibo Land Tenure*, 48.

192. Uchendu, *Igbo of Southeast Nigeria*, 59, 88.

193. Lynn, *Commerce and Economic Change*, chap. 2; Lovejoy, *Transformations in Slavery*, 187–88.

194. Hutton, quoting an 1835 letter in *Report from the Select Committee on the West Coast of Africa*, pt. 1, 224.

195. Dike and Ekejiuba, *The Aro*, 257.

196. Ohadike, *Ekumeku Movement*, 31.

197. Dalrymple-Smith, "Commercial Transition," chap. 6.

198. Rao, "Changing Position," 12; see also Korieh, *The Land Has Changed*, chap. 1.

199. Lynn, *Commerce and Economic Change*, 111.

200. Syfert, "A History of the Liberian Coasting Trade, 1821–1900," 207–11.

201. Buxton, *The African Slave Trade*, 220.

202. Burton, *A Mission to Gelele*, 248–49.

203. In Johnston, "British West Africa," 119.

204. *Report from the Select Committee on Africa (Western Coast)*, 212; see also Ofonagoro, *Trade and Imperialism*, 13.

205. Mann, *Lagos*, 92–94.

206. *Report from the Select Committee of the House of Lords, Appointed to Consider the Best Means Which Great Britain Can Adopt for the Final Extinction of the African Slave Trade*, 547.

207. Robert S. Smith, *The Lagos Consulate, 1851–1861*, 27; Mann, *Lagos*, 44–50; Akinjogbin, *War and Peace in Yorubaland, 1793–1893*.

208. Mann, *Lagos*, 99–102.

209. Palmerston, "Minute on Protection of Trade," 22 April 1860, in Newbury, *British Policy 1786–1874*, 120; Hopkins, "Economic Imperialism in West Africa," 589.

210. Biobaku, *Egba*, 67.

211. Olabimtan, "Townsend, Henry"; Tucker, *Abbeokuta; or, Sunrise within the Tropics*.

212. See, for example, Campbell, *A Few Facts*, 14–15; Vincent, "Cotton Growing in Southern Nigeria."

213. Campbell, *A Pilgrimage to My Motherland*, 83; Ogunremi, "Economic Development and Warfare."

214. Johnson and Johnson, *The History of the Yorubas*, 490–91; Biobaku, *Sources of Yoruba History*, 245.

215. See Falola, "Yoruba Toll System"; Agiri, "Aspects of Socio-Economic Changes"; Awẹ, "Militarism and Economic Development."

216. Hopkins, "Economic Imperialism in West Africa," 595; but see Law, "Crisis of Adaptation"; Falola, *Political Economy*.

217. Moody, *Palm Tree*, 144.

218. Crowther, *The River Niger*, 18; Mann, *Lagos*, 164–67, 213.

219. Getz, *Slavery and Reform*, 125–27; Mann, *Lagos*.

220. Shields, "Palm Oil and Power," 111–12.

221. Oroge, "Institution of Slavery," 202, 390.

222. Ohadike, *Ekumeku Movement*, 75; Ohadike, *Anioma*, 197.

223. Mackeown, *Twenty-Five Years in Qua Iboe*, 95–96.

224. Hubbard, *The Sobo*, 283.

225. Quoted in Dike, *Trade and Politics*, 85.

226. Stilliard, "Legitimate Trade," 55–56.

227. Dike, *Trade and Politics*, 87.

228. Nair, *Politics and Society*, 90–94.

229. Nair, *Politics and Society*, 93.

230. Lynn, "Factionalism, Imperialism and the Making and Breaking of Bonny Kingship c. 1830–1885," 179.

231. Lynn, "Factionalism, Imperialism and the Making and Breaking of Bonny Kingship c. 1830–1885," 183; Dike, *Trade and Politics*, 144.

232. Pepple, *Brief Statement*.

233. G. Jones, *Trading States*, 121–27.

234. Dike, *Trade and Politics*.

235. Dike, *Trade and Politics*, 107; Nair, "Trade in Southern Nigeria," 429.

236. Flint, *Sir George Goldie*, 20–21.

237. Quoted in Nair, "Trade in Southern Nigeria," 429.

238. Dike, *Trade and Politics*, 115; Lynn, *Commerce and Economic Change*, 109–16, 146.

239. Flint, *Sir George Goldie*, 29.

240. Lynn, *Commerce and Economic Change*, 147–50.

241. A good review of the scholarly debate can be found in Cain and Hopkins, *British Imperialism*.

242. See Flint, *Sir George Goldie*; Crowder, *West Africa under Colonial Rule*.

243. Hopkins, "Economic Imperialism in West Africa," 601–3.

244. Ayandele, *Ijebu of Yorubaland*, chap. 1; Crowder, *West Africa under Colonial Rule*, 126.

245. Akintoye, "The British and the 1877–93 War," 287–88.

246. Gueye and Boahen, "African Initiatives and Resistance," 127–29.

247. Rudin, *Germans in the Cameroons*, 77, 259; see also Austen and Derrick, *Middlemen of the Cameroons Rivers*.

248. Dike, *Trade and Politics*, 183–84; but see Hargreave's revisionist account of Jaja's origins, "Political Economy," 102.

249. Flint, *Sir George Goldie*, 27.

250. Dike, *Trade and Politics*, 188–98; Hargreaves, "Political Economy."

251. Dike, *Trade and Politics*, 215–16; Lynn, *Commerce and Economic Change*, 141, 180; Hargreaves, "Political Economy," 289.

252. Crowder, *West Africa under Colonial Rule*, 120–21.

253. Alagoa, *Small Brave City-State*, 91–116.

254. Quoted in Anene, *Southern Nigeria*, 171.

255. Falola, *Colonialism and Violence*, 9–11.

256. Belloc and Blackwood, *Modern Traveller*, 41.

257. Stilliard, "Legitimate Trade," 58; see similar dismissive views in McPhee, *Economic Revolution*, 32; Hopkins, "An Economic History of Lagos, 1880–1914," 18.

258. "A Mine of Palm-Oil," *The Colonies and India*, 20 Sept 1879, 9.

259. Jamieson, *Western Central Africa*, 24.

260. Gertzel, "Commercial Organization," 289.

261. Mockler-Ferryman and Day, *Up the Niger*, 257.

262. Hopkins, "Economic Imperialism in West Africa," 591–92.

Chapter 4

1. Freight bill for *Hawke*, 1777, D/DAV/11/5/1, Merseyside Maritime archives; slavevoyages.org voyage 91739.

2. Pory's introduction to Africanus, *Geographical Historie*, 42; Boyle, *General Heads*, 74. Several sources mention soap as an article of commerce, but it's not clear if this was for export or for regional trade in Africa. See, for example, Mentelle, *Cosmographie Élémentaire*, 314.

3. W. Smith, *A New Voyage to Guinea*, 241.

4. Barbot, *Description*, 106.

5. Purdy, *Memoir*, 106.

6. Naismith, *Observations on the Different Breeds*, 44; Wadström, *Essay on Colonization*, 31; J. Burke, *British Husbandry*, vol. 2, 476–77; Norris, *Memoirs of the Reign of Bossa Ahádee*, 146.

7. Gittins, "Soapmaking in Britain, 1824–1851," 35.

8. The recipe called for 1,120 pounds of tallow and 336 pounds of rosin, with 20 pounds of palm oil added at the end. Carmichael, *Treatise on Soap-Making*, 57; see also recipe in Aikin and Aikin, *Dictionary of Chemistry*, vol. 2, 331. Today's rosin is purer than the harsh-smelling stuff available in the 1800s. See Ure, *Dictionary*, 1144.

9. Gittins, "Soapmaking in Britain, 1824–1851," 31.

10. "Palm Soap," advertisement for Samuel Berry, *Morning Post*, 5 January 1804.

11. Morfit, *Chemistry Applied*, 220, 98; "Johnstone's Royal Patent Windsor Soap," *La Belle assemblée; or, Bell's Court and Fashionable Magazine* 16 (1817), 21.

12. Lewkowitsch, *Cantor Lectures*, 18; Leicester, *Background of Chemistry*, 174. Carl Scheele identified glycerin as a constituent of fats in 1779, but he died prematurely (likely from chemical poisoning). Lennartson, *The Chemical Works of Carl Wilhelm Scheele*, 79.

13. Dijkstra, "How Chevreul (1786–1889) Based His Conclusions on His Analytical Results"; Wisniak, "Edmond Frémy." Chevreul's "margaric acid" turned out to be mostly palmitic acid. See Parnell, *Applied Chemistry*, 311.

14. Zallen, *American Lucifers: The Dark History of Artificial Light*, 51; Robins, "Oil Boom"; Musson, *Enterprise in Soap and Chemicals*, 23.

15. Lynn, *Commerce and Economic Change*, 28; Cameron et al., "Improved Method"; Kurten, *Art of Manufacturing Soaps*, 99–100; Morfit, *Chemistry Applied*.

16. J. Anderson, *Barilla Question*, 24.

17. "Report of a Survey and Enquiry into the Manufacture of Soap in the Country directed to be made by the Honorable Boards order of 5th September 1837," CUST

119/409, TNA. I am indebted to Nancy Shoemaker for providing notes of this document.

18. Gittins, "Soapmaking in Britain, 1824–1851," 34.

19. Lynn, "Liverpool and Africa."

20. African Institution, *3rd Report*, 12, 26–27, 31–32.

21. Sherlock and Towne, *Duties, Drawbacks, and Bounties*, 43; *Customs Tariffs of the United Kingdom.*

22. The duty on tallow fell from 4s. (3s. 2d. on British ships) in 1823 to 1s. 6d. (1d. from British possessions) in 1846, to no duty in 1860.

23. "Substitute for Tallow," *Glamorgan, Monmouth, and Brecon Gazette and Merthyr Guardian*, 26 December 1840. On merchants' rhetoric, see Sherwood, *After Abolition.*

24. Association of London and Country Soap Manufacturers, *Case of the Soap Duties*, 18; "Diplomacy and Commerce No. IV," 34.

25. Association of London and Country Soap Manufacturers, *Case of the Soap Duties*, 1.

26. The real figure was probably much higher, due to tax evasion. Prison inmates got an eleven-pound annual soap ration! G. Wilson, *On the Stearic Candle*, 66; see also Association of London and Country Soap Manufacturers, *Case of the Soap Duties*; Gittins, "Soapmaking in Britain, 1824–1851."

27. Metropolitan Working Classes' Association for Improving the Public Health, *Bathing and Personal Cleanliness*; *A Circular Addressed by the Soap Manufacturers*, 11; Musson, *Enterprise in Soap and Chemicals*, 26.

28. Soames, "The Soap Duty," 10.

29. McClintock, *Imperial Leather*, 208.

30. *Report from the Select Committee on the West Coast of Africa*, pt. 1, 214.

31. Crowther, *Charge Delivered*, 15.

32. Simmonds, *Commercial Products of the Vegetable Kingdom*, 527.

33. "Cheerily, Cheerily," 26.

34. See Vos, "Coffee, Cash, and Consumption."

35. The palm kernel oil was interchangeable with coconut oil. C. Wilson, *History of Unilever*, vol. 1, 31.

36. Musson, *Enterprise in Soap and Chemicals*, 84; Lewis, *So Clean*, chap. 2.

37. Fouque, *Histoire raissonée du commerce de Marseille*, 154–61; Masson, *Marseille et la colonisation française*, 372; Schnapper, *La politique et commerce*, 121–24; Soumonni, "Compatibility of the Slave and Palm Oil Trades"; on industrialization in France, see M. Smith, *The Emergence of Modern Business Enterprise in France, 1800–1930*, 283–84.

38. "Memorial of the Citizens of New Bedford Praying That the Duty on Foreign Tallow, Olive, and Palm Oil, &c., May Not Be Repealed." 22nd Congress first session, HR doc 123, 21 February 1832.

39. The tariff was reimposed in 1846 as a revenue measure, at 10 percent ad valorum.

40. Schisgall, *Eyes on Tomorrow*, 11; Lief, *It Floats*, 28, 50; Greeney, "When Laundry Detergent Was Edible."

41. G. Wilson, *On the Stearic Candle*, 5.

42. Forbes, *Early Petroleum History*, 146; Irwin, "Spermaceti Candle," 48; Zallen, *American Lucifers: The Dark History of Artificial Light.*

43. Manicler, "Patent 6121, 2 June 1831," 240.

44. Unsaturated fatty acids make candles soft and runny, and burning glycerin produces unpleasant smells.

45. Whorton, *The Arsenic Century*, chap. 7; to get a sense of why people cared about smoke, "snuffing" and "guttering," see Bowers, *Lengthening the Day*; Dillon, *Artificial Sunshine.*

46. Advertisement, *Cardiff and Merthyr Guardian*, 26 November 1859, 1; see also "Candles, Slavery and War," 279.

47. "Manufactures of Price's," *Chemist and Druggist*, 67.

48. G. Wilson, "Manufactures," 141; G. Wilson, *On the Stearic Candle*; "Composition and Spermaceti Candles."

49. Fraser, "Candlemaking," 161; the company offered rewards for the prosecution of patent-infringers in 1844: "Fifty Pounds Reward," *Justice of the Peace*, 8, no. 31 (3 August 1844), 528. Fouqet and Pearson, "Seven Centuries"; "A History of Price's Candles."

50. G. Wilson, "Manufactures," 150; Lamborn, *Modern Soaps, Candles and Glycerin*, 446.

51. "New Price's Patent Candle Co.," *The Standard* (London), 1 May 1852, 1; Hatton, "Price's Patent Candle Company," 407.

52. See Latham, "Palm Oil Exports from Calabar."

53. "Investigations in Connection with the African Palm Oil Industry," 389; Musson, *Enterprise in Soap and Chemicals*, 90; for a detailed chart of impurities, melting points, and free fatty acid content of various palm oils, see Benedikt, *Chemical Analysis*, 429.

54. Price's Patent Candle Company, "Letter . . . to the Board of Directors," 18.

55. "The Training Schools of Price's Patent Candle Company," 106.

56. Zallen, "American Lucifers: Makers and Masters," 53.

57. Bede, "Literature," 190; compare with Teltscher's reading in *Palace of Palms*, 257–62.

58. Fraser, "Candlemaking," 170.

59. G. Wilson, *On the Stearic Candle*, 34.

60. G. Wilson, "Manufactures," 154; Clapp, *An Environmental History of Britain since the Industrial Revolution*, 228.

61. Musson, *Enterprise in Soap and Chemicals*, 215.

62. See, for example, Mathe-Shires, "Lagos Colony"; contrast with Austen, "Distorted Theme," 274; Butt, "James Young," 51.

63. Quote from Murfitt, "English Patent System," 216; "To Thomas Motley"; Booth, "Patent Axle-Grease and Lubricating Fluid, No. 6814"; Smiles, *Memoir of the Late Henry Booth*, 43.

64. Browne, "Car Axle-Boxes," 389; "Unguents," 410; Laucks, *Commercial Oils*, 87.

65. "Railway and Waggon Grease."

66. McPhee gives 13,000 tons as a figure for railways in 1865 (McPhee, *Economic Revolution*, 31), but I was unable to locate the edition of Spon's *Dictionary* that he cited. I use figures cited in other versions of Spon, derived from T. Richardson and Watts, *Chemical Technology*, 744–47.

67. Russell and Hudson, *Early Railway Chemistry and Its Legacy*, 18; "Design and Construction," 286.

68. Gibbins, *Britain's Railways*, 50.

69. "Railway Grease," 134; see also "The Bubble of the Age," 11.

70. Watson, *The Reasoning Power in Animals*, 294; see the nearly identical account in "Railway Theft Extraordinaire," *Liverpool Mercury*, 18 March 1853.

71. Dickens, "Our Oil-Flasks," 262.

72. Maclean, *British Railway System*, 77.

73. There is a growing literature on this theme, but see especially MacFadyen, *Flax Americana*; Cushman, *Guano*; Evans, *Bound in Twine*; Moore, *Capitalism*.

74. See Zeide, *Canned*.

75. "The Palm-Oil Districts of Africa," 26.

76. Minchinton, *British Tinplate*, chap. 1.

77. Dunbar, *Tin-Plate Industry*, 10–11.

78. William Williams and Thomas Hay, 1834 patent for "improvements in preparing certain metals . . . ," *London Journal of Arts and Sciences* 7 (1836), 130–38; Lynn, *Commerce and Economic Change*, 117, 130.

79. Flower, *A History of the Trade in Tin*, 169.

80. Bailey, *Industrial Oil*, 433.

81. Letter to editor, *Western Mail*, 17 March 1874, 5.

82. Swank, *All about Tinplates*, 14.

83. *Report of the Departmental Committee on Compensation*, 3849–3917; Minchinton, *British Tinplate*, 123.

84. *Forty-Fourth Annual Report on Alkali*, 30; Dunbar, *Tin-Plate Industry*, 45; J. Jones, *Tinplate Industry*, 259–60.

85. Lynn underestimated the continuing use of palm oil in tinplating in *Commerce and Economic Change*, 117; J. Jones, *Tinplate Industry*, 17–18; Hammond, "Manufacture of Tin-Plates."

86. Smaller mills often continued using palm oil and older machinery, and some larger mills switched back from flux during periods of poor trade. "American Tariff Bill," *South Wales Daily News*, 18 August 1894, 6; "The Quality of Tin Roofs."

87. Swansea Chamber of Commerce and Jones, *Commercial Year Book*, 104; Camp, "Relation of the Iron and Steel Industries," 702.

88. Pantzaris, *Pocketbook*, 148; Stepina and Vesely, *Lubricants and Special Fluids*, 619; Nurul, Syahrullail, and Teng, "Alternative Lubricants."

89. Lynn, *Commerce and Economic Change*, 111–14.

90. For a concise treatment of the voluminous literature on petroleum, see Ross, *Ecology and Power*, chap. 6.

91. Headrick, "Botany, Chemistry, and Tropical Development"; C. Jones, *Routes of Power*, 115; Lynn, *Commerce and Economic Change*, 117.

92. Lewkowitsch, *Chemical Technology*, vol. 3, 285; McGucken, *Biodegradable*, chap. 1.

93. PPCC advertisement, 1884, Evanion Collection of ephemera, Evan.4318, British Library (http://www.bl.uk/onlinegallery/onlineex/evancoll/a/014eva000000000u04318000.html).

94. Fouqet and Pearson, "Seven Centuries," 171; Zallen, *American Lucifers: The Dark History of Artificial Light*.

95. Wynter, "Candle Making," 80; according to Newman, 12 percent of PPCC candles used paraffin in 1870, but 90 percent did by 1900: Newman, *Battersea's Global Reach*, 28.

96. W. Smith, "Report on Section III," 633; Malden, *The Victoria History of the County of Surrey*, 408; Pantzaris, *Pocketbook*, 145.

97. Burton, *Abeokuta and the Camaroons Mountains*, 324.

98. Moloney, Druce, and Shelley, *Sketch*, 35.

99. Simmonds, *Commercial Products of the Vegetable Kingdom*, 525.

100. Read, *Negro Problem Solved*, 169.

101. "Liberian Invention and Manufacture," *The Friend* 23, no. 50 (1850): 397.

102. Olukoju, *"Liverpool" of West Africa*, 51.

103. Aitken, "Feeding Stuffs," 37; Coleman, *Cattle*, 54; "Palm Kernel Meal."

104. Henriques and Warde, "Fuelling the English Breakfast"; Lampe and Sharp, *A Land of Milk and Butter*, chaps. 6–7.

105. Snodgrass, *Margarine*, 125; Commercial Manufacturing Company, *Brief History*; see also Clayton, *Margarine*, 1–2; Howard, "Margarine Industry"; Stuyvenberg, *Margarine*.

106. Spiekermann, "Redefining Food," 11; Cohen, *Pure Adulteration*, chap. 4.

107. Lewkowitsch, *Cantor Lectures*, 18.

108. Talbot, *Oil Conquest*, 214–15; Unilever Art, Archives & Records Management, "Loders & Nucoline Ltd."

109. Talbot, *Oil Conquest*, 208.

110. H. Smith and Pape, *Consols of the East*, 169.

111. W. Williams, "Chemistry of Cookery," 104; Brannt, *Animal and Vegetable Fats*, 334; KG Berger, "Lewkowitsch Memorial Lecture," box 2/1987, Lewkowitsch papers, Hagley Museum and Library, Wilmington, DE.

112. "Adulteration of Butter," *Liverpool Daily Post*, 28 April, 1862.

113. Skertchly, *Dahomey As It Is*, 34; Simmons and Mitchell, *Edible Fats and Oils*, 92, 126. Chemists developed tests to identify palm oil in margarine. See, for example, Lewkowitsch, *Chemical Technology*.

114. E. R. Bolton in Rutgers, LePlae, and Tingey, *Oil Palms and Their Fruit*, 34; Berger and Martin, "Palm Oil," 405. The United Kingdom imported about 3,000 tons of refined palm oil from Europe in 1909. Simmons and Mitchell, *Edible Fats and Oils*, 137–38.

115. Hunt, "Raw Materials," 38; Boldingh, "Margarine."

116. Robins, "Oil Boom."

117. List and Jackson, "Battle over Hydrogenation"; Schleifer, "The Perfect Solution"; Snodgrass, *Margarine*.

118. Hoover, "The Popular Cooking Fat in Austria."

119. Quoted in USDA, *Inventory of Seeds and Plants Imported*, 5.

120. "Palm Oil Cannot Be Used," *Omaha Daily Bee*, 12 August 1902; Howard, "Margarine Industry," 124, 271; "Palm Oil in Margarine"; M.P.C., "The Palm Oil Question," *Country Gentleman*, 25 May 1905.

121. Howard, "Margarine Industry," 124; see Cliff v. United States, 195 U.S. 159 (1904) and Moxley v. Hertz, 216 U.S. 344 (1909); "Butterine Company in Trouble," *American Food Journal* 9 no. 9 (1914): 571; Eddy, "Something New."

122. Theodoridis, Warde, and Kander, "Trade and Overcoming Land Constraints"; Warde, "Trees, Trade and Textiles."

123. Chambers, "The Light Question," 295.

124. G. Wilson, "Manufactures," 153.

125. Spinage, *African Ecology*, 152; see discussion in Mccann, "Climate and Causation"; Rönnbäck, *Labour and Living Standards.*

126. Theodoridis, "Ecological Footprint"; this estimate is 800 pounds per acre. Compare to 108 gallons (about 950 pounds) cited in AJ Walker (DC Birim) to Comm. Eastern Province, 22 August 1921, ADM 11/1/1241, PRAAD-Accra.

127. Based on figures given in Clifford, "London's Soap Industry."

128. Lund, "Fats and Oils," 35.

129. Kiple, *Movable Feast*, 216–23.

130. Jones to Kew, 3 August 1938, "Oil Palm and Oil Producing Nuts and Seeds," PRO 5/0/1, RBGK.

131. C. Wilson, *History of Unilever*, vol. 2, 177.

132. Lifebuoy soap advertisement in *Liberty Magazine*, 17 April 1926, 39.

133. Advertisement, *Women's Home Companion*, July 1912, 47.

134. Palmolive advertisement, 1938, Duke University Libraries digital repository, https://repository.duke.edu/dc/adaccess/BH1277; see also "A Day in the Palmolive Factory," Palmolive, Milwaukee, MI, ca. 1925. PAM 2009.162, Hagley pamphlet collection, Wilmington, DE.

135. The figure is for 1827–50. Inikori excluded a short-lived guano boom (1844–45) in calculating these figures: "West Africa's Seaborne Trade," 58.

136. Austen, "Distorted Theme," 274.

Chapter 5

1. C. Wilson, *History of Unilever*, vol. 2, 107.

2. Calvert, *Togoland*, 40.

3. Quoted in Billows and Beckwith, *Consols of the West Coast*, 7.

4. "The African Oil Palm Industry. III. Machinery," 63; Fickendey, "Über Die Verwendungsfähigkeit Des Palmöls Als Speisefett."

5. Robins, "Smallholders and Machines," 73.

6. Chinery, *West African Slavery*, 5.

7. See Robins, "Smallholders and Machines"; and examples in Shields, "Palm Oil and Power."

8. Veeser, "Forgotten Instrument," 1147.

9. For the Royal Niger Company and the Nigerian conquest, see Flint, *Sir George Goldie*; for French and Belgian Congo, see Coquery-Vidrovitch, *Congo*, vol. 1.

10. Boss, "Scramble for Palms"; Cushman, *Guano*, 101–8; Ramamurthy, *Imperial Persuaders*, 53.

11. Adam, *Le palmier à huile*, 258.

12. See guidebooks published for investors, including Adam, *Le palmier à huile*; Billows and Beckwith, *Consols of the West Coast*; Milligan, *Oil Palm*; Belfort and Hoyer, *Coconuts*; Newland, *Planting, Cultivation and Expression*; and see correspondence in CO 879/115/8, TNA.

13. Daniel, *Le palmier à huile au Dahomey*, 6, 26.

14. Hubert, *Le palmier à huile*, 267–73.

15. Oloruntimehin, "French Estate"; Clozel, "Land Tenure among the Natives of the Ivory Coast."

16. "La Question Des Palmeraies," 8.

17. "La Question Des Palmeraies," 3–6.

18. "The Dahomey Oil Palm," 28; Henry, "L'exploitation Du Palmier á Huile," 312–14; and see another case documented in Manning, *Dahomey*, 174.

19. Henry, "L'exploitation Du Palmier á Huile," 318–22; Brocard, "Afrique Occidentale Francaise," 240.

20. Committee on Edible and Oil-Producing Nuts and Seeds, *Minutes of Evidence*, 5417–18, 5597–98.

21. Committee on Edible and Oil-Producing Nuts and Seeds, *Minutes of Evidence*, 5609, 5633–35.

22. Committee on Edible and Oil-Producing Nuts and Seeds, *Minutes of Evidence*, 5432, 5438–41, 5567.

23. Buell, *Native Problem*, vol. 2, 23.

24. See Coquery-Vidrovitch, *Congo*, vol. 1, 10–11.

25. C. Wilson, *History of Unilever*, vol. 1, 179; Coquery-Vidrovitch gives a higher figure of £92,000. Coquery-Vidrovitch, *Congo*, vol. 2, chap. 14, quote on p. 333.

26. "History of Compagnie Propretaire du Kouilou Niari CPKN," UAC /1/11/14/1/37/49, UARM.

27. Henry, "L'exploitation Du Palmier á Huile," 368.

28. See Van Pelt, "Oil Palm"; Chevalier, "Le Palmier à huile à la Côte d'Ivoire."

29. W. D. Smith, *German Colonial Empire*, chaps. 5–6.

30. Müllendorff, "Development of German West Africa," 74.

31. Rudin, *Germans in the Cameroons*, 260–61; Müllendorff, "Development of German West Africa," 77; "An Imperial Industry," 329–30.

32. Chevalier, *Documents sur le palmier à huile*, 91.

33. Rudin, *Germans in the Cameroons*, 260.

34. Rudin, *Germans in the Cameroons*, 261.

35. Bederman, "Plantation Agriculture in Victoria Division, West Cameroon," 354.

36. Ven, "Jurgens [Hzn.], Antonius Johannes (1867–1945)"; C. Wilson, *History of Unilever*, vol. 2, 108.

37. Lawrance, *Locality, Mobility, and "Nation,"* 31; see also A. Jones and Sebald, *African Family Archive*; Bay and Mann, *Rethinking the African Diaspora*.

38. Zimmerman, "German Alabama," 1391.

39. "The African Oil Palm Industry. III. Machinery," 68; Lewkowitsch, *Chemical Technology*, 536; a full description of the mill can be found in ADM 39/5/6, PRAAD-Accra.

40. Agu Plantation to DC Ho, 15 July 1919, ADM 39/5/8, PRAAD-Accra.

41. Letter, "King Kofi and his people, Leleklele and his people," 24 August 1914, and DC Agu to Lome, 30 October 1914, ADM 36/5/6, PRAAD-Accra.

42. Agu Manager to Controlling Officer German Firms, ca. 1919, ADM 39/5/8, PRAAD-Accra.

43. Committee on Edible and Oil-Producing Nuts and Seeds, *Minutes of Evidence*, 1784–92.

44. The idea that Lever was forced to leave British West Africa was first constructed by Lever himself, and later retold by Nworah, "Lever's West African Concessions"; for more recent examples of this narrative, see Phillips, *Enigma of Colonialism*; S. Martin, *Palm Oil and Protest*, 60–66; Berger and Martin, "Palm Oil," 398.

45. Davies, *Trade Makers*, 140–43; Boss, "Scramble for Palms."

46. Lever Bros. to CO, 7 July 1908, in *West Africa: Correspondence Respecting the Grant of Exclusive Rights*.

47. Nworah, "Lever's West African Concessions," 250.

48. Gov. Sierra Leone to Lever Bros., 2 January 1909, in *West Africa: Correspondence Respecting the Grant of Exclusive Rights*.

49. Clifford to Sec. State for Colonies, 17 February 1913, CO 879/115/8, TNA.

50. The firms were APOL, Brunner Mond, Crosfield, and Gossage. These estates (at Winneba, Butre [or Sese], Dixcove, Akwida, and Elmina) were amalgamated in 1914 as Palm Oil Estate Managers Ltd. See BT 31/41073 and BT 31/41074, TNA, and POEM property correspondence, UAC/2/34/AY/2/1/1, UARM.

51. Committee on Edible and Oil-Producing Nuts and Seeds, *Minutes of Evidence*, 2493.

52. Davies, *Trade Makers*, 140–43.

53. Lever Bros. to CO, 3 October 1911, in *West Africa: Correspondence Regarding the Grant of Exclusive Rights*.

54. Sec. State to Gov. Gold Coast, 6 January 1912, *West Africa: Correspondence Respecting the Grant of Exclusive Rights*.

55. "Elmina," *Gold Coast Leader*, 28 July 1911.

56. "Frank Friend," *Sierra Leone Guardian*, 14 April 1911.

57. See Trevor's testimony cited above, Leighton correspondence in CO 879/115/8, TNA and POEM records in BT 31/41073-41074, TNA.

58. Correspondence in CO 879/115/8, TNA and Boss, "Scramble for Palms," 45; Musson, *Enterprise in Soap and Chemicals*, 256.

59. TH Maskey to Gov. Sierra Leone, 15 May 1914 and Merewether to Sec State, 30 March 1914, CO 879/115/8, TNA.

60. Slater to Amery, 28 September 1925, CO 267/631/13, TNA.

61. Anonymous, "Oil Palm," 76; but see rebuttal in Addison, "Palm Nut Tree," 13.

62. Report, enclosed in Slater to Amery, 28 September 1925, CO 267/631/13, TNA.

63. Correspondence and enclosures in ADM 11/1/1034 and in ADM 11/1/1751, PRAAD-Accra.

64. Anonymous, "Oil Palm," 76.

65. Addison, "Palm Nut Tree," 13.

66. *West Africa. Palm Oil and Palm Kernels*, 64.

67. Phillips, *Enigma of Colonialism*, 95; Udo, "Plantation Agriculture"; Committee on Improved and Increased Production of Palm Oil and Palm Kernels in West Africa, *Palm Oil*.

68. Hancock, *Commonwealth Affairs*, vol. 2, pt. 1, 192–94; see Carland, *Colonial Office and Nigeria*; Hyam, *Understanding the British Empire*.

69. Harding, memorandum, 20 September 1924, in West Africa Confidential Print no. 1113, CO 879/122/4, TNA.

70. Quoted in Phillips, *Enigma of Colonialism*, 97.

71. Clifford suggested a "palm oil association," modeled on the British Cotton Growing Association. Meredith, "Government and Decline," 315. The Association was a "semi-philanthropic" body that built and operated cotton gins in the British Empire. See Robins, *Cotton and Race*.

72. G. G. Auchinleck, "*The Plantation Oil-Palm Industry in the East*," in CO 96/670/4, TNA.

73. Ormsby-Gore to Flood, 18 January 1927, minute on Auchileck's report, CO 96/670/4, TNA.

74. "Report on the Eastern Province of the Gold Coast Colony for 1937–38," 264, Provincial Development Book (Eastern Province), ADM 29/5/2, PRAAD-Accra.

75. [illegible] to Snelling, Niger Company, 18 March 1929, CO 96/690/15, TNA.

76. Phillips, *Enigma of Colonialism*, 102.

77. Enclosures and correspondence, 1929, CO 96/690/15, TNA.

78. Hancock, *Commonwealth Affairs*, vol. 2, pt. 1, 192.

79. Agricultural Department report for 1930–31, ADM 5/1/92, and 1937–38, ADM 5/1/101, PRAAD-Accra.

80. Correspondence and memorandums, "Sierra Leone Oil Palm Industry, 1927," CO 267/619/6, TNA.

81. Gov. to Sec. State, 8 October 1940, CO 267/675/4.

82. Aubrey Cooke, minute on "Palm oil: Nigeria," 12 March 1929, CO 583/168/11; Faulkner, confidential report to Chief Secretary, Lagos, 6 September 1929, CO 583/168/11; Correspondence in "Palm Oil Industry: Nigeria, 1930," CO 583/170/5, TNA.

83. Correspondence, 1934–35, in CO 267/623/4, TNA; the poor performance of Sumtran "Deli" seed led to speculation that the variety "probably did not originate in West Africa." OPRS annual report, 13 February 1942, CO 852/437/4, TNA.

84. Correspondence, 1934–35, in CO 267/623/4, TNA; and Hancock, *Commonwealth Affairs*, vol. 2, pt. 1, 191–92.

85. Nicolaï, "Le Congo," 3.

86. Vansina, *Being Colonized*, 8–9.

87. Nicolaï, "Le Congo," 3.

88. See Hochschild, *King Leopold's Ghost*; Pavlakis, *British Humanitarianism and the Congo Reform Movement, 1896–1913*.

89. Fieldhouse, *Unilever Overseas*, 500.

90. Sidney Edkins, "Notes on the History of the H.C.B.," UAC/1/11/14/1/11, UARM. Lever did ask the Royal Botanical Gardens at Kew to recommend a tree surveyor in 1910 to get a better inventory of palms, but he didn't hire the man Kew recommended. Lever Bros. to RBGK, 7 February 1910, Southern Nigeria: Agricultural Department 1899–1917, MR/499, f97, RBGK.

91. Lever to Kew, 7 February 1910, MR/499, RBGK.

92. LePlae, "Oil Palm Groves," 9.

93. Quoted in "La Question Des Palmeraies," 8.

94. Fieldhouse, *Unilever Overseas*, 502.

95. Marchal, *Lord Leverhulme's Ghosts*, 3.

96. Quoted in Knox, *Coming Clean*, 174.

97. Sidney Edkins, "Notes on the History of the H.C.B.," UAC/1/11/14/1/11, UARM.

98. Leverhulme, *Leverhulme*, 173.

99. LePlae, "Oil Palm Groves," 14.

100. Sidney Edkins, "Notes on the History of the H.C.B.," UAC/1/11/14/1/11, UARM.

101. Martin, "Visit to Yangambi," 12 March 1942, UAC/2/36/5/1/2/1, UARM.

102. LePlae, "Oil Palm Groves," 10.

103. Letter from C. Malet, *Planter* 11, no. 2 (September 1930), 37–38.

104. Fieldhouse, *Unilever Overseas*, 523.

105. Ministère des colonies, *Le palmier a huile*, 6.

106. Sidney Edkins, "Notes on the History of the H.C.B.," UAC/1/11/14/1/11, UARM.

107. LePlae, "Oil Palm Groves," 13–14.

108. Clement, "The Land Tenure System in the Congo," 91.

109. Fieldhouse, *Unilever Overseas*, 503–6.

110. Sidney Edkins, "Notes on the History of the H.C.B.," UAC/1/11/14/1/11, UARM.

111. Sidney Edkins, "Notes on the History of the H.C.B.," UAC/1/11/14/1/11, UARM.

112. The historical demography of Congo is highly contested; see Sanderson, "Le Congo belge"; Marchal, *Lord Leverhulme's Ghosts*.

113. Marchal, *Lord Leverhulme's Ghosts*, 10.

114. C. Wilson, *History of Unilever*, vol. 1, 176.

115. Fieldhouse, *Unilever Overseas*, 512.

116. Marchal, *Lord Leverhulme's Ghosts*, 63–64.

117. Marchal, *Lord Leverhulme's Ghosts*, 12.

118. Marchal, *Lord Leverhulme's Ghosts*, 6–7.

119. Nelson, *Congo Basin*, 141; Northrup, *Bend in the River*, 99–100.

120. Quoted in Buell, *Native Problem*, vol. 2, 542.

121. Fieldhouse, *Unilever Overseas*, 514.

122. Marchal, *Lord Leverhulme's Ghosts*, 140.

123. Fieldhouse, *Unilever Overseas*, 512.

124. Marchal, *Lord Leverhulme's Ghosts*, 175.

125. Vanderyst, "La Récolte," 26–28; Harms, *River of Wealth*, 55.

126. The text reads "53 feet," but this almost certainly a typo. Committee on Edible and Oil-Producing Nuts and Seeds, *Minutes of Evidence*, 5560–61.

127. Vanderyst, "La Récolte."

128. Vanderyst, "La Récolte," 30–31.

129. Marchal, *Lord Leverhulme's Ghosts*, 67.

130. Nelson, *Congo Basin*, 141.

131. Jewsiewicki, "African Peasants," 62.

132. One typical quota was thirty-six bunches per week. Nelson, *Congo Basin*, 142.

133. Marchal, *Lord Leverhulme's Ghosts*, 142.

134. Quoted in Jewsiewicki, "African Peasants," 62.

135. Marchal, *Lord Leverhulme's Ghosts*, 29, 31; Houben and Seibert, "(Un)Freedom," 185.

136. Quoted in Jewsiewicki, "African Peasants," 62.

137. LePlae, "Agriculture in the Belgian Congo," 63.

138. Leplae, quoted in Nelson, *Congo Basin*, 125.

139. Nelson, *Congo Basin*, 162.

140. Vansina, *Being Colonized*, 215.

141. Sidney Edkins, "Notes on the History of the H.C.B.," UAC/1/11/14/1/11, UARM.

142. Sidney Edkins, "Notes on the History of the H.C.B.," UAC/1/11/14/1/11, UARM.

143. Marchal, *Lord Leverhulme's Ghosts*, 158–59, 163.

144. Fieldhouse, *Unilever Overseas*, 514.

145. Nelson, *Congo Basin*, 187; Henriet, "'Elusive Natives.'"

146. See Seibert, "More Continuity Than Change"; and Nelson, *Congo Basin*.

147. Leverhulme, *Leverhulme*, 173.

148. Nelson, *Congo Basin*, 20; on anthropogenic groves, see also Munier, *Le palmier a huile*, 17.

149. Vanderyst, "Origine Des Palmeraies"; Vanderyst, "Les Palmeraies"; Zeven, *Semi-Wild Oil Palm*, 30; see also M. Vanderwyen in "Benin 1949 Conference," CO 852/1156/6, TNA.

150. Weeks, *Among the Primitive Bakongo*, 97; Duchesne, "Du Droit Des Indigènes Sur Les Palmeraies Naturelles."

151. Bakulu, "History of Oil Palm Use."

152. Vanderyst, "Origine Des Palmeraies," 75.

153. See Fairhead and Leach, *Reframing Deforestation*, especially the chapters on Liberia and Ghana. On climate and agriculture, see Fleur, *Fusion Foodways*.

154. G. Thompson, *Palm Land*, 219.

155. Juhé-Beaulaton, "La palmeraie du Sud Bénin"; and see Fairhead and Leach, *Reframing Deforestation*.

156. See Dudgeon, *Agricultural and Forest Products*; Farquhar, *Oil Palm*; Unwin, *West African Forests and Forestry*.

157. Quoted in Vanderyst, "Les Palmeraies," 122.

158. Chevalier, *Documents sur le palmier à huile*, 84–85.

159. Chevalier, *Documents sur le palmier à huile*, 91.

160. Zeven, *Semi-Wild Oil Palm*, 21, 49–51.

161. Marvin Miracle collected numerous examples of palm planting from the ethnographic literature, and he noted that patterns of palm cultivation changed in response to markets for palm oil, palm wine, and other products. Miracle, *Agriculture in the Congo Basin*.

162. Hellermann, "Reading Farm and Forest"; H. Thompson, "The Forests of Southern Nigeria," 129; for pre-1900 examples from Igboland, see Isichei, *Igbo Worlds*, 204, 208, 210, 230, 243, 274.

163. McEwan, "Representing West African Forests," 16.

164. Giles-Vernick, *Cutting the Vines of the Past*, chap. 6.

165. Uzozie, "Tradition and Change," 161.

166. Watkins, "Afro-Brazilian Landscape," 12.

167. Vanderwyen, quoted in "Benin 1949 Oil Palm Research Conference," typescript, CO 852/1156/6, TNA.

168. Marchal, *Lord Leverhulme's Ghosts*, 83.

169. Morel, *Black Man's Burden*, 125–26, 186.

Chapter 6

1. W. F. Hutchinson, "Report of the Commission on Economic Agriculture in the Gold Coast," 54–55, 1889, ADM 5/3/7, PRAAD-Accra.

2. Farquhar, *Oil Palm*.

3. Tingey, "Transporting Oil Palm Fruit," 24.

4. W. F. Hutchinson, "Report of the Commission on Economic Agriculture in the Gold Coast," 54–55, 1889, ADM 5/3/7, PRAAD-Accra; Billows and Beckwith, *Consols of the West Coast*, 25; Mba, *Nigerian Women Mobilized*, 74.

5. Quoted in Ukegbu, "Nigerian Oil Palm Industry," 182.

6. Farquhar, *Oil Palm*, 29; Meredith, "Government and Decline," 321–22.

7. Scott, "Production for Trade," 228; Faulker and Mackie credit hard oil with 55–65 percent extraction vs. 55 percent for soft. *West African Agriculture*, 101.

8. R. Coull, report enclosed in Agricultural Department report 1923–1924, ADM 5/1/81, PRAAD-Accra.

9. Esuman-Gwira is incorrectly spelled as "Gwina" or "Gvira" in several sources.

10. Sarbah, "Oil Palm," 243; Committee on Edible and Oil Producing Nuts and Seeds, *Minutes of Evidence*, 192 see also Sarbah to Secretary for Native Affairs, 28 November 1908, ADM 11/1/223, PRAAD-Accra.

11. Robins, "Smallholders and Machines," 81–83; Meredith, "Government and Decline," 322; for a detailed account of the various machines, see Ukegbu, "Nigerian Oil Palm Industry," 210–23.

12. Committee on Edible and Oil-Producing Nuts and Seeds, *Minutes of Evidence*, 192.

13. LePlae, *Palmier à Huile*, 66.

14. Faulkner and Mackie, *West African Agriculture*, 6.

15. Robins, "Smallholders and Machines"; S. Martin, *Palm Oil and Protest*, 64; a skilled cracker could make one bushel (forty-eight to sixty-eight pounds, depending on the measure) in one day. Addison, "Palm Nut Tree (*Elaeis guineensis*)," 18.

16. Agricultural Department Report for 1916, ADM 5/1/73, PRAAD-Accra.

17. O. T. Faulkner to Ormsby-Gore, 8 February 1928, CO 583/155/14, TNA.

18. Zeven, *Semi-Wild Oil Palm*, 93; Forde, "Rural Economies," 51.

19. Leis, "Palm Oil, Illicit Gin, and the Moral Order of the Ijaw," 830; Hofstra, "Social Significance," 108.

20. Uzozie, "Tradition and Change," 425.

21. Sarbah, "Oil Palm."

22. Nwokeji, "The Slave Emancipation Problematic," 330–31; see discussion in Imbua, "Anti-Pawning Campaign." S. Martin notes that although the law was repealed in 1915, slave trading continued into the 1930s in eastern Nigeria. *Palm Oil and Protest*, 43.

23. Manning, "Economic History," 54, 139–40; Wartena, "Making a Living."

24. See oral evidence in Ukegbu, "Nigerian Oil Palm Industry," 90–92; Isichei, *Igbo Worlds*, 83.

25. Uku, *Seeds in the Palm*, 62.

26. Otite, "Rural Migrants as Catalysts in Rural Development," 228; Hubbard, *The Sobo*, 44; Ekeh, *History of the Urhobo People*.

27. "Dika Nuts," 22; Usoro, *Nigerian Oil Palm Industry*, 93.

28. Pratten, *Man-Leopard Murders*, 124.

29. Committee on Edible and Oil-Producing Nuts and Seeds, *Minutes of Evidence*, 2080.

30. Korieh, *The Land Has Changed*, 114.

31. Scott, "Production for Trade," 236; Ukegbu, "Nigerian Oil Palm Industry," 231.

32. A. Martin, *Oil Palm Economy*, 12.

33. Barnes, *Mechanical Processes for the Extraction of Oil*, 6, 19.

34. On cookware, see Osborn, "Bauxite to Cooking Pots."

35. Chevalier, "Le Palmier à huile à la Côte d'Ivoire," 221.

36. Chuku, *Igbo Women*, 93; Chima Korieh recalled cracking kernels by hand as a boy, calling it "one of the tasks I most hated as a child." *The Land Has Changed*, 83.

37. Agricultural Department Report for 1937–1938, 8, ADM 5/1/101, PRAAD-Accra. The Duchscher press turned a handle down a threaded shaft, while other designs turned a shaft to press on a loose plate.

38. Report, Agricultural Investigation Station, Cape Coast, 21 September 1936, ADM 31/1/11, PRAAD-Accra; W. Miller, "Economic Analysis," 47; Meredith, "Government and Decline," 323.

39. Leis, "Palm Oil, Illicit Gin, and the Moral Order of the Ijaw," 831–32; see also Ukegbu, "Nigerian Oil Palm Industry," 235.

40. Suret-Canale, *French Colonialism*, 226; Mondon, "Côte d'Ivoire," 84–85.

41. Strickland, *Report on the Introduction of Co-Operative Societies into Nigeria*, 233; Pratten, *Man-Leopard Murders*, 185n68.

42. Strickland, *Report on the Introduction of Co-Operative Societies into Nigeria*, 234.

43. Chuku, *Igbo Women*, 91.

44. 7d. per tin was more than 10 percent the export price of palm oil in 1937 (about 4s. 4d. for a tin). S. Martin, *Palm Oil and Protest*, 65, 146; and see A. F. B. Bridges, "Report of Oil Palm Survey, Ibo, Ibibio and Cross River Areas," June 1937, Mss. Afr. S 697 (1), Rhodes House, Oxford. Enrique Martino kindly provided a digital copy of this document.

45. Ukegbu, "Nigerian Oil Palm Industry," 69–70; but see W. Miller, "Economic Analysis," chap. 3, which argues for the efficiency of head-portage.

46. Chuku, *Igbo Women*, 153–56.

47. S. Martin, *Palm Oil and Protest*, 85; Chuku, *Igbo Women*; Korieh, *The Land Has Changed*, 116–17.

48. Bauer, *West African Trade*, 24–25; Mba, *Nigerian Women Mobilized*, 47.

49. Usoro, *Nigerian Oil Palm Industry*, 26; Dyke, *Report*, 4.

50. Company of African Merchants, Liverpool, to RBGK, 25 February 1910, ECB/8/1, RBGK.

51. Chevalier, *Documents sur le palmier à huile*, 44–45.

52. "Investigations in Connection with the African Palm Oil Industry," 365.

53. Sarbah, "Oil Palm," 236.

54. "The Varieties of the Oil Palm in West Africa. (, Jacq.)"; see updates in the 1914 edition, "The Varieties of Oil-Palm in West Africa. (*Elaeis guineensis*, Jacq.)"; and "The West African Oil Palm. (*Elaeis guineensis*, Jacq.)."

55. Tudhope to Imperial Institute, 7 May 1909, ADM 11/1/107, PRAAD-Accra; see also Janssens, "Le Palmier a Huile Au District Du Kasai," 32.

56. Hofstra, "Social Significance," 116; Houard, Castelli, and Lavergne, *Contribution à l'étude Du Palmier à Huile*, 5.

57. Rudin, *Germans in the Cameroons*, 260.

58. Gascon, Noiret, and Meunier, "Oil Palm," 481.

59. Chevalier, *Documents sur le palmier à huile*, 45.

60. See comparison of sugarbeets, cotton, and oil palm in Zimmerman, *Alabama in Africa*.

61. Praid to Sec. State for Colonies, 18 February 1908, ADM 11/1/1144, PRAAD-Accra.

62. See reports from Gold Coast, Dahomey, and Cameroon in "The Varieties of the Oil Palm in West Africa. (*Elaeis guineensis*, Jacq.)."

63. Unwin, *West African Forests and Forestry*, 466–67.

64. Vaderwyen in "Benin 1949 Oil Palm Conference," CO 852/1156/6, TNA.

65. Zeven, *Semi-Wild Oil Palm*, 20.

66. Unwin, *West African Forests and Forestry*, 466–67; Zeven, *Semi-Wild Oil Palm*, 31.

67. "The Varieties of the Oil Palm in West Africa. (*Elaeis guineensis*, Jacq.)," 45.

68. Farquhar, *Oil Palm*, 29.

69. Chevalier, *Documents sur le palmier à huile*, 73.

70. Beccari, *Contributo Alla Conoscenza Della Palma a Olio*; Burkill and Birstwistle, *Dictionary*, 911; Schultes, "Taxonomic, Nomenclatural and Ethnobotanic Notes

on Elaeis," 174. Europeans were especially interested in "mantled" palm fruit, once called *E. poissonii* but now recognized as a mutation of *E. guineensis*. Africans didn't consider the mantled palm very useful.

71. Unwin, *West African Forests and Forestry*, 467; for a detailed analysis of rainfall and palm produce exports, see Manning, *Dahomey*, 101–3.

72. D. Prain to Under Sec for Colonies, 24 July 1917, MR/446, RBGK; Dir. of Agriculture to Sec. Native Affairs, 3 March 1910, ADM 11/1/1144, PRAAD-Accra.

73. The French firms protested the use of their money on palm oil, given their reliance on peanut oil. The government wound up providing much of the funding. Bonneuil, "Mise En Valeur," 47.

74. Royal Botanic Gardens, Kew, *Useful Plants of Nigeria*, 736.

75. In "The Emergence of an Export Cluster," Giacomin overstates the "compartmentalization" of science in Africa, compared with Southeast Asia. Researchers in Africa published in journals and organized several trans-colonial conferences.

76. Beirnaert, "Que Pouvons-Nous Attendre Des Palmeraies Ameliorees Du Congo Belge?," 49; Beirnaert and Vanderweyen, *Contribution à l'étude genetique*.

77. Marchal, *Lord Leverhulme's Ghosts*, 199–200.

78. Chevalier, "Le Palmier à huile à la Côte d'Ivoire," 228–29.

79. Suret-Canale, *French Colonialism*, 226; "Dahomey," 94–95.

80. Dept of Agr. Pamphlet No. 7, 30 April 1926; memo, 20 February 1926, ADM 36/1/11, PRAAD-Accra; "Scheme of Plantation Work in Connection with Oil Palms to be Undertaken by the Government to Stimulate Production of Palm Products," August 1923, MR1440, RBGK.

81. *Despatches Relating to the Sierra Leone Oil Palm Industry*, 28, 39.

82. See Richards, *Indigenous Agricultural Revolution*; Mouser et al., "Commodity and Anti-Commodity."

83. West Africa Commission, "Technical Report," 37; A. Martin, *Oil Palm Economy*, 10.

84. Kilby, *Industrialization in an Open Economy*, 142.

85. N. C. Hollins, report on Masanki, 5 September 1935, CO 267/651/11, TNA.

86. Faulkner and Mackie, *West African Agriculture*, 100; Udom, "Nigerian Government Policy."

87. LePlae, *Palmier à Huile*, 8, 61–63.

88. Dyke, *Report*, 4; for earlier examples, see Committee on Edible and Oil-Producing Nuts and Seeds, *Minutes of Evidence*, 764–66; and "Recent Progress in Agriculture" (1914), 130.

89. This section is based on Robins, "Imbibing the Lesson of Defiance."

90. Daniel, *Le palmier à huile au Dahomey*, 3; G. Thompson, *Palm Land*, 219.

91. "Guinée Française," 70; Brouwers, "Rural People's Response," 71.

92. Forde, "Land and Labour in a Cross River Village, Southern Nigeria," 42; Robins, "Imbibing the Lesson of Defiance."

93. Adam, *Le palmier à huile*, 227–29.

94. Knoll, *Togo under Imperial Germany*, 143; Rudin, *Germans in the Cameroons*, 259.

95. "Recent Progress in Agriculture" (1911), 297.

96. Agriculture Department annual report for 1922–1923, 10, ADM 5/1/79, PRAAD-Accra.

97. Robins, "Imbibing the Lesson of Defiance."

98. Slater, "Gold Coast," 347.

99. Antheaume, "La palmeraie du Mono"; Dissou, *Économie de la culture*, 47; Manning, *Dahomey*, 238.

100. Heap, "Cooking the Gins"; see also Korieh, "Alcohol and Empire."

101. See Akyeampong, *Drink, Power, and Cultural Change*; see also the discussion in Bersselaar, *King of Drinks*.

102. Robins, "Imbibing the Lesson of Defiance."

103. Ukegbu, "Nigerian Oil Palm Industry," 156.

104. Quoted in Usoro, *Nigerian Oil Palm Industry*, 45; see also Oil Palm Research Station, 2nd Annual Report, 13 February 1942, CO 852/437/4, TNA.

105. On adulteration and commodity forms, see Cohen, *Pure Adulteration*.

106. Moloney, Druce, and Shelley, *Sketch*, 72.

107. Stilliard, "Legitimate Trade," 115; Clerk's account in Pedler notes, UAC /1/11/14/1/25, UARM.

108. G. Jones, *Trading States*, 144.

109. See photo Af,A48.15 in British Museum collection https://www.britishmuseum.org/collection/object/EA_Af-A48-15. The photo caption misidentifies the fraudulent puncheon as sculpture.

110. Moloney, Druce, and Shelley, *Sketch*, 73; S. Martin, *Palm Oil and Protest*, 111.

111. Farquhar, *Oil Palm*, 30.

112. J. Smith, *Trade and Travels*, 200–201.

113. Ofonagoro, *Trade and Imperialism*, 115; Hopkins, "Economic Imperialism in West Africa," 594.

114. Olukoju, "Government, the Business Community," 100.

115. Knoll, *Togo under Imperial Germany*, 141.

116. Adam, *Le palmier à huile*, 231–42.

117. Leubuscher, "Marketing Schemes for Native-Grown Produce in African Territories," 167.

118. "Report by Mr. F. A. Stockdale on his Visit to Nigeria, Gold Coast and Sierra Leone, October 1935–February 1936." Colonial advisory council of agriculture and animal health, C.A.C. 270, ADM 5/3/30, PRAAD-Accra.

119. Mba, *Nigerian Women Mobilized*, 47; Chuku, *Igbo Women*, 229.

120. S. Martin, *Palm Oil and Protest*, 58, 112; Olukoju, *"Liverpool" of West Africa*, 89–91, 183; also see Gardner, *Taxing Colonial Africa*.

121. Quoted in Korieh, *The Land Has Changed*, 136.

122. See, for example, Lawrance, "En Proie à La Fièvre Du Cacao," 152.

123. Coquery-Vidrovitch, "French Colonization," 171.

124. Frankema, "Raising Revenue in the British Empire, 1870–1940," 463; see also Frankema and Waijenburg, "Metropolitan Blueprints of Colonial Taxation?"; Frankema, "Colonial Taxation and Government Spending in British Africa, 1880–1940"; Gardner, *Taxing Colonial Africa*; Afolabi, "Colonial Taxation Policy."

125. Mellor notes, UAC/1/11/14/1/22, UARM.

126. Hailey, *An African Survey*, 590.

127. S. Martin, *Palm Oil and Protest*, 113–17; Korieh, *The Land Has Changed*, chap. 4.

128. Uku, *Seeds in the Palm*, 94–95.

129. Ukegbu, "Nigerian Oil Palm Industry," 168.

130. Falola, *Colonialism and Violence*, 112.

131. Mba, *Nigerian Women Mobilized*, 75.

132. For a concise narrative of the Women's War, see Falola, *Colonialism and Violence*, 117–23; for more detailed analysis, see Korieh, *The Land Has Changed*, 126; Chuku, *Igbo Women*; S. Martin, *Palm Oil and Protest*; Mba, *Nigerian Women Mobilized*; Van Allen, "Sitting on a Man."

133. Uku, *Seeds in the Palm*, 95.

134. S. Martin, *Palm Oil and Protest*, appendix 14; Korieh, *The Land Has Changed*, 134; Pratten, *Man-Leopard Murders*, 114–21. The continuing circulation of older, slave-trade era currency was also a problem, as sellers could not easily convert this money to British currency. Naanen, "Tax from the Dead."

135. Leith-Ross, *African Women*, 167.

136. The average Urhobo family consisted of a husband, two wives, and three working-age children. Olukoju, "Confronting the Combines," 54.

137. See examples in Pratten, *Man-Leopard Murders*, 124; Korieh, *The Land Has Changed*.

138. Olukoju, "Confronting the Combines," 55.

139. Olukoju, "Nigeria or Lever-Ia?"

140. Leith-Ross, *African Women*, 62.

141. Clauson, minute, 1935, CO 852/17/9, TNA; see also CO 554/95/16.

142. On the nature of colonial archives, see Stoler, *Along the Archival Grain*.

143. The oil-to-soap yield varies greatly, depending on the type of soap. This estimate is based on 50 percent water content, which was typical for soft soaps. Hard soaps contained less water.

144. Fieldhouse, *Unilever Overseas*, 340–57.

145. Manning, *Dahomey*, 102.

146. Ekundare, *An Economic History of Nigeria, 1860–1960*, 166, 213.

147. Maier, "Precolonial Palm Oil"; Adam, *Le palmier à huile*, 261; Gold Coast bluebook for 1900, ADM 7/1/34; F. M. Purcell manuscript, ADM 11/1/1294, PRAAD-Accra.

148. Unwin, *West African Forests and Forestry*, 476; *"Recent Progress in Agriculture"* (1914), 130; Daniel, *Le palmier à huile au Dahomey*.

149. Hupfeld, "Palm Oil," 762; Daniel, *Le palmier à huile au Dahomey*.

150. W. G. A. Ormsby-Gore, "Report by The Hon. W. G. A. Ormsby-Gore, MP, on his visit to West Africa during the year 1926," ADM 5/3/24, PRAAD-Accra.

151. Agricultural Report for 1922–1923, ADM 5/1/79, PRAAD-Accra. Assuming a 500 milliliter bottle, this works out to about sixty-five grams per family per day.

152. Phillips, *Enigma of Colonialism*, 103.

153. Memo, 13 February 1934, CSO 8/3/3, PRAAD-Accra; Slater to Lord Passfield, 6 September 1929, CO 96/690/15, TNA; and Gold Coast Agriculture Department annual report for1936–37, ADM 5/1/94, PRAAD-Accra.

154. Correspondence in POEM: Krobo mill file, UAC/2/4/1/2/4, UARM.

155. Suret-Canale, "L'industrie Des Oléagineux En A.O.F.," 286.

156. Hinds, "Government Policy and the Nigerian Palm Oil Export Industry, 1939–49," 460; *Report of the Mission to Enquire into the Production and Transport of Vegetable Oils*, 35.

157. Faulkner and Mackie, *West African Agriculture*, 44, 95; West Africa Commission, "Technical Report," 24–25; Korieh, *The Land Has Changed*, 85–90.

158. Korieh, *The Land Has Changed*, 90.

159. Compare with S. Martin, *Palm Oil and Protest*, 53.

160. Lynn, *Commerce and Economic Change*, 34–35; S. Martin, *Palm Oil and Protest*, 46; Fenske, "Imachi Nkwu," 64.

161. Korieh, *The Land Has Changed*, 81.

162. A. F. B. Bridges, "Report of Oil Palm Survey, Ibo, Ibibio and Cross River Areas," June 1937, Mss. Afr. S 697 (1), Rhodes House, Oxford; A. Martin, *Oil Palm Economy*, 7; Wickizer, "The Smallholder in Tropical Export Crop Production," 81–82.

163. Toovey in "Benin 1949 Oil Palm Research Conference," CO 852/1156/6, TNA.

164. Chubb, *Ibo Land Tenure*, 50.

165. Faulkner and Mackie, *West African Agriculture*, 97; Uzozie, "Tradition and Change."

166. Ukegbu, "Nigerian Oil Palm Industry," 167.

167. Chubb, *Ibo Land Tenure*, 48.

168. Fenske, "Imachi Nkwu," 53.

169. Green, *Ibo Village Affairs*, 133.

170. A. Martin, *Oil Palm Economy*, 5; Chubb, *Ibo Land Tenure*, 49–50.

171. Zeven, *Semi-Wild Oil Palm*, 90.

172. League of Nations, *International Statistical Yearbook*, 1926 and 1939.

173. Dyke, *Report*, 15.

174. Phillips, *Enigma of Colonialism*, 106.

175. Hailey, *An African Survey*, 1404.

176. Chevalier, "Le Palmier à huile à la Côte d'Ivoire," 228–29.

177. J. E. W. Flood, minute, 9 May 1927, CO 583/146/9, TNA.

178. G. Jones, *Earth Goddess*, 9; see Corey Ross's critical appraisal of Jones and other writers in *Ecology and Power*.

179. Faulkner and Mackie, *West African Agriculture*, 5–6.

180. Olukoju, "United Kingdom," 125; Buchanan-Smith to Sec. State for Colonies, 23 November 1931, CO 583/199/8, TNA.

181. Chevalier, "Le Palmier à huile à la Côte d'Ivoire," 228–29.

Chapter 7

1. Colombijn, "Ecological Sustainability"; Kathirithamby-Wells, *Nature and Nation*, chap. 1; Peluso and Vandergeest, "Genealogies," 766–67; Dove, *Banana Tree at the Gate*.

2. See Barlow, *Natural Rubber Industry*, 20; Tate, *RGA History*, 451.

3. See Michael Dove's analysis in *Banana Tree at the Gate*, 24; see also Haraway, "Anthropocene, Capitalocene, Plantationocene, Chthulucene"; Ross, "Plantation Paradigm."

4. See Jorgensen, Jorgensen, and Pritchard, *New Natures*; Uekötter, *Comparing Apples, Oranges, and Cotton.*

5. Aso, *Rubber and the Making of Vietnam.*

6. Brass and Bernstein, "Introduction," 9.

7. Curtin, *Rise and Fall.*

8. Drabble, *Rubber in Malaya*; Barlow, *Natural Rubber Industry*; Dean, *Brazil and the Struggle for Rubber*; Tully, *The Devil's Milk.*

9. For doctrinaire accounts of rubber in Malaya, see Tate, *RGA History*; Barlow, *Natural Rubber Industry*; for more critical accounts of Malaya and beyond, see Drabble, *Rubber in Malaya*; Tully, *The Devil's Milk*; and Aso, *Rubber and the Making of Vietnam*, chap. 1.

10. Ross, *Ecology and Power*, 104; Grandin, *Fordlandia.*

11. Clarence-Smith, "Rubber Cultivation."

12. Lees, *Planting Empire*, 43; Breman, *Taming the Coolie Beast*, 16.

13. Barral, "Paternalistic Supervision," 242.

14. Nonini, *British Colonial Rule*, 53; White, *Post-Colonial Malaysia*, 85.

15. See Peluso and Vandergeest, "Genealogies," 787–89; Stoler, *Capitalism and Confrontation*, 22–23; Breman, *Taming the Coolie Beast*, 20–28.

16. Stoler, *Capitalism and Confrontation*, 16; the 999-year lease was offered by Perak in 1879; the lease in perpetuity became typical of FMS districts in the late 1890s. Raja, *Economy of Colonial Malaya*, 170, 189; G. Allen and Donnithorne, *Western Enterprise*, 115.

17. Jomo Kwame Sundaram, *A Question of Class*, 85.

18. Harper, "The Politics of the Forest in Colonial Malaya," 11; Aiken, "Losing Ground," 171.

19. Boomgaard, "Land Rights," 484–85.

20. Pelzer, *Planter and Peasant*, 50–51.

21. See Hall, Hirsch, and Li, *Powers of Exclusion.*

22. Cleary, "Plantation Agriculture"; Aiken, "Losing Ground," 161, 170.

23. See, for example, Burkill's Malayan diary, 1914, BUR/1/5, RBGK.

24. Jackson, *Planters and Speculators*, 224.

25. Jackson, *Planters and Speculators*, 9–10, 224.

26. Pelzer, *Planter and Peasant*, 42–43.

27. Pelzer, *Planter and Peasant*, 52–53.

28. Barnard, *Nature's Colony*, 142.

29. Agriculture report for Lower Perak, 1906, quoted in Jackson, *Planters and Speculators*, 234.

30. Suppiah and Raja, *Chettiar Role*, 35; Ross, *Ecology and Power*, 107–8.

31. Nonini, *British Colonial Rule*, 69; Jomo Kwame Sundaram, Chang, and Khoo, *Deforesting Malaysia*, 66.

32. Graham, Floering, and Fieldhouse, *Modern Plantation*, 41.

33. Examples of remorse include Fauconnier, *Malaisie*; Székely, *Tropic Fever*; see discussion in Ross, *Ecology and Power*, 111–14.

34. Harper, "The Politics of the Forest in Colonial Malaya," 7.

35. Baxendale, "Plantation Rubber," 186.

36. "Notes on Mr. Muir's and Mr. Mellor's conversation with Mr. Lunn. Discussion of a Government policy with regard to Palm Plantations and Land tenure." UAC /1/11/14/1/38/113, UARM.

37. Gordon, "Towards a Model of Asian Plantation Systems," 314; Engerman, "Servants to Slaves," 266.

38. Li, "The Price of Un/Freedom"; Stoler, *Capitalism and Confrontation*, 8–9; see also discussion of "coolie" discourse in Lees, *Planting Empire*, 55–59; Bosma, *Making of a Periphery*.

39. In particular, see Sandhu, *Indians in Malaya*; Ramasamy, "Labour Control and Labour Resistance"; Breman, *Taming the Coolie Beast*; Kaur, "Plantation Systems"; Engerman, "Servants to Slaves."

40. Examples are from "Notes on Sapong Enquiry," CO 874/697, TNA.

41. Breman, *Taming the Coolie Beast*, 32.

42. Shakespeare Junior, "The Labour Question Solved," *The Planter* 1, vol. 1 (1920): 9. 9.

43. Reprint from *Financial Times*, 26 September 1912, in Straits Plantation papers, MS37737, LMA.

44. Chanderbali, "Indian Indenture," 308.

45. Sandhu, *Indians in Malaya*, 171.

46. Breman, *Taming the Coolie Beast*, 59. Klaveren argues that figures were lower, but his own data show alarmingly high mortality before 1900, for example, 60.2 per 1,000 on average and 86.2 among Chinese men in 1897. Klaveren, "Death among Coolies."

47. Thee, "Colonial Extraction"; Thee, *Plantation Agriculture*, 40.

48. Kaur, "Tappers and Weeders"; Ramasamy, "Labour Control and Labour Resistance"; Jain, "Tamilian Labour and Malayan Plantations, 1840–1938."

49. Sandhu, *Indians in Malaya*, 96.

50. Barlow, *Natural Rubber Industry*, 46; Gordon, "Contract Labour in Rubber Plantations."

51. Stoler, *Capitalism and Confrontation*, 25–30; Breman, *Taming the Coolie Beast*, chap. 4.

52. Editorial, *Singapore Free Press and Mercantile Advertiser*, 3 January 1913, 5.

53. Lees, *Planting Empire*, 47, 189–204; Sandhu, *Indians in Malaya*, 89–93.

54. See Lim Teck Ghee, *Peasants*; Lees, *Planting Empire*, 47, 204–14; Nonini, *British Colonial Rule*, 52; Suppiah and Raja, *Chettiar Role*.

55. Quoted in Parmer, *Colonial Labor Policy*, 244.

56. Golden Hope Minute book, annual general meeting, 27 September 1949, MS37703, LMA.

57. R. Smith, "Oil Palms," 221.

58. S. Martin, *UP Saga*, 82; Lamb, "A Time of Normalcy."

59. Kuala Selangor Rubber Co. Annual General Meeting, 23 June 1925, MS37828-001, LMA.

60. A. A. Cowan to J. H. Better, 19 April 1926, PRO 5/0/1, RBGK.

61. Hunger, *Oliepalm*, 266–67; Pelzer, *Planter and Peasant*, 55.

62. Rutgers, "Cultivation of the Oil Palm," 2–3.

63. Rutgers, "African Oil Palm," 595.

64. For a history of the garden at Buitenzorg, see Maat, *Science Cultivating Practice*; Goss, *The Floracrats*.

65. Hunger, *Oliepalm*, 8.

66. The Bourbon pair were "male," while the Amsterdam pair were "female," based on early flowering. Teysmann later observed blooms of both sexes on all the palms. Hunger, *Oliepalm*, 11; and see Cramer, "Oliepalm." Corley and Tinker suggest that all four palms sprouted in Amsterdam from African seed in the 2008 edition of *Oil Palm*, 6; but they walk back from this claim in the 2015 edition (p. 6). See another version in Hartley, *Oil Palm*, 3rd ed., 20.

67. Rutgers et al., *Investigations on Oilpalms*, 1; see also Jagoe, "Deli Oil Palms"; Hardon and Thomas, "Breeding and Selection of Oil Palm in Malaya." Cramer was sure that Teysmann's "Bourbon" was Reunion, and that the addition of "or Mauritius" was erroneous. PJS Cramer to Evans, 25 September 1940, PRO 5/0/1, RBGK.

68. Hunger erroneously calls him "D. F. Pryce" in the 1917 edition, but he corrected the mistake in his expanded 1924 edition and includes a photograph. Hunger, *Oliepalm*; for biographical information on Pryce, see *Bye-Gones*, 289.

69. De Bruijn Kops, "Notulen van de een en twintigste vergadering der Nederlandsch-Indische Maatschappij van Nijverheid," 389; Vriese, *Tuinbouw-flora van Nederland en zijne overzeesche bezittingen*, 125–28; International Exhibition of Australia, *Official Record*, 59.

70. Cramer, "Oliepalm," 448, ruled out Mauritius, but he was unaware of an 1837 catalog of the gardens (Bojer, *Hortus Mauritianus*, 304). No oil palms are listed in 1801 and 1822 accounts of Mauritius: Vaux, *History of Mauritius*; *A Catalogue of the Exotic Plants Cultivated in the Mauritius*. Hunger notes an unpublished 1854 catalog from Bogor describing the palms as "ex hort. bot. Amst. et Pryce hort. Bourb.," apparently confirming separate origins. Hunger, *Oliepalm*, 10, 34.

71. For example, Voigt, *Hortus Suburaanus Calcuttensis*, 643.

72. NEI officials wrote to Elmina for palm oil advice in the 1850s. Hunger, *Oliepalm*, 21–27.

73. Corley and Tinker, *Oil Palm*, 135; Hayati et al., "Genetic Diversity of Oil Palm (*Elaeis guineensis* Jacq.) Germplasm Collections from Africa"; Cochard et al., "Geographic and Genetic Structure of African Oil Palm Diversity Suggests New Approaches to Breeding."

74. Hunger, *Oliepalm*, 37, 73. Teysmann's pollinating may well have favored one or two parent palms or even a single self-pollinated palm, thinning the gene pool further; see also Hardon and Thomas, "Breeding and Selection of Oil Palm in Malaya."

75. "Java," *Straits Times*, 21 March 1863, 1.

76. Quoted in Hunger, *Oliepalm*, vii.

77. "Uses of Palm Oil in Cooking," *Straits Times*, 21 January 1933, 6.

78. Hunger, *Oliepalm*, 102, 124–25.

79. Burkill and Birstwistle, *Dictionary*, 912.

80. Arthur Hill to Under Sec for Colonies, 27 August 1908, ECB/8/1, RBGK; Bunting, Eaton, and Georgi, "The Oil Palm in Malaya."

81. "Palm Cultivation in Sarawak" in ECB/8/1, RBGK, and Ooi Keat Gin, "Economic History of Sarawak," 194, 330n22. The project was patronage for her friend James Brooke, the first "white raja" of Sarwak. When he died in 1868, she lost interest in the country.

82. Corely and Tinker state that the palms came from Ghana, but the correspondence at Kew states that a collector in Monrovia was the supplier. Corley and Tinker, *Oil Palm*, 6; see correspondence in ECB/8/1, RBGK.

83. Jagoe, "Deli Oil Palms," 5; see correspondence printed in "Oil Palm in Labuan."

84. "Palm Cultivation in Sarawak" August 1891, ECB/8/1, RBGK.

85. "The Oil Palm in Sarawak," *Singapore Free Press and Mercantile Advertiser*, 9 June 1891, 10; *Straits Times*, 16 November 1892, 6; "Sarawak in 1893," *Singapore Free Press*, 23 January 1894, 11.

86. Barnard credits the Singapore garden with great work acclimating oil palm, but the tree grew perfectly well in the region, and the Singapore garden's staff added nothing useful to working knowledge of the tree. Barnard, *Nature's Colony*, 87. See also "Notes on Products and Soils," *Straits Times*, 13 March 1895, 3.

87. "Mr. Von Donop in North Borneo," *Straits Times*, 11 June 1884, 10.

88. G. Allen and Donnithorne, *Western Enterprise*, 140.

89. Egerton to Thiselton-Dyer, 7 June 1904, Director's Correspondence, RBGK. http://plants.jstor.org/stable/10.5555/al.ap.visual.kadc1446.

90. Ridley, "Oil Palm," 37.

91. Egerton to Lord Crewe, 22 June 1908, 1957/0591061, ANM; "Oil Palms in British Guiana," *Straits Times*, 10 December 1912, 10.

92. "Oil Palm seeds," ca. 1909, 1957/0615552, ANM; J. B. Carruthers to Kew, 9 January 1909, ECB/8/1, RBGK.

93. Arthur W. Hill, to Undersec for Colonies, 27 August 1908, 1957/0591061, ANM. See also "African Oil Palm in Malaya," *Malaya Tribune*, 10 February 1914, 10.

94. *FMS Annual Report for 1909*, 11, CO1071/236, TNA; *Federated Malay States Agricultural Bulletin*, 1, no. 1 (August 1912), 31.

95. "Palm Oil from Malaya: Ten Years' Progress," *Straits Times*, 27 June 1929, 9.

96. Cunyngham-Brown, *The Traders*, 252.

97. Hunger, *Oliepalm*, 266–67.

98. T. Fleming to Nickalls, 13 July 1987, MS37394-005, LMA.

99. *FMS Annual Report for 1911*, 9, CO1071/37, TNA; Pakiam, "Smallholder Involvement," chap. 3; Boss, "Scramble for Palms."

100. Hallet, "Note Sur Le Palmier a Huile," 279; Fauconnier, "Essais de Culture," 21.

101. Clarence-Smith, "Rivaud Hallet Plantation Group."

102. Hallet said he had every intention of following up rubber planting with oil palms when he started working in Southeast Asia; Hallet, "Note Sur Le Palmier a

Huile," 279. The exact date of Hallet's car breakdown is the subject of some controversey: Rival and Levang say 1905 (*Palms of Controversies*); they are likely misreading LePlae, *Palmier à Huile*. Hallet's own account suggets the breakdown occurred in 1911 (reproduced in Hunger, *Oliepalm*, 268). Fauconnier's account does not mention the breakdown, but he insists that he was with Hallet in June 1911, when they tested the oil palm fruit.

103. Fauconnier, "Essais de Culture," 21; compare with Tate's account in *RGA History*, 452–53.

104. Van Pelt puts the date of Hallet's first oil palm plantings as 1910, not 1911, but most sources point to 1911. See Van Pelt, "La Culture et l'Exploitation moderne."

105. Hunger, *Oliepalm*, 268–69.

106. Hallet, "Les Plantations d'Elaeis En Malaisie," 47.

107. For example, see "The Palm Nut Tree," *Straits Times*, 13 December 1913, 7.

108. See Boss, "Scramble for Palms," 46–50, 54.

109. Van Pelt, "La Culture et l'Exploitation moderne," 162.

110. Rutgers, "Crop Records of the Oil Palm"; see bibliography in Hunger, *Oliepalm*.

111. Lewis Smart in Rutgers, LePlae, and Tingey, *Oil Palms and Their Fruit*, 37.

112. Milligan, *Oil Palm*.

113. L. Henderson, "African Oil Palm in Malaya," 53.

114. "Oil Palms," *The Planter* 1, no. 2 (1920), 43.

115. Guthrie to London, 19 November 1920, G/COR/5, SOAS.

116. W. Wilson, "A Slump Soliloquy," 21.

117. Minute, 9 October 1920, 1957/0609646, ANM; one official with "some considerable knowledge of West Africa" said he was "not at all inclined to believe that the plantation of the African Palm Oil tree will ever be able to compete" with Africa and its "teeming population." "Mr. James, re: residents conference," 30 September 1920, 1957/0213249, ANM.

118. "Mr. James's report on Resident's Conference, 30.9.1920," 1957/0213249, ANM.

119. In the 1910s, estates interplanted rubber and coconuts: "it was usually the coconuts that were condemned." Baxendale, "Plantation Rubber," 185; White, *Post-Colonial Malaysia*, 85; for interplanting cases, see Brooklands Estate to DO Talok Datok, 30 June 1925, 1957/0237140; Shand, Halldane and Co correspondence 1926–27, 1957/0242280, ANM.

120. Fauconnier to Selangor Resident, 21 May 1917, 1957/0195159, ANM.

121. LePlae, *Palmier à Huile*, 29.

122. E. L. B. to Resident, Selangor, 28 August 1917, 1957/0604273; African Oil Palms Syndicate to Resident, Kuala Lumpur, 6 July 1917, 1957/0195775, ANM.

123. Guthrie to Sec. Res. Selangor, 9 June 1920, 1957/0210077, ANM.

124. Duff Development Co to British Adviser, Kota Bharu, 30 May 1921, 1957/0501458, ANM. The district official agreed to pay the bill, though he noted "that this payment must not be considered as creating a precedent."

125. See, for example, *Federated Malay States Agricultural Bulletin* 4, no. 8 (May 1916).

126. C. W. Harrison to Sec. Res. Selangor, 9 October 1923, 1957/0227176, ANM.

127. Tate, *RGA History*, 451; S. Martin, *UP Saga*, 6.

128. Grist, "Agricultural Education," 127; Lim Teck Ghee, *Peasants*; Harper, *The End of Empire and the Making of Malaya*, 26; Pelzer, *Planter and Peasant*, 50.

129. White, *Post-Colonial Malaysia*, 172; A. Booth, *Colonial Legacies*, 58.

130. Malet, "Palm Oil Planting," 324.

131. D. O. to Sec. Res Selangor, 20 July 1923, 1957/0227176, ANM.

132. Auchinleck, *The Plantation Oil-Palm Industry in the East*, 12. The extent to which this claim was true was hotly debated in the 1920s–30s. Manufacturers preferred low-acid oil, because they could extract more glycerine from it and because they appreciated the flexibility it provided, but soaps and other nonfood uses did not need such a high level of quality. In the end, soap firms wanted what was cheapest. See Olukoju, "United Kingdom."

133. Quote from DO Kuala Langat, 15 January 1920, 1957/0206789, ANM; see also E. Bateson, "Report on Oil Palm Cultivation, British North Borneo," June 1925, CO 874/158, TNA; Rutgers, "Crop Records of the Oil Palm," 248.

134. DO Kuala Langat to Sec Res Sel, 7 March 1924, 1957/0230161, ANM.

135. Sec Res Sel to Undersecretary FMS, 22 June 1927, 1957/0242280, ANM.

136. Pugh, *Great Enterprise*, 156.

137. Mr. Laurent to D. C. Selangor, 2 August 1918, 1957/0200946; and application file, 1957/0186683W, ANM.

138. The workers hadn't been paid in six months and complained of harsh treatment leading up to the strike. Ramasamy, "Labour Control and Labour Resistance," 101.

139. *Straits Times*, 9 May 1913, 8.

140. Van Pelt, "La Culture et l'Exploitation moderne," 163–64.

141. Van Pelt, "La Culture et l'Exploitation moderne," 163.

142. Ooi Jin-Bee, *Peninsular Malaysia*, 104.

143. Ministère des colonies, *Le Palmier à Huile*, 6.

144. Chevalier, "Le Palmier à huile à la Côte d'Ivoire," 225.

145. DO Kuala Langat to Secretary to Resident Selangor, 15 January 1920, 1957/0206789, ANM.

146. Van Pelt, "La Culture et l'Exploitation moderne," 163–64.

147. Draft letter, A. H. Lemon to Chief Secretary, 7 April 1920, 1957/0206789, ANM. See the case of Gladys Frugtniet, whose estate was deemed too swampy even for oil palms. DO Ulu Langat to Sec Res Sel, 16 June 1930, 1957/0267968, ANM.

148. The major drive into peat soils did not come until the late 1970s, however. Ooi Jin-Bee, *Peninsular Malaysia*, 81–82.

149. H. J. Simpson, State Agricultural Officer, Pahang, "Report by S.A.O. on Dura Oil Palm Estate," 1939/32, 1957/0534106, ANM. Swidden cultivation was often used as justification for displacing communities in the colonial era. Ross, *Ecology and Power*, 291–92.

150. DO to Sec for Res Selangor, 21 August 1924, United Sua Betong Rubber—Application for Land, 1957/0232106, ANM.

151. The applications I examined mostly concerned Selangor and Pahang. On Semai land use and the strategy of retreat, see Dentan, *Semai*, 43; and more broadly, Aiken, "Losing Ground," 172; Gomes, *Looking for Money*, 179.

152. Hartley, *Oil Palm*, 2nd ed., 13–14; see also S. Martin, *UP Saga*, 49–50.

153. Rutgers et al., *Investigations on Oilpalms*, 15, 107. Rutgers noted that on the first estates, some bunches were 70 percent fruit by weight; on others, they were only 32 percent.

154. Rutgers et al., *Investigations on Oilpalms*, 103; Rutgers claimed that Deli palms gave 50 percent more oil than typical African palms, but Fauconnier downgraded that estimate to 30 percent. He was more impressed with how early young palms bore fruit in Southeast Asia. LePlae, *Palmier à Huile*, 26.

155. See presentation in "Cultivation of the African Oil Palm."

156. Henry, "Documents Sur Le Palmier à Huile à Sumatra," 201–4, 213.

157. L. Henderson, "African Oil Palm in Malaya," 23.

158. Rutgers admitted that data on Sumatra palms highlighted the best performers, and that undeveloped fruit were discarded from bunches. Many reports on African palms showed yields equivalent to the best Deli palms. One report from Ivory Coast found "wild" palms that gave sixty-five kilograms of fruit at eight years old, versus a maximum of sixty-eight kilograms for twenty-year-old Deli palms. LePlae, *Palmier à Huile*, 33.

159. Hunger stated that the Singapore theory was "certainly based on the wrong information." Hunger, *Oliepalm*, 262–63; Rutgers agreed; see Rutgers et al., *Investigations on Oilpalms*, 5; Corley and Tinker, *Oil Palm*, 139.

160. Maas, "Het Planten van Oliepalmen," 499–500.

161. Ferrand, "L'Avenir Du Palmier à Huile," 224.

162. DO for Ulu Selangor to Sec res Sel, 21 June 1920, 1957/0211073, ANM. The commercial lifespan of the tree was, by the 1950s, redefined at thirty years, the point at which a plantation palm grows too tall to harvest without a ladder.

163. Minute, 29 June 1920, 1957/0211073, ANM.

164. Fauconnier, "Essais de Culture," 21.

165. Application for land at Mukim Sengei Tiggi, Selangor, 1915–1918, 1957/0186683W, ANM.

166. Letter from L. P. Jorgenson, 8 September 1920, 1957/0211073, ANM.

167. R. Garnier to DO, Kuala Selangor, 20 September 1920, 1957/0211073, ANM. Jagoe, citing unpublished correspondence, differentiates between the "bungalow garden" at Tennamaram, which he says was planted with "the best jungle species" from Africa, and the plantation proper, which Fauconnier said was sourced from his Rantau Panjang palms. Jagoe, "Deli Oil Palms," 7–8.

168. Oliver Marks to A. B. Voules, 12 November 1920, 1957/0209921, ANM.

169. Ferrand, "L'Avenir Du Palmier à Huile," 219; Van Pelt complained in 1920 that there was no useful "ground-work" on oil palm breeding, and he warned about the limits rubber planters faced when they tried to develop a better tree from a small genetic pool. Van Pelt, "Oil Palm," 102.

170. Hartley, *Oil Palm*, 2nd ed., 196–97.

171. Henry, "Documents Sur Le Palmier à Huile à Sumatra," 214.

172. Rutgers to RBG Kew, 5 February 1921, MR/446, RBGK.

173. Van Pelt, *La Culture Du Palmier à Huile*, 57.

174. L. Henderson, "African Oil Palm in Malaya," 21.

175. Sparnaaij, "Bunch Production," 12. A. A. Cowan complained in 1926 that while he had been given a good tour of AVROS, the plantations often refused to let him see their palm oil factories. Cowan to J. H. Better, 10 April 1926, PRO 5/0/1, RBGK.

176. L. Henderson, "African Oil Palm in Malaya," 53.

177. Cramer, "Comparison between Oilpalms and Coconuts," 342.

178. Eaton, "Annual Report of the Chemical Division for 1924," 183; Heurn, "De Bereiding van Palmvet," 523.

179. Milsum, "African Oil Palm," 103.

180. Maas, "La Culture et la Sélection," 189; and Alex Cowan to Better, 10 April 1926, PRO 5/0/1, RBGK.

181. L. Henderson, "African Oil Palm in Malaya," 9.

182. Georgi, "Malayan Palm Kernels," 60; Leubuscher, *Processing*, 37. In the 1930s, Asian plantations averaged 4.85 tons oil for each ton of kernels, slipping to 5.3:1 in 1939 and 6:1 in 1940 (League of Nations, *Statistical Yearbook*, 1944).

183. L. Henderson, "African Oil Palm in Malaya," 23; see also letter in *Straits Times*, 12 June 1928, 8.

184. P. J. Bliek to Lewis Smart, 3 November 1924, MR446 RBGK.

185. Auchinleck, *The Plantation Oil-Palm Industry in the East.*

186. Marsh, "Oil Palm," 315.

187. One planter fretted about a new plant using carbon disulfide solvent ("beastly stuff"): (illegible) To J. Crowe, 17 July 1908, MS 37742, LMA.

188. Milsum, "African Oil Palm," 99.

189. Blommendaal, "Manufacture of Palm Oil"; "The Palm Oil Factory at Elaeis Estate, Johore"; Georgi, "Comparison," 117; S. Martin, *UP Saga*, 53–54.

190. Clarence-Smith, "Rivaud-Hallet Plantation Group," 127–32.

191. L. Henderson, "African Oil Palm in Malaya," 21.

192. L. Henderson, "African Oil Palm in Malaya," 23.

193. "The Palm Oil Tree in Indo-China," *Straits Times*, 9 June 1914, 12.

194. For example, Malet, letter to editor, *The Planter* 11, no. 2 (September 1930): 37, and Milligan, *Oil Palm*, 45.

195. Rutgers et al., *Investigations on Oilpalms*, 41; LePlae, *Palmier à Huile*, 78.

196. See Ruf, "Agroforests."

197. Luytjes, "La Situation de La Culture Du Palmier à Huile Sur La Côte Orientale de Sumatra et Dans La Province d'Atjeh," 241.

198. *FMS Annual Report for 1920*, 7, CO 1071/236, TNA.

199. Georgi, "The Removal of Plant Nutrients in Oil Palm Cultivation," 484; the most critical assessment came from Henry, "Documents Sur Le Palmier à Huile à Sumatra," 205. Rebuttals aimed at Henry's report did not refute his assessment of soil fertility, and they candidly admitted that oil palm was a cash crop like any other and would need fertilizers. Ferrand, "L'Avenir Du Palmier à Huile," 226; Luytjes, "La Situation de La Culture Du Palmier à Huile Sur La Côte Orientale de Sumatra et Dans

La Province d'Atjeh," 241; also see Beirnaert, "Que Pouvons-Nous Attendre Des Palmeraies Ameliorees Du Congo Belge?"

200. West Africa Commission, "Technical Report," 29.

201. Handover, "The Eradication of Lalang"; Ross, *Ecology and Power*, 111–16; Aso, *Rubber and the Making of Vietnam*, 70–71.

202. Spring, "Cover Crops," 169; Rutgers et al., *Investigations on Oilpalms*, 39; and numerous articles and notes in the *Malayan Agricultural Journal* and *The Planter*.

203. See Ross, *Ecology and Power*.

204. Cushman, *Guano*; Teaiwa, *Consuming Ocean Island*.

205. Rutgers, "African Oil Palm," 595; Rutgers et al., *Investigations on Oilpalms*, 65; also see L. Henderson, "African Oil Palm in Malaya." Donna Haraway calls plantations an "epidemic-friendly" way of arranging nature in Mitman, "Reflections on the Plantationocene."

206. Milsum, "African Oil Palm," 103; Rutgers et al., *Investigations on Oilpalms*, 70; S. Martin, *UP Saga*, 60–61.

207. Abdullah, *Planter's Tales*, 24.

208. Abdullah, *Planter's Tales*, 27.

209. Cumberbatch to Sec Res Pahang, 28 July 1932, 1957/0455331, ANM; *Johore Annual Report for 1931*, 11–12, CO1071/219, TNA.

210. "Report of visit to Mentara Oil Palm Estate, and two rubber estates in Kelantan, 24.7.39," 1957/0526051, ANM.

211. Kathirithamby-Wells, *Nature and Nation*, 190–92.

212. Gater, "Les Insectes," 195.

213. Gater, "Insects on African Oil-Palms," 250.

214. Hunger, *Oliepalm*, 37; Rutgers et al., *Investigations on Oilpalms*, 3.

215. Milsum, "African Oil Palm," 94; Van Pelt, "Oil Palm," 103.

216. Gater, "Insects on African Oil-Palms," 254–55.

217. Rutgers, "Crop Records of the Oil Palm," 251.

218. Rutgers et al., *Investigations on Oilpalms*; see description in Abdullah, *Planter's Tales*, 46.

219. Jagoe, "Observations."

220. S. Martin, *UP Saga*, 103.

221. Malet, letter to editor, *The Planter* 10, no. 12 (July 1930): 343; Alston, "Fruit-Rot," 360.

222. Common Spear Rot can kill young palms, but most recover after a year or two. Rutgers et al., *Investigations on Oilpalms*, 67–68; Hartley, *Oil Palm*, 3rd ed., 593–94.

223. Lanagan, *The Palm Oil Industry in West Malaysia*, 4.

224. Lever Bros to Curator, Kew, 10 May 1915, ECB/8/1, RBGK; "Diseases of the Oil Palm in West Africa."

225. Dijk, Onguene, and Kuyper, "Knowledge and Utilization of Edible Mushrooms by Local Populations of the Rain Forest of South Cameroon."

226. Caption on oil palm photo, Wellcome collection no. 568298i, https://wellcomecollection.org/works/tqah29vc; Kwa, "Environmental Change."

227. Byerlee, "Fall and Rise."

228. Cowan to Better, 10 April 1926, PRO 5/0/1, RBGK.

229. Olukoju, "United Kingdom"; Bacon et al., *World Trade in Agricultural Products*, 300, 304–5.

230. *Johore Annual Report for 1935*, 20, CO 1071/219, TNA.

231. "Notre Carnet Financier," *L'Indochine: revue économique d'Extrême-Orient*, no. 140 (July 1932): 144.

232. S. Martin, *UP Saga*, 63.

233. Malet, letter to editor, *The Planter* 11, no. 2 (September 1930): 38.

234. Tate, *RGA History*, 465n23.

235. Reprint from *Investor's Chronicle*, 14 July 1933, MS37737, LMA.

236. Allied Sumatran Plantations Ltd., 13th annual report, 21 June 1939, MS 37946/1, LMA.

237. South, "Position of Coconut," 423; and "Sorry Predicament of Oil Industries," *Straits Times*, 17 September 1934, 6.

238. vE. "Expanding N. I. Palm Oil Industry Depends on U.S.," 154; Freas et al., *Demand and Price Structure*, 12; United States Tariff Commission, *Summaries of Tariff Information*, 103; S. Martin, *UP Saga*, 209.

239. Buelens and Frankema, "Colonial Adventures in Tropical Agriculture," table 3.

240. Stoler, *Capitalism and Confrontation*, 88; Huff, "Entitlements, Destitution, and Emigration in the 1930s Singapore Great Depression," 310; Jomo Kwame Sundaram, *A Question of Class*, 192.

241. Correspondence, July–September 1934, in 1957/0283442, ANM.

242. Dyke, *Report*, 6.

243. Stockdale, "Visit to Malaya," 373.

244. Milsum and Georgi, "Small Scale Extraction of Palm Oil," 53.

245. Pakiam, "Smallholder Involvement."

Chapter 8

1. Staples, *Birth of Development*, 71–81; for U.S. wartime fat policies, see Prodöhl, "Dinner to Dynamite."

2. Minute, ca. 1946, CO 852/603/8, TNA.

3. Jasper Knight to Eric Roll, 27 January 1947, MAF 83/2195, TNA.

4. Chuku, "'Crack Kernels, Crack Hitler.'"

5. T. M. Knox, "The Oil Palm Industry," paper to be read at the British Association, Dundee, 4 September 1939, UAC/1/11/14/1/17, UARM.

6. Creech Jones, "Agricultural Productivity in Africa," 22 February 1947, CO 852/1003/3, TNA.

7. Pim, *Colonial Agricultural Production*, 128.

8. Pim, *Colonial Agricultural Production*, 139.

9. Richards to Trenchard, 28 September 1944, CO 852/604/1, TNA; see also G. H. C. Amos minute, 21 June 1946, MAF 83/1295, TNA.

10. *Report of the Mission to Enquire into the Production and Transport of Vegetable Oils*; Hodge, *Triumph of the Expert*, chap. 7.

11. G. H. C. Amos, minute, 21 June 1946, MAF 83/2195, TNA.

12. Kisch, minute, 18 April 1946, CO 852/604/3, TNA; draft telegram CO 852/604/3; see also CO 852/314/4.

13. See Azu Mate Kole, Ten-Year Plan for Manya Krobo state 1946–1947, CSO 21/22/181, PRAAD-Accra; see similar warnings in CO 852/604/3, TNA.

14. Cowen, "Early Years of the Colonial Development Corporation"; Baring, "Aspects"; V. Thompson and Adloff, "French Economic Policy," 131–32.

15. Rist, *History of Development*; Leys, *Rise & Fall*.

16. Kratoska, *The Japanese Occupation of Malaya*, 235; Pakiam, "Why Don't Some Cuisines Travel?"

17. United Plantations was a notable exception, immediately replacing its machinery at the company's own expense. S. Martin, *UP Saga*, 100–101.

18. White, *End of Empire*, 72–73; see correspondence about Japanese Surrendered Personnel (JSP) in MAF 83/2178, TNA.

19. White, *End of Empire*, 275.

20. Draft memo in Guthries to Unilever, 23 December 1947, UAC/2/10/M6/1/2/1, UARM.

21. Elaeis Plantations journal 1949–1957, G/EP/2, SOAS.

22. Memo from Guthrie & Co., 18 November 1953; Draft memo on "decontrol," 3 December 1953, MAF 83/2178, TNA; White, *Post-Colonial Malaysia*, 169.

23. Pelzer, *Planter and Peasant*, 123–25.

24. Pelzer, *Planter and Peasant*, 135; see also Stoler, *Capitalism and Confrontation*, chap. 4.

25. See Pelzer, *Planters against Peasants*, chap. 2.

26. Stoler, *Capitalism and Confrontation*, 129.

27. Pelzer, *Planters against Peasants*, 70–73; Stoler, *Capitalism and Confrontation*, 153–57.

28. Stoler, *Capitalism and Confrontation*, 140.

29. Stoler, *Capitalism and Confrontation*, 140; for an account of the path to nationalization, see Lindblad, *Bridges*.

30. Mr. Hopegood to Nickalls, nd. MS37394/004, LMA.

31. W. A. Faure to Sir Geoffrey Evans, 10 July 1946; Evans to Faure, 9 July 1946, PRO 5/0/1, RBGK; see also Imperial Mycological Institute to Mr. Cotton, 21 January 1941, File 4/L/12A Lever/Unilever, RBGK.

32. Wardlaw, "Notes on a Visit"; on fusarium and bananas, see Soluri, *Banana Cultures*; for oil palms, see Hartley, *Oil Palm*, 3rd ed., 602–7.

33. White, *End of Empire*, 200–201.

34. White, *End of Empire*, chap. 5.

35. S. Martin, *UP Saga*, 295–304; Mohd Tayeb Dolmat, *Technologies for Planting Oil Palm on Peat*.

36. Cumberbatch & Co. (agents for Elmina Estate) to Resident Commissioner for Selangor, 14 September 1946, 1957/0303479, ANM.

37. Giacomin, "The Transformation of the Global Palm Oil Cluster."

38. S. Martin, *UP Saga*, 103–4; Ferwerda, "Questions Relevant to Replanting in Oil Palm Cultivation"; and "Report on Oil Palm Estates in Malaya," 1955, 1957/0534468, ANM.

39. Shell advertisements in *Malayan Agricultural Journal*, 1969. The ad was for "lalang oil" mixed with Gramevin, a powerful herbicide.

40. Stoler, *Capitalism and Confrontation*, 164.

41. Tinker, *Life's Adventures*, 49.

42. Toovey in Benin 1949 Oil Palm Conference, typescript, CO 852/1156/6, TNA.

43. Tinker, *Life's Adventures*, 49.

44. M. Cammaerts to Faure, 10 September 1942; De Blanck to J. M. Lenahan, 8 April 1942, UAC/2/36/5/1/2/1, UARM.

45. S. Martin, *UP Saga*, 149–51 and letters to Far Eastern Economic Review, 4 February and 17 March 1988, in MS37394/004, LMA.

46. Compare with McCook's "creole science," *States of Nature*, 5.

47. Gascon, Noiret, and Meunier, "Oil Palm," 489.

48. S. Martin, *UP Saga*, 142.

49. Gray, in Benin 1949 Oil Palm Conference, typescript, CO 852/1156/6, TNA.

50. G. F. Clay, "Report on a Visit to the Belgian Congo and Sierra Leone," 1948, CO 852/1377, TNA.

51. See "Oil Palm" notes in *Malayan Agricultural Journal* 1951–1955, and Hartley, *Oil Palm*, 3rd ed., 199–201; Tate, *RGA History*, 461.

52. T. Parker, report on centrifuges, 8 May 1941, 1957/0679898, ANM; see also files in 1957/0681146.

53. S. Martin, *UP Saga*, 185–87. I use "expeller" here instead of "screw" to avoid confusion with hand-operated screw presses like the Duchscher press.

54. Committee minutes, 27 June 1956, and "The Colour and Bleaching of Palm Oil," Oil Palm paper no. 8, AY 3/31, TNA.

55. G. F. Clay, "Report on a Visit to the Belgian Congo and Sierra Leone," 1948, CO 852/1377, TNA.

56. G. F. Clay, "Report on a Visit to the Belgian Congo and Sierra Leone," 1948, CO 852/1377, TNA.

57. Benin 1949 Oil Palm Conference, typescript, CO 852/1156/6; quote from minutes, 22 September 1953, Oil palm subcommittee, AY 3/29, TNA.

58. D. L. Martin in "Consultative Committee on Oils and Oilseeds: Oil Palm subcommittee," 25 June 1952, AY 3/28, TNA.

59. Benin 1949 Oil Palm Conference, typescript, CO 852/1156/6, TNA.

60. *FAO Yearbook of Food and Agricultural Statistics*, Rome: FAO, 1951, 1961. After 1960, significantly more kernel crushing took place in producing countries, making kernel exports less useful as a figure.

61. Minutes, 22 September 1953, Oil palm subcommittee, AY 3/29, TNA.

62. *Report of the Mission to Enquire into the Production and Transport of Vegetable Oils*, 10.

63. Hanney to Imperial Institute, 14 October 1952, AY 3/29, TNA; and correspondence in AY 3/28.

64. "Note on the visit of Dr. Azikiwe," Colonial Products Laboratory, 20 Sept 1955, CO 554/1204, TNA.

65. Graham, Floering, and Fieldhouse, *Modern Plantation*, 99–100.

66. Dove, *Banana Tree at the Gate*, 28.

67. "The Future of the Nigerian Oil Industry," UAC, Sep. 1944, CO 852/604/1, TNA; Fieldhouse, *Merchant Capital and Economic Decolonization*, 212–14.

68. A. Cowan, extract from *West African Review*, December 1933, and untitled paper, 7 February 1938, UAC/1/11/14/1/78, UARM.

69. United Africa Company, "The Future of the Nigerian Oil Industry," 50.

70. Quoted in Ukegbu, "Nigerian Palm Oil Industry," 381–82; see discussion in Hinds, "Government Policy and the Nigerian Palm Oil Export Industry, 1939–49."

71. Compare to Fieldhouse, *Merchant Capital and Economic Decolonization*, 214–15.

72. E. Hallett to Faure, 7 April 1948, UAC/2/19/1/5/2/1, UARM.

73. Hinds, "Chiefs and the Making of Industrial Policy," 475.

74. Morgan, "Farming Practice," 332–33.

75. Floyd, *Eastern Nigeria*, 191.

76. Hartley, "Improvement," 67.

77. Udo, *Geographical Regions of Nigeria*, 94–96; Udo, "Plantation Agriculture."

78. Ukegbu, "Nigerian Palm Oil Industry," 404.

79. Gov. Nigeria to Sec. State for Colonies, 22 August 1951; Lyttelton to Johnson, 29 September 1953, CO 554/369, TNA.

80. Ukegbu, "Nigerian Palm Oil Industry," 405.

81. Kisch memo, 27 November 1959, CO 852/1614, TNA; see Udo, "Plantation Agriculture"; Ukegbu, "Nigerian Palm Oil Industry," 257–58; Tignor, *Capitalism and Nationalism*, 249; Falola, *Economic Reforms*, 122.

82. John Barlow to Oliver Lyttleton, 2 May 1952, CO 554/657, TNA.

83. Ukegbu, "Nigerian Palm Oil Industry," 371.

84. Meredith, "State Controlled Marketing and Economic 'Development,'" 89; Hinds, "Chiefs and the Making of Industrial Policy"; on the Eastern Region boards, see Falola, *Economic Reforms*, 122–40.

85. The Marketing Boards bought British securities with the savings, effectively using Nigerian farmers' money to prop up the UK financial system. Helleiner, "The Fiscal Role of Marketing Boards," 128, 132; Tignor, *Capitalism and Nationalism*, 219.

86. Tignor, *Capitalism and Nationalism*, 251; IBRD, *Economic Development of Nigeria*; the most prominent critique of the boards is Bauer, *West African Trade*.

87. Nwanze, "Economics"; Kaniki, "Economical Technology"; S. Martin, *Palm Oil and Protest*, 128–29.

88. "Proposals for the Expansion of the Department of Commerce and Industries, Nigeria," 1949, CO 852/1156/7, TNA; Falola, *Economic Reforms*, 137.

89. E. V. Rochfort Rae, 14 April 1945, CO 852/604/2; Report from Nigerian Governor to Stanley, Sec State for Colonies, 3 April 1945, CO 852/604/2, TNA.

90. "Mr. Wilkes' notes," CO 852/514/7, TNA.

91. Hinds, "Government Policy and the Nigerian Palm Oil Export Industry, 1939–49," 473–74; Ukegbu, "Nigerian Palm Oil Industry," 393–95.

92. The "unrefined oil" refers to liquid and fibers left over from the first oil-skimming operation. In this community, men owned the first collection of oil, but women owned whatever they could extract from the leftover material. Fibers from

the pulp and kernel shells were also valued as fuel and were considered part of a wife's share. Nwabughuogu, "Oil Mill Riots," 75; for Annang/Ibibio examples, see Pratten, *Man-Leopard Murders*, 329.

93. Nwabughuogu, "Oil Mill Riots," 72; Chuku, *Igbo Women*, 102, 233; Mba, *Nigerian Women Mobilized*, 109–10; see also Oriji, "Igbo Women"; Aghalino, "British Colonial Policies."

94. Oriji, "Igbo Women," 16.

95. Mba, *Nigerian Women Mobilized*, 110.

96. Pratten, *Man-Leopard Murders*, 329.

97. Usoro, *Nigerian Oil Palm Industry*, 100; Nwabughuogu, "Oil Mill Riots," 77.

98. J. Macpherson to Sec. State for Colonies, 26 March 1953, CO 554/658, TNA; Oriji, "Igbo Women," 20; Falola, *Economic Reforms*, 127.

99. Chuku, *Igbo Women*, 102, 233; Mba, *Nigerian Women Mobilized*, 109–10; S. Martin, *Palm Oil and Protest*, 132.

100. Floyd, *Eastern Nigeria*, 191; Kilby, *Industrialization in an Open Economy*, 150.

101. Kilby, *Industrialization in an Open Economy*, 153–55; Helleiner, *Peasant Agriculture*, Table IV-A-8; Onyioha, *ENDC Oils.*

102. Kilby, "Reply," 200–201; see data in W. Miller, "Economic Analysis."

103. Tichit, "L'amélioration," 81.

104. Ukegbu, "Nigerian Palm Oil Industry," 411–12; Kilby, *Industrialization in an Open Economy*, 155.

105. Ukegbu, "Nigerian Palm Oil Industry," 414–21.

106. Munier, *Le palmier a huile*, 18, 34, 40.

107. Leroy, "Rapport sommaire sur une mission en A.O.F. (10 janvier–10 mars 1957)," 173.

108. Carrière De Belgarric, "Les huileries de palme du plan," 381; Suret-Canale, "L'industrie Des Oléagineux En A.O.F."

109. Carrière De Belgarric, "Souhaits de Bienvenu," 10.

110. V. Thompson and Adloff, *French West Africa*, 387.

111. Carrière De Belgarric, "Les huileries de palme du plan," 381.

112. Grivot, *Réactions Dahoméennes*, 47.

113. Carrière De Belgarric, "Les huileries de palme du plan," 381.

114. Cognard, "La Palmeraie Du Dahomey," 93, 96.

115. Dissou, "Développement et mise en valeur," 489; Service agriculture du Côte-d'Ivoire, "Les Travaux d'aménagement," 118; and see Wartena, "Making a Living."

116. Munier, *Le palmier a huile*, 36, 40–41; for Nigerian examples, see Ukegbu, "Nigerian Palm Oil Industry," 429.

117. *Conférence Franco-Britannique*, 27; Dumont, *Afrique Noire*, 105.

118. Rapley, *Ivoirien Capitalism*, 31, 44.

119. Hartley, *Oil Palm*, 3rd ed., 25; Surre, *L'Institut de Recherches pour les Huiles et Oléagineux*, 41.

120. Skinner, *Agricultural Economy*, 19–20; Due, "Agricultural Development in the Ivory Coast and Ghana"; Daddieh, "Contract Farming."

121. For the critique, see Amin, *Neo-Colonialism*; for a rebuttal, see Settié, *L'ère de l'économie Des Plantations En Côte d'Ivoire.*

122. Minute by Green, 26 September 1950; minute by Emmanuel, 7 October 1950; Willis to Winterbottom, 21 November 1950, CO 717/208/5, TNA.

123. Minute by Green, 26 September 1950; minute by Emmanuel, 7 and 11 October 1950; Willis to Winterbottom, 21 November 1950, CO 717/208/5, TNA.

124. Rendell, *History of the CDC*, 7.

125. Rendell, *History of the CDC*, 26, 37.

126. Minutes and correspondence, March–June 1952, CO 1022/436, TNA.

127. Willis to R. L. Sharp, draft letter, 1952, CO 1022/436, TNA.

128. "Two women stole fruit," *Straits Times*, 6 May 1953, 7.

129. CDC Application for Capital Sanction in Winterbottom to Willis, 14 March 1952; CDC "Schemes in Operation," 24 April 1954, CO 1022/436, TNA.

130. For a brief account focused on the Emergency's impact on plantations, see White, *End of Empire*, 97–124.

131. "Notes of a talk delivered by Mr. T. M. Walker, Chairman of the board of directors, Guthrie Agency . . . 11 Oct 1962 to Persatuan Ekonomi Malaya," 1957/0694520, ANM.

132. Winterbottom to Willis, 15 June 1951; "Labour" memo (n.d.); Willis to Winterbottom, 3 January 1952, CO 717/208/6, TNA.

133. Winterbottom to Willis, 15 June 1951; minute by Anthony Gann, 5 November 1951; "Labour" memo (n.d.); quote from Willis to Winterbottom, 3 January 1952, CO 717/208/6, TNA.

134. "Troops hunt killers," *Singapore Free Press*, 26 July 1954, 1. Rendell, *History of the CDC*, 37.

135. "Dead men hit in first burst," *Straits Times*, 26 July 1954, 1.

136. "Reds raid estate, kill clerk," *Singapore Free Press*, 5 October 1954, 1; "Planter is shot in estate battle," *Straits Times*, 13 May 1957, 5.

137. "600 workers back after 2½ months," *Straits Times*, 15 July 1956, 15; "Oil palm labourers end strike," *Straits Times*, 23 September 1959, 9.

138. Kaur, "Plantation Systems"; Menon and Leggett, "The NUPW in the Nineties."

139. E. J. H. Berwick, "Preliminary Suggestions for a RIDA Project in Selangor. Oil Palm Smallholdings Run on a Plantation Basis," 4 April 1952, 1985/0021421, ANM.

140. Minutes by Othman [Mohamad?], 27 April 1952; Raja Uda, 2 May 1952; and following minutes from colonial officials through 1954, 1985/0021421, ANM.

141. White, "'Ungentlemanly Capitalism,'" 114–15; White, *Post-Colonial Malaysia*, 173; but see Pakiam, "Smallholder Involvement."

142. White, *Post-Colonial Malaysia*, 98.

143. "Notes of a talk delivered by Mr. T. M. Walker, Chairman of the board of directors, Guthrie Agency . . . 11 Oct 1962 to Persatuan Ekonomi Malaya," 1957/0694520, ANM.

144. Minutes by Othman [Mohamad?], 27 April 1952; Raja Uda, 2 May 1952; and following minutes from colonial officials through 1954, 1985/0021421, ANM.

145. IBRD, *Economic Development*, 260.

146. Lim and Dorall, "Contract Farming," 72–74; Shamsul Bahrin and Lee, *FELDA*.

147. Chadwick to R. C. C. Hunt (draft, n.d.), in DO 35/9993, TNA.

148. Minute by E. N. Larmour, 9 June 1958, DO 35/9993, TNA; Little, *Social Cost*, 57.

149. "Pioneers start clearing jungle," *Straits Times*, 4 August 1959, 16.

150. "The Oil Palm Smallholders' Scheme," ca. 1959, in DO 35/9995, TNA.

151. "Factory shares offer for land scheme settlers," *Straits Times*, 7 November 1962, 6.

152. J. A. E. M. to MacNaghten, 6 September 1961, CLC/B/207/MS40718/001, LMA.

153. Little, *Social Cost*, 57.

154. Rendell, *History of the CDC*, 67.

155. White, "Surviving Sukarno," 1292.

156. Allied Sumatran Plantations Ltd., Report of the 33rd Annual General Meeting, 24 September 1959, MS37946/1, LMA.

157. Roadnight, *United States Policy.*

158. Mackie, *Konfrontasi*, 202.

159. Simpson, *Economists with Guns*, 207–48.

160. White, "Surviving Sukarno," 1312.

161. Yacob and Khalid, "Adapt or Divest," 462.

162. "Company Profile: Harrisons & Crosfield Ltd," June 1977, MS 37397, LMA.

163. Memo H&C Berhad, 1 April 1967, MS 37736, LMA.

164. H&C and H&C ANZ correspondence and memos, 1963–1964, MS 37855, LMA.

165. Longayroux, "Hoskins Oil Palm Project: An Introduction," 1; see the brief historical summary in Koczberski, Gibson, and Curry, *Improving Productivity*; for political context, see Downs, *Australian Trusteeship*, 331–35, 512–16.

166. Minutes, 23 May 1950 and following, CO 852/1160/3; and Clay to Hansard, 22 June 1951, CO 852/1160/4, TNA.

167. Beckman, "Ghana, 1951–78."

168. POEM property correspondence, UAC/2/34/AY/2/1/1, UARM.

169. Korieh, *The Land Has Changed*, 200.

170. See British policy discussions in C. J. Tredwell to Acting High Commissioner, Lagos, 8 September 1965 and Lagos telegram (unsigned), 31 December 1965; Farm Settlements Scheme, Eastern Nigeria, 29 April 1966, OD 30/56, TNA.

171. Purvis, "Sources of Growth," 275–76.

172. McWilliam, *The Development Business*, 91.

173. Korieh, *The Land Has Changed*, 203, 205.

174. IBRD, "Economic Growth of Nigeria, Vol. IV," 14, 41.

175. Purvis, "Sources of Growth," 272.

176. Purvis, "Sources of Growth," 271.

177. IBRD, "Economic Growth of Nigeria, Vol. IV," 14.

178. Laurent, *Tree Crops*, 12, 18.

179. Stopler, *Inside Independent Nigeria*, 111–12.

180. IBRD, "Current Economic Position," 21–22.

181. Onwueme, *Like a Lily among Thorns*, 182.

182. Korieh, *The Land Has Changed*, 219–24; Western Africa Regional Office, "Rivers State," 2.

183. De Blank to F. Ferguson, PAMOL, Nigeria, 22 April 1948, UAC/2/36/5/1/2/7; see also correspondence in boxes 4–6, UARM.

184. Clement, "The Land Tenure System"; Draschouossoff, "Agricultural Change."

185. Fieldhouse, *Unilever Overseas*, 521–24.

186. HCB correspondence and memos, UAC 2/36/5/1/2/9, UARM.

187. Fieldhouse, *Unilever Overseas*, 538–39; Fieldhouse, *Merchant Capital and Economic Decolonization*, 476–77.

188. C. Wilson, *Unilever 1945–1965*, 215.

189. According to the manager, Mulele "went to Moscow and Peking to be instructed in the terrorist activities in Malaya with the idea of creating trouble in the Belgian Congo. He started a reign of terror in the south-westerly part of the Congo, beating up his compatriots, destroying our cattle ranches and creating Communist conditions." L. C. Beaumont, notes in "Africanisation—Belgian Congo," 1972, UAC/1/11/14/1/35, UARM.

190. Fieldhouse, *Unilever Overseas*, 543.

191. See Orstom, *Unilever*; Konings, *Unilever Estates*.

Chapter 9

1. Byerlee, Falcon, and Naylor, *Tropical Oil Crop Revolution*, 6.

2. *Diet, Nutrition and the Prevention of Chronic Disease*, 17–20; and Food and Agriculture Organization (2020), FAOSTAT database, http://www.fao/org/faostat.

3. Barani, *Palm Oil*; Moore, *Capitalism*; but see the clearer explication in Patel and Moore, *History of the World in Seven Cheap Things*, 140.

4. Winson, *Industrial Diet*, 27–28.

5. Calliauw, "Dry Fractionation"; see also Hayes and Pronczuk, "Replacing Trans Fat."

6. Tempel, *Raw Materials and Pricing*, 14–15.

7. The cost was about £5 per ton by 1920, and was down to 5–8 percent of the oil's finished price in the 1940s. Corlett, *Economic Development*, 67; Waterman, *Hydrogenation*, 235–36; for palm oil as an alternative, see *Production, Properties and Uses*, 15; S. Martin, *UP Saga*.

8. S. Martin places the shift toward invisible, malleable fats in the 1950s (*UP Saga*, 199), but as I argue here and in Robins, "Oil Boom," this shift was underway by the 1920s.

9. Schleifer, "The Perfect Solution"; R. Allen, "Hydrogenation"; on convenience food, see Warner, *Pandora's Lunchbox*.

10. Abbott, "Accomplishments of the Margarine Industry"; see Petrick, "Larding the Larder"; Robins, "Oil Boom."

11. For Unilever examples, see G. Jones, *Renewing Unilever*, 121–51; see the discussion of consumer choice and commodification in Soluri, *Banana Cultures*, 220–23.

12. Cohen, "Analysis as Border Patrol."

13. Nicholls, "Some Economic Aspects of the Margarine Industry"; Hand, "Marketing Health Education"; Hunt, "Raw Materials"; Riepma, "Margarine in Western Europe."

14. Eckey, *Vegetable Fats and Oils*, 335.

15. "A Handbook on the Jewish Dietary Laws," Union of Orthodox Jewish Congregations of America, 1954, in box 5, Home Economics pamphlet collection (8006), Cornell University Archives; Horowitz, *Kosher USA*.

16. Gaskell, "Palm Oil Revolution," 62.

17. Pantzaris, *Pocketbook*, 88–89, 97; Gaskell, "Palm Oil Revolution."

18. "Proposals for the Organisation of Palm Fruit Processing by FLDA Smallholders in Malaya," Tropical Products Institute, AY 4/2928, TNA.

19. Pakiam, "Why Don't Some Cuisines Travel?," 58.

20. Gaskell, "Palm Oil Revolution," 32.

21. Gaskell, "Palm Oil Revolution," 39–40.

22. Faure to Raymond, 18 April 1963, AY 4/2928, TNA; Blume, *Organisational Aspects*, 161.

23. USDA Foreign Agricultural Service (2020), Production, Supply, and Distribution Database, https://apps.fas.usda.gov/psdonline.

24. Gaskell, "Palm Oil Revolution," 73–75.

25. Ray, "The Body and Its Purity"; see also Hardgrove, "Politics of Ghee"; Knox, *Coming Clean*, 108.

26. Choudhury, "A Palatable Journey through the Pages"; Pinto, "40 Years Ago . . . And Now"; Tandon, *Punjabi Saga*, 342–45.

27. Gandhi, *Diet and Diet Reform*; see also Tandon, *Punjabi Saga*, 391; Mathur, *Vanaspati Industry in India*, 92.

28. Claiborne, "Dispute over Animal Fat in Cooking Oil Heats Up Indian Politics," *Washington Post*, 18 November, 1983; see also G. Jones, *Renewing Unilever*, 170–71.

29. Editorial, *Times of India*, 29 July 1949, 6.

30. In 1980, consumption was as low as 10 grams of fat per capita per day. National Nutrition Monitoring Bureau, *Report*; Meenakshi, "Coconuts and Oil Palm in the Indian Oilseeds Economy."

31. Byerlee, Falcon, and Naylor, *Tropical Oil Crop Revolution*, 118; S. Martin, *UP Saga*, 246.

32. Sagar et al., "India in the Oil Palm Era"; Schleifer and Sun, "Emerging Markets and Private Governance."

33. T. Burke, *Lifebuoy Men, Lux Women*; Ferme, *The Underneath of Things*, 194.

34. FAO, "African Small-Scale Palm Oil," 3.

35. See discussion in Watkins, "Afro Brazilian Landscape," 433.

36. Cheyns, "La consommation urbaine de l'huile de palme rouge en Côte d'Ivoire."

37. Adjei, "Making of Quality."

38. Gallegos, "Palm Oil Tensions."

39. Boutwell et al., *Analysis*, 17–19.

40. M. Martin, "U.S. Palm Oil Imports."

41. Joint Conference on H.R. 5262, HRG-1977-CFS-0004, 19 July 1977; see the voluminous record of diplomatic correspondence over the proposed ban in the Wikileaks "Kissenger Cables" and "Carter Cables" (http://www.wikileaks.org).

42. H.R. 12952 (94th Congress); H.R. 14921 (94th Congress); see Yacob, "Government, Business and Lobbyists"; S. Martin, "Edible Oil."

43. Kevin Lanagan, "Palm Oil Production Costs," presentation to the Agricultural Technical Advisory Council, January 1976, author copy courtesy Kevin Lanagan.

44. S.Res. 444 (94th Congress).

45. David Broder, "Palm oil dilutes the issue," *Boston Globe*, 24 July 1977, A7; H.R. 5262 (95th Congress).

46. Deare to FCO 19 October 1976, FCO 24/2296, TNA; Kevin Lanagan, "Palm Oil," Contribution to *International Economic Report of the President*, 1976, author copy.

47. "Palm Oil Import Review," Hearings before the subcommittee on oilseeds and rice and subcommittee on cotton of the committee of agriculture, House of Representatives, 94th Congress, 18 March and 15 May 1976.

48. See Nestle, *Food Politics*; for early works, see Van Itallie, "Nutritional Research in Atherosclerosis; a Progress Report"; Stamler, "Diet and Atherosclerotic Disease"; for critical reevaluations, see Werkö, "End of the Road for the Diet-Heart Theory?"; Ramsden et al., "Re-Evaluation of the Traditional Diet-Heart Hypothesis"; for a contrarian view, see Steinberg, *The Cholesterol Wars*.

49. Dr. Barnes quoted in clipping, 11 April 1958, "Cardiac Problems told to 100 Nurses here," box 3, Tompkins County Public Health records (2921), Cornell University Archives.

50. See advertisements in the *Journal of the American Dietetic Association* 1950s–1970s, and Davids, "Technology as the New Frontier."

51. K. L. Milstead, "Food Fats and Nutritional Quackery," 30 October 1961, in box 190, American Home Economic Association records, Cornell University Archives.

52. Wartella, Lichtenstein, and Boon, *Front-of-Package Nutrition*; McNamara, "Palm Oil and Health." One 1978 headline read, "Palm Oil: Unfriendly Fat No. 1." "Few of us realize," the article warned, "when ordering French fries that they were cooking in palm oil, probably." Lawrence Power, "Palm Oil: Unfriendly Fat No. 1," *Los Angles Times*, 1 October 1978, H8.

53. George V. Mann, "The Saturated vs. Unsaturated Fats Controversy," 1972, box 2 folder 16, 29-4-2733 Nutritional Sciences Cooperative Extension, Cornell University Archives.

54. M. Martin, "U.S. Palm Oil Imports," 25.

55. Kwang, "Interchangeability," 9.

56. "Hearing on Tropical Fats." The ASA's president said that manufacturers "have promoted vegetable oil as healthful, boasting 'no cholesterol, low in saturated fat, and high in polyunsaturates' . . . They were talking about soybean oil. . . . The terms vegetable oil and soybean oil are one and the same in the mind of most consumers." "Escape From Tropical Fats," *Soybean Digest*, June/July 1987, supplement, 2.

57. "Meet the Man Who's Trying to Put You Out of Business," *Soybean Digest*, March 1986, 49–52.

58. Quoted by Kurt Berger, Joint Hearing, First Session on H.R. 3232, 10 September 1987, 179.

59. Porter and Vogt, "Labeling of Tropical Oils," 28–29; McNamara, "Palm Oil and Health"; Yacob, "Government, Business and Lobbyists."

60. Berger, Joint Hearing, First Session on H.R. 3232, 10 September 1987, 180; S. Martin, "Edible Oil," 224–26.

61. Kris-Etherton and Seligson, "The Heart of the Tropical Oils Issue."

62. S. Martin, *UP Saga*, 276.

63. Othman, Houston, and McIntosh, "Health Issue," 215, 222.

64. Schleifer, "Fear of Frying."

65. Schleifer, "The Perfect Solution"; Schleifer, "Categories Count." See also McNamara, "Palm Oil and Health," 242–43.

66. Schleifer, "We Spent a Million Bucks and Then We Had To Do Something."

67. Shankar and Hawkes, "India Has a Problem with Palm Oil."

68. Author interview with K. C. Hayes, 6 September 2019; author interview with Daniel Perlman, 4 September 2019; Malaysian Palm Oil Board, *Going for Liquid Gold*, 130.

69. Tullis, "Hooked."

70. Belt, "Recipe Ruined?"; Meehan, "Berger Cookies."

71. Coombes, "Trans Fats"; W. Loh, "Trans Fat Reformulation," 99; Klonoff, "Replacements for Trans Fats—Will There Be an Oil Shortage?"; Prodöhl, "Dinner to Dynamite." Interesterified fats offer one alternative, but they require complex catalytic or enzymatic processes. Their effect on human health is still being researched. Hayes and Pronczuk, "Replacing Trans Fat."

72. "Project Proposal: Business and Economic Intelligence Centre for the Palm Oil Industry," February 2003, MIER, 2010/0015978, ANM; Malaysian Palm Oil Board, *Going for Liquid Gold*, 129–31.

73. Kadandale, Marten, and Smith, "Palm Oil Industry."

74. See Malaysian Palm Oil Board, *Going for Liquid Gold*.

75. Lam et al., "Malaysian Palm Oil."

76. Yacob, "Government, Business and Lobbyists," 919.

77. "Palm oil to help pay for fighter jets as Malaysia barters with Russia," Reuters, 7 June 1994; Adrian David, "Russia offers palm oil defense offset," *New Straits Times*, 29 March 2019.

78. Khoo Boo Hock, RAM Consultancy, "Financing Projects Overseas with Palm Oil," 2010/0015971, ANM.

79. Levitt, *Oil, Fat, and Soap*, 141.

80. Levitt, *Oil, Fat, and Soap*, 151–52.

81. Palm oil was widely used in ore "beneficiation" in the early twentieth century, but it was replaced with cheaper substitutes in most places. Peša, "Between Waste and Profit"; Shengo et al., "Review of the Beneficiation."

82. Kozlik and Boomgaard, *The Demand for Fats and Oils in the Soap Industry*, 429; Sills and Doty, *Detergents*, 2; C. Wilson, *History of Unilever*, vol. 2, 349.

83. Sills and Doty, *Detergents*, vi; Corlett, *Economic Development*, 38–45.

84. Wubs, *International Business and National War Interests*, 130; Baptista and Travis, "I. G. Farben in America"; Puplett, *Synthetic Detergents*, 20–25.

85. White, *Post-Colonial Malaysia*, 168; G. Jones, *Renewing Unilever*, 22–23. Tide was made with coconut oil or petroleum, depending on the price of the former; McGucken, *Biodegradable*, 17.

86. Fochtman et al., *Animal Fats in Hot Dip Tinning*, i; Yacob, "Government, Business and Lobbyists," 911; *Stockpile Report to the Congress July–December 1956*, 9; *Stockpiling—Palm Oil*; CONGOPALM, *Production, Properties and Uses*, 17.

87. Schisgall, *Eyes on Tomorrow*, 249–58; McGucken, *Biodegradable*.

88. Rupilius and Salmiah, "The Changing World of Oleochemicals."

89. Rabiu, Elias, and Oyekola, "Oleochemicals from Palm Oil for the Petroleum Industry."

90. See S. Martin, *UP Saga*, 233–34.

91. MPOB policy paper, 23 November 2001, 2012/0005393, ANM.

92. Castaneda and Giodano, *Palm Oil Prospects for 2005*, 11; Rasiah, "Explaining Malaysia's Export Expansion."

93. Malaysian Palm Oil Board, *Going for Liquid Gold*, 147.

94. Byerlee, Falcon, and Naylor, *Tropical Oil Crop Revolution*, chap. 6.

95. LePlae, "Oil Palm Groves," 16; Eaton, "A New Fuel"; LePlae, *Palmier à Huile*, 36; De Wildeman, *Remarques a Propos de La Foret Équatoriale Congolaise*, 80–81.

96. Knothe, "Historical Perspectives"; Knothe, "George Chavanne and the First Biodiesel"; Pioch and Vaitilingom, "Palm Oil and Derivatives."

97. Brandão and Schoneveld, *The State of Oil Palm Development in the Brazilian Amazon*, 8.

98. S. Ooi, Nascimento, and Da Silva, "Oil Palm Industry in Brazil," 370.

99. Taussig, *Palma Africana*, 189; on flex crops see Byerlee, Falcon, and Naylor, *Tropical Oil Crop Revolution*, 135–58; Borras et al., "Rise of Flex Crops."

100. Neslen, "Leaked Figures Show Spike in Palm Oil Use for Biodiesel in Europe."

101. Houtart, *Agrofuels*, ix; also see Meijaard and Sheil, "The Moral Minefield of Ethical Oil Palm and Sustainable Development"; Pye, "Commodifying Sustainability."

102. Dauvergne and Neville, "The Changing North-South and South-South Political Economy of Biofuels," 1090–91.

103. Energy Post, "Biofuels Turn Out to Be a Climate Mistake"; DeCicco et al., "Carbon Balance Effects of U.S. Biofuel Production and Use."

104. Jiwan, "Political Ecology," 61.

105. Dauvergne, "The Global Politics of the Business of 'Sustainable' Palm Oil," 36.

106. Dauvergne, *Environmentalism of the Rich*; Dauvergne, "The Global Politics of the Business of 'Sustainable' Palm Oil"; Pye, "Commodifying Sustainability."

107. Varkkey, *Haze Problem*.

108. Tsing, *Friction*, 43.

109. Quoted in Gallegos, "Palm Oil Tensions," 23.

110. "Deadly Palm Oil in Your Shopping Trolley," Palm Oil Action, 2007 (archived at https://web.archive.org/web/20070624222831/http://www.palmoilaction.org.au/).

111. For example, Breyer, "25 Sneaky Names for Palm Oil."

112. "Watch the Moment an Orangutan Tries to Defend Its Jungle Home Being Destroyed by a Digger," *The Independent*, 7 June 2018, https://www.independent.co

.uk/news/world/asia/orangutan-defends-jungle-home-video-digger-ape-borneo-indonesia-deforestation-a8387836.html.

113. See, for example, Glickman, *Buying Power.*

114. Dauvergne, *Environmentalism of the Rich*, 120–21.

115. Barani, *Palm Oil*, 64.

116. Pye, "Commodifying Sustainability."

117. Tullis, "Hooked."

Chapter 10

1. "A thousand workers and farmers in NGPI" to Lord Kindersley, 21 November 1982, OD 71/104; typescript in OD 71/115; Thorne, "File notice," 18 August 1986, OD 71/116, TNA.

2. HC Deb., 14 February 1984, col. 237. NGPI (also called NDC-Guthrie) and NGEI were two phases of the project; I refer to both as NGPI here.

3. Kroef, "Private Armies and Extrajudicial Violence in the Philippines," 9.

4. HC Deb., 7 December 1982, col. 784.

5. Sacerdoti and Ocampo, "Guthrie and the Angels"; Priests of the Diocese of Butuan to CDC, 22 November 1982, OD 71/104, TNA.

6. Untitled clipping and memo by Father Ton Zwart, OD 71/105, TNA.

7. Marten to Buss, 2 December 1982; Clipping, *Sunday Times*, 28 November 1982, OD 71/104, TNA.

8. Brian Eads, "£6 million aid dilemma for UK," *Observer*, 28 November 1982, clipping in OD 71/104, TNA; Sacerdoti and Ocampo, "Guthrie and the Angels."

9. Coote, "Behind the Palm Trees."

10. Morgan to Rednall, 14 May 1984 OD 71/107, TNA.

11. HC Deb., 7 December 1982, col. 777; see also translated letter, October 1982, in OD 71/104, TNA.

12. Father Ton Zwart, "Development for Whom?" 15 September 1983, OD 71/105, TNA.

13. "Notes for NGPI meeting," ca. 1982, OD 71/104, TNA.

14. Memo, Derek Nesbit, 19 May 1983, OD 71/105, TNA.

15. "Brief for Minister for OD to Catholic Herald," 8 October 1986, OD 71/116, TNA.

16. HC Deb., 7 December 1982, col. 776–777.

17. Father Ton Zwart, "Development for Whom?" 15 September 1983; "A study of a British financing corporation's investments," 25 August 1983, OD 71/105, TNA.

18. Father Ton Zwart, "A study of a British financing corporation's investments" 25 August 1983, OD 71/105, TNA.

19. Report by Peter Meinertzhagen, 15 November 1982, OD 71/104, TNA.

20. Morgan to Rednall, 10 August 1984, OD 71/107, TNA; Coote, "Behind the Palm Trees."

21. Thorne to Jones, 31 October 1984, OD 71/108; Graham to Embassy (Manilla), 24 September 1984, OD 71/107, TNA.

22. A. Thorne to Rednall, 5 October 1983, OD 71/105; Note on HC Debate reprint in OD 71/106, TNA.

23. HC Deb 14 February 1984, cols. 238–239.

24. Doyle to Kindersley, 8 September 1983; Meinertzhagen to Doyle, 19 September 1983, OD 71/105, TNA.

25. "Rebels raid palm oil factory," *Malaya*, 18 December 1984; Thorne to Jones, 21 December 1984, OD 71/108, TNA.

26. Thorne to Jones, 10 April 1985, OD 71/115, TNA.

27. Thorne to Carter, 9 November 1984; Morgan to Jones, 29 November 1984, OD 71/108, TNA.

28. Typescript in OD 71/115; Thorne, "File notice," 18 August 1986, OD 71/116, TNA.

29. Tyler and Dixie, "Investing in Agribusiness," 24; the Bank's report says nothing about the body count at NGPI, referring only to "allegations" of coercion.

30. Figures for "oil palm fruit, area harvested" 1961–2018, FAOSTAT.

31. CDC Group. *Report and Accounts*, 1999. https://assets.cdcgroup.com/wp-content/uploads/1999/06/25150804/Annual-Report-and-Accounts-1999.pdf.

32. Cramb and McCarthy, "Characterising Oil Palm Production," 29; S. Martin, *UP Saga*.

33. S. Martin, *UP Saga*, 294; Tullis, "Hooked." See chapter 7 for the history of the *E. kamerunicus* and oil palm pollination.

34. See Yacob and White, "Unfinished Business."

35. Kaur, "Plantation Systems," 210; see also De Koninck, "La Paysannerie."

36. R. A. Pranam to Ministry of Rural Development, 18 July 1962, 1974/0000026; also "Extracts . . . Assistant Minister's speech," July 1962, 1974/0000026, ANM.

37. Graham, Floering, and Fieldhouse, *Modern Plantation*, 110.

38. Perumal, "Smallholder Crop," 395; Pletcher, "Regulation with Growth," 629. As Drabble notes, development aid for settlers "constituted a package of material largesse which the political elites found extremely useful as a means of dispensing patronage to supporters;" Drabble, *Economic History*, 220. For a study of settler resistance, see Salleh, "Peasants, Proletarianisation and the State: FELDA Settlers in Pahang"; for comparison of the Malaysian case to Indonesia, see Barral, "Paternalistic Supervision."

39. Manshard and Morgan, *Agricultural Expansion and Pioneer Settlements in the Humid Tropics*.

40. Shamsul Bahrin and Lee, *FELDA*, 38.

41. A. C. Maby to Mr. West, 12 February 1965, OD 25/83, TNA.

42. IBRD, "Review of the economic situation," in *Agriculture in West Malaysia*, 10 July 1967, in OD 39/102, TNA.

43. Talib, *Raja Muhammad Alias: The Architect of Felda*, 91.

44. The final figure was 100,000 acres of plantation and 50,000 acres of managed timber. Tippetts-Abbett-McCarthy-Stratton, and Hunting Technical Services Ltd., "Jengka Triangle," 14.

45. World Bank, "The Jengka Triangle," 52–53.

46. "Big land scheme survey by foreign experts," *Straits Times*, 2 July 1965, 6; quote from "Battle plans for Jengka," *Straits Times*, 1 March 1966, 9.

47. "Razak makes a new promise at rally," 7 May 1969, 9.

48. World Bank, "The Jengka Triangle," viii, 51.

49. World Bank, "The Jengka Triangle," viii.

50. World Bank, "Malaysia: Jengka Triangle," 20; World Bank, "The Jengka Triangle," 10.

51. World Bank, "The Jengka Triangle," viii.

52. Shamsul Bahrin, Lee, and Dorall, "The Jengka Triangle: A Report on Research in Progress."

53. World Bank, "The Jengka Triangle," 29–30; Salleh, "Peasants, Proletarianisation and the State: FELDA Settlers in Pahang."

54. Graham, Floering, and Fieldhouse, *Modern Plantation*, 109–10; Cramb and Curry, "Oil Palm and Rural Livelihoods in the Asia-Pacific Region," 228.

55. Pye, "Commodifying Sustainability"; Lim Teck Ghee and Dorall, "Contract Farming," 110–12; Fold, "Oiling the Palms."

56. Perumal, "Smallholder Crop," 400.

57. Report, 8 April 1975, MS 37812, LMA.

58. "The Guthrie Segama River Estate"; BBT to Bombay Burmah Trading Co., 26 May 1960, CLC/B/207/MS40718/001, LMA; Winterbotham to Willis, 15 June 1951, CO 717/208/6; Minutes for February–May 1952, CO 1022/95; TNA.

59. Lai, *Glance of Tawau*, 227.

60. McWilliam, *The Development Business*, 156; Tyler and Dixie, "Investing in Agribusiness"; see also minutes in OD 39/109, TNA.

61. Cramb and McCarthy, "Characterising Oil Palm Production," 208–9, 229–30; Kaur, *Economic Change in East Malaysia*, 186.

62. Cramb, "Agrarian Transition," 74.

63. Sutton, "Agribusiness," 96; World Bank, "Malaysia—Sabah Land Settlement and Environmental Management Project."

64. Sutton and Buang, "A New Role for Malaysia's FELDA"; Fold, "Oiling the Palms."

65. Bissonnette and Bernard, "Oil Palm Plantations in Sabah," 130; Cramb, "Re-Inventing Dualism"; Bissonnette and Koninck, "The Return of the Plantation?"

66. Report to directors, Sabah Plantations Ltd., 28 November 1969, MS37812; Chairman's report, 24 June 1969, MS 37813, LMA; Majid Cooke, "In the Name."

67. See especially Ngidang, "Contradictions in Land Development Schemes"; Majid Cooke, Toh, and Vaz, "Making an Informed Choice"; McCarthy and Cramb, "Policy Narratives"; V. King, "Models and Realities"; Cramb and McCarthy, "Characterising Oil Palm Production"; Brookfield, Byron, and Potter, *In Place of the Forest*.

68. Gunggut et al., "Where Have All the Forests Gone?," 369; Aiken and Leigh, *Vanishing Rain Forests*, 68; Dauvergne, *Shadows in the Forest*, 105.

69. See Badrun, *Milestone of Change*, 48; Barani, *Palm Oil*.

70. There are no reliable estimates of African palm grove coverage, but see the rough figures in Carrere, "Oil Palm in Africa," 6, which amount to less than 7 million hectares. Indonesia may now have more than 15 million hectares of oil palm.

71. As Jelsma et al. argue, "smallholder" means different things to different people. Many smallholders don't hold legal title to their land. Formal contracts and informal ties between smallholders and oil mills vary greatly, depending on location and the age of the project. Indonesia has tried several variations on the NES model. Jelsma et al., "Unpacking."

72. Cramb and McCarthy, "Characterising Oil Palm Production," 444.

73. Teoh, "Key Sustainability Issues," 11.

74. World Bank, "Nucleus Estate and Smallholders II Project," 6.

75. The Dutch began transmigration programs during the colonial period. For an overview, see Hardjono, *Transmigration*.

76. Operations Evaluation Department, "Project Performance Audit Report: Indonesia First and Second North Sumatra Estates Projects," 16–17; see also World Bank, "Project Performance Audit Report: Indonesia, Nucleus Estate and Smallholders I, II, and III Projects"; for a thorough review of Indonesia's changing smallholder policies (and a guide to the alphabet soup of acronyms for state agencies, firms, and programs), see Zen et al., "Interventions."

77. Tyson, Varkkey, and Choiruzzad, "Deconstructing the Palm Oil Industry Narrative in Indonesia," 425; McCarthy and Cramb, "Policy Narratives," 116.

78. Gaskell, "Palm Oil Revolution," 36–42; Casson, *Hesitant Boom*; Larson, "Indonesia's Palm Oil Subsector."

79. "Note for Record," MC Wood, 11 November 1988. The Bank's representative is identified as Baudelaire. FCO 15/5373, TNA; Peluso, "Plantation and the Mine," 40.

80. Hardjono, *Transmigration*, 39–40; Secrett, "Environmental Impact," 84.

81. Whitten, "Indonesia's Transmigration," 243.

82. See discussion in Gellert, "Palm Oil Expansion."

83. Lord Avebury to KGW Frost, 29 April 1986, FCO 15/4635, TNA.

84. Tsing, *Friction*, ix.

85. J. Malcolm, "Brown envelopes," n.d., FCO 15/5378, TNA.

86. Reports from White (16 February 1989) and Benjamin (14 February 1989), OD 120/86, TNA.

87. Colchester, "Struggle for Land," 105.

88. Quoted in Colchester, "Unity and Diversity," 89.

89. Szczepanski, "Land Policy"; Acciaioli and Dewi, "Opposition"; P. Anderson, "Free, Prior, and Informed"; for a broader analysis, see Tsing, "Land as Law: Negotiating the Meaning of Property in Indonesia."

90. World Bank, "Indonesia Impact Evaluation Report: Transmigration I, II, and III," xiii; see also Persoon, "The Kubu and the Outside World (South Sumatra, Indonesia). The Modification of Hunting and Gathering," 516. "Kubu" is now seen as a pejorative.

91. Kurniawati, "Losing the Forest."

92. Fearnside, "Transmigration in Indonesia," 567.

93. Potter and Lee, *Tree Planting*, 36.

94. See, for example, Secrett, "Environmental Impact"; Colchester, "Struggle for Land."

95. Secrett, "Environmental Impact," 79.

96. Pye, "Introduction," 11; Pye, "Biofuel Connection."

97. German colonizers introduced oil palms in the 1890s, but nothing came of them. Grieve, "Oil Palm Industry"; Longayroux, "Hoskins Oil Palm Project: An Introduction," 1.

98. Fleming, "The Company View," 10.

99. Ploeg, "Sociological Aspects," 29, 36–40.

100. Guy Nickalls interview with Philip McLachan, MS 37394/006, LMA.

101. Ploeg, "Sociological Aspects," 29, 36–40, 91.

102. "300 Highlanders Attack Athlete," *Port Moresby Post-Courier*, 19 March 1971; Ploeg, "Sociological Aspects," 91; also see account in Nickalls interview with McLachan, MS 37394/006, LMA.

103. "Shots in Kimbe Payback," *Papua New Guinea Post-Courier*, 14 February 1973; Noel Pascoe, "A Second Body Found," *Papua New Guinea Post-Courier*, 15 February 1973.

104. Woolford, "Trouble Smoulders," 29; Koczberski, Gibson, and Curry, *Improving Productivity*, 124.

105. Nancy Sullivan & Associates Ltd., "Welpam Woris," 27.

106. Hulme, "Economic Appraisal," 332; for management, see Nickalls interview with McLachan, MS37394/006, LMA.

107. Ploeg, "Sociological Aspects," 91; Operations Evaluation Department, "New Britain Smallholder," 12–14.

108. Ploeg, "Sociological Aspects," 100.

109. Newton, "Feasting for Oil Palm," 71.

110. Koczberski, Gibson, and Curry, *Improving Productivity*, 12; West, *Dispossession and the Environment*, 82.

111. Koczberski, "Loose Fruit Mamas," 1177.

112. McCann, *American Company*, 41.

113. Edelman and León, "Cycles of Land Grabbing," 1704.

114. McCann, *American Company*, 33.

115. Named after the region in France where it was first used to protect grape vines from fungal attacks, the blend of copper sulfate and slaked lime has been used for well over a century. It is still widely used today, even on "organic" vegetables.

116. Marquardt, "Pesticides, Parakeets, and Unions in the Costa Rican Banana Industry, 1938–1962," 28; Soluri, *Banana Cultures*, 116–27.

117. Rogers, *Abaca, Cacao, and the African Oil Palm in Costa Rica*, 6.

118. The strategic interest was focused on palm oil's role in steel-rolling and tinplating, rather than food uses. Crawford, *African Oil Palm*, 2–3; D. Richardson, "Historia."

119. Patiño, "Informacion Preliminar."

120. May, *Costa Rica*, 112, 88–89.

121. May and Plaza Lasso, *United Fruit*, 105.

122. May and Plaza Lasso, *United Fruit*, 145.

123. Fieldhouse, *Unilever Overseas*, 547; R. Escobar and Peralta, "Oil Palm Industry," 17.

124. Soluri, *Banana Cultures*, 195–96.

125. Rhoades, *Pacifico Costarricense*, 174; McCann, *American Company*, 42; Lundberg, *Adventure in Costa Rica*, 203–4.

126. Rhoades, *Pacifico Costarricense*, 35.

127. The literature on bananas and sugar is huge; for a start, see Soluri, "Bananas before Plantations. Smallholders, Shippers, and Colonial Policy in Jamaica, 1870–1910"; Curry-Machado, "In Cane's Shadow."

128. For an outline of the Aguan project, see *Fomento del Cultivo de la Palma Africana en el Valle del Aguan*; Péhaut, "Le cocotier"; Moll, *The Economics of Oil Palm*, 274–75; for critical analysis, see Kersson, *Grabbing Power*, 19–20; Fontenay, "Dual Role," 4; León, "Rebellion," chap. 3.

129. Moll, *The Economics of Oil Palm*, 275; Jung, "African Palm," 63–64.

130. Edelman and León, "Cycles of Land Grabbing," 1709.

131. Rhoades, *Pacifico Costarricense*, 77.

132. Rhoades, *Pacifico Costarricense*, 36, 93–129, 281.

133. *Report of the FAO Oilseed Mission for Venezuela*; Markley, *Vegetable Oils*, 9–12.

134. Taussig, *Palma Africana*, 9.

135. Leal, *Landscapes of Freedom*.

136. Patiño, "Informacion Preliminar," 11–13; A. Escobar, *Territories of Difference*, 70–71; Díaz, "Palma Africana En La Costa Caribe," 104–6; quote from Ospina Bozzi and Ochoa Jaramillo, *La palma africana en Colombia*, 29.

137. Ridler, "Import Substitution," 25.

138. Quoted in A. Escobar, *Territories of Difference*, 70; D. Adams and Herron, *Produccion y Consumo*, 8–10; for details on Gutt's proposals and his links to IRHO and the Ferrand visit, see "Three propositions," 27 July 1960, Inchape papers, MS 27352, LMA.

139. A. Escobar, *Territories of Difference*, 78.

140. Ollagnier, "Coopération française en Colombie"; Colombia's government also adopted new rules in the 1950s that encouraged investment in domestic agricultural development, which helped finance the Colombian oil palm boom. See Tovar, "Décadas 1960 y 1970," 88.

141. Forero Rueda, "La Lucha"; quote from Counter, "Land on Trial," chap. 2.

142. Ollagnier, "Coopération française en Colombie"; Péhaut, "Le cocotier," 80–82.

143. Moll, *The Economics of Oil Palm*, 255; for more on Peru's fisheries, see Cushman, *Guano*.

144. Ridler, "Import Substitution," 26.

145. Indupalma, "Nuestra historia"; Ospina Bozzi and Ochoa Jaramillo, *La palma africana en Colombia*, 115–23; for the workers' account, see FUNDESVIC, *Empezó Nuestro Sueño*; FUNDESVIC, *De Siervos a Obreros*.

146. Forero Rueda, "La Lucha"; Counter, "Land on Trial," chap. 2.

147. Figures from Byerlee, "Learning," 33; for different takes on "failure" and "success," see Rival and Levang, *Palms of Controversies*, 16; Rasiah, "Explaining Malaysia's Export Expansion"; Giacomin, "The Transformation of the Global Palm Oil Cluster."

148. See, for example, Kajisa, Maredia, and Boughton, "Transformation."

149. Lorenzini, *Global Development*, 144 and passim.

150. Byerlee, Falcon, and Naylor, *Tropical Oil Crop Revolution*, 45.

151. Olomola, *Financing Oil Palm*, 2.

152. Moll, *The Economics of Oil Palm*, 216.

153. Traugott, "The Economic Origins of the Kwilu Rebellion"; De Witte, "The Suppression of the Congo Rebellions and the Rise of Mobutu, 1963–5"; Weigert, "Congo/Zaire"; see also the religious explanation emphasized in Fox, de Craemer, and Ribeaucourt, "The Second Independence."

154. Nicolaï, "Le Congo."

155. See Grossman, "Deep in the Jungle, Scientists Explore the Links between the Congo and Climate Change."

156. World Bank, "Zaire—Oil Palm Development Project," 1978, 1.

157. World Bank, "Zaire—Oil Palm Development Project," 1991; G. Jones, *Renewing Unilever*, 198–202. Unilever sold its remaining stake in the northern plantations to Feronia in 2009.

158. FAO has no data for the 1980s and early 1990s, but recorded imports of palm oil began in 1996 and reached 88,000 tons by 2016. FAOSTAT.

159. FAOSTAT gives figures as high as 1.6 million tons imported in 2014, but the USDA-Production, Supply and Distribution database estimate is closer to 500,000 tons. As Byerlee, Falcon, and Naylor, *Tropical Oil Crop Revolution*, note, there are notable discrepancies between FAO, USDA, and OECD figures.

160. Burr to Patterson at Treasury, 17 September 1965, OD 20/365, TNA.

161. IBRD, "Economic Growth of Nigeria, Vol. IV," 19.

162. Western Africa Regional Office, "Rivers State," 1.

163. Dada, "African Export Industry," 9.

164. World Bank, "Project Completion Report, Nigeria"; Olomola, *Financing Oil Palm*; on movements of scientists from Africa to Asia, see Giacomin, "The Emergence of an Export Cluster"; and S. Martin, *UP Saga*, 148–54.

165. Williams, "The World Bank in Rural Nigeria, Revisited," 51–52.

166. Wire netting to protect young palms was a suspiciously costly part of the smalholder project budget, amounting to nearly 32 percent of the smallholder services outlay in the Ondo State project. The seedlings themselves amounted to 55 percent of the budget, and fertilizer only 8 percent. Olomola, *Financing Oil Palm*, 25.

167. Western Africa Regional Office, "Rivers State," annex I.

168. World Bank, "Supplementary Annexes to the Nigerian Oil Palm Projects Appraisal Reports," 11.

169. World Bank, "Project Completion Report, Nigeria," vii.

170. World Bank, "Project Completion Report, Nigeria."

171. Orewa, "Designing Agricultural Development Projects for the Small Scale Farmers," 298.

172. Gyasi, "RISONPALM," 31–32; Korieh, *The Land Has Changed*, 211–12; see also World Bank, "Implementation Completion Report: Nigeria Tree Crops Project."

173. Gyasi notes that landowners hired Ogoni and Ibibio workers to do most of the harvesting in palm groves. Gyasi, "RISONPALM," 35.

174. World Bank, "Project Completion Report, Nigeria."

175. Okonta and Douglas, *Where Vultures Feast.*

176. Orewa, "Designing Agricultural Development Projects for the Small Scale Farmers"; S. Martin, *Palm Oil and Protest*, 135.

177. Smallholder figures from RSPO, "Cooperatives in Cote D'ivoire," http://rsep.rspo.org/index.php/oil-palm-smallholder-initiatives-worldwide/item/cooperatives-in-cote-d-ivoire. Calculating Ivory Coast's exports is complicated by the fact that a sizable reexport trade seems to have developed, taking advantage cheap Malaysian palm oil. See data in MIT's *Observatory of Economic Complexity*, https://atlas.media.mit.edu/en/.

178. Moll, *The Economics of Oil Palm*, 191–92; Cheyns, Colin, and Ruf, "Relations Entre Agro-industries," 4.

179. Daddieh, "Contract Farming," 197.

180. Moll, *The Economics of Oil Palm*, 191–92; Cheyns, Colin, and Ruf, "Relations Entre Agro-industries," 4.

181. Jannot, "Emplois, économie, environnement," 394.

182. Cheyns, Colin, and Ruf, "Relations Entre Agro-industries," 7–8.

183. Daddieh, "Contract Farming," 198.

184. Daddieh, "Contract Farming," 208; Cheyns, Colin, and Ruf, "Relations Entre Agro-industries," 28.

185. Cheyns, Colin, and Ruf, "Relations Entre Agro-industries," 31n2.

186. Daddieh, "Contract Farming," 206.

187. Cheyns, Colin, and Ruf, "Relations Entre Agro-industries," 20, 22–23, 31; Béhi et al., "Vin de Palme."

188. Cheyns and Rafflegeau, "Family Agriculture," 116; Béhi et al., "Vin de Palme"; for detailed analyses of the food crop-palm wine cycle in Benin under similar conditions (a state-backed smallholder scheme), see Brouwers, "Rural People's Response."

189. Apeh and Opata, "Oil Palm Wine Economy."

190. On Liberia's troubled plantation history, see Mitman, "Forgotten Paths of Empire"; Atkinson, "Palm Oil in Liberia."

191. World Bank, "Liberia: Decoris Oil Palm" quotes on pp. iii, iv, 2, and 11.

192. World Bank, "Sierra Leone Integrated Agricultural Development Project."

193. World Bank, "Sierra Leone Integrated Agricultural Development Project," annex 8, 2.

194. The Bank estimated that farmers could earn seventy Leones per ton by making oil, compared with twenty-five Leones per ton offered by the mill. The Bank estimated that the "tipping point" was forty Leones per ton, though it never got a chance to test this. World Bank, "Sierra Leone Integrated Agricultural Development Project," annex 1, 12.

195. Leach, *Rainforest Relations*, 145, 190–91.

196. World Bank, "Sierra Leone Eastern Integrated Agricultural Development Project III," 8.

197. Daddieh, "Contract Farming," 193; Obeng-Odoom, "Understanding Land Reform."

198. Daddieh, "Contract Farming," 191; Gyasi, "Oil Palm Belt."

199. Ghana: other group interests, UAC/1/11/14/1/38/98, UARM.

200. Daddieh, "Contract Farming," 209.

201. Nkongho, Ndjogui, and Levang, "History of Partnership between Agro-industries and Oil Palm Smallholders in Cameroon"; Nkongho, Feintrenie, and Levang, *The Non-industrial Palm Oil Sector in Cameroon.*

202. Konings, *Unilever Estates*, 24–33; Moll, *The Economics of Oil Palm*, 174; Feintrenie, "Transfer of the Asian Model of Oil Palm Development"; Potter, *Managing Oil Palm Landscapes*, 106–12; Nkongho, Ndjogui, and Levang, "History of Partnership between Agro-industries and Oil Palm Smallholders in Cameroon"; Orstom, *Unilever*; World Bank, "Cameroon—Rubber and Oil Palm Projects."

203. Konings, "Unilever, Contract Farmers and Co-operatives in Cameroon," 120; Konings, *Unilever Estates.*

204. G. Jones, *Renewing Unilever*, 197–203.

Chapter 11

1. CDC Group, "A Joint statement from CDC Group, BIO, DEG, and FMO."

2. Human Rights Watch, "A Dirty Investment"; World Rainforest Movement, "FERONIA in the Democratic Republic of the Congo"; CDC Group, "A Joint Statement from CDC Group, BIO, DEG and FMO."

3. On the recent oil palm boom, see Cramb and McCarthy, *Oil Palm Complex*; Potter, *Managing Oil Palm Landscapes.*

4. Khor, "Felda Case Study"; Fold, "Oiling the Palms"; Byerlee, "Fall and Rise."

5. Varkkey, *Haze Problem*; McCarthy and Cramb, "Policy Narratives."

6. World Bank, "Indonesia Impact Evaluation Report: Transmigration," 31.

7. Fitzherbert et al., "How Will Oil Palm Expansion Affect Biodiversity?"; Abood et al., "Relative Contributions of the Logging, Fiber, Oil Palm, and Mining Industries to Forest Loss in Indonesia."

8. Zen and McCarthy, "Agribusiness," 121–23; Gatto, Wollni, and Qaim, "Oil Palm Boom and Land-Use Dynamics in Indonesia"; Potter and Lee, *Tree Planting*, 21.

9. Saravanamuttu, "Political Economy"; Li, "Intergenerational Displacement in Indonesia's Oil Palm Plantation Zone"; Li, "The Price of Un/Freedom."

10. Li, "The Price of Un/Freedom"; Saravanamuttu, "Political Economy"; Majid Cooke and Mulia, "Migration and Moral Panic."

11. Mason and McDowell, "Palm Oil Labor Abuses Linked to World's Top Brands, Banks."

12. See Saravanamuttu, "Political Economy"; Majid Cooke and Mulia, "Migration and Moral Panic"; for technology examples, see Malaysian Palm Oil Board, *Going for Liquid Gold.*

13. Baglioni and Gibbon, "Land Grabbing."

14. Dada, "African Export Industry"; Schoneveld, "The Politics of the Forest Frontier," 156.

15. Hellermann, "The Chief, the Youth and the Plantation"; Potter, *Managing Oil Palm Landscapes*, 99–102.

16. Dada, "African Export Industry," 5. Calculating the region's imports is complicated by reexports and smuggling of imported Southeast Asian palm oil.

17. Kay, "Reflections on Latin American Rural Studies in the Neoliberal Globalization Period"; Pietilainen and Otero, "Power and Dispossession," 1142.

18. Rhoades, *Pacifico Costarricense*, 36, 93–129, 281; Beggs and Moore, *Social Landscape of African Oil Palm.*

19. A portion of the 1996 sale went into a trust fund for landowners. PNG's government later changed its mind about privatization and took back a controlling stake in the company. Curtin, "Privatization Policy," 357; Filer, "Asian Investment in the Rural Industries of Papua New Guinea."

20. Majid Cooke, "Maps and Counter-Maps"; Cramb and McCarthy, "Characterising Oil Palm Production," 234; Bujang, "Doppa Disagrees with Kok's Statement to Stop Expansion of Oil Palm Plantations."

21. Start with McCarthy and Cramb, "Policy Narratives"; and see Majid Cooke, "Vulnerability, Control and Oil Palm in Sarawak"; Majid Cooke, "Maps and Counter-Maps," 276; Cramb and Sujang, "The Mouse Deer and the Crocodile"; F. Loh, "Where Has All the Violence Gone?," 34–37; and see relevant chapters in Cramb and McCarthy, *Oil Palm Complex*.

22. Santika et al., "Does Oil Palm Agriculture Help Alleviate Poverty?"

23. Li, "Situating Transmigration," 368.

24. Putzel, *Captive Land*, 340–41; Hambloch, "Governance and Land Reform"; B. Miller, "'Killed, Forced, Afraid'"; Panganiban, "Tension Grips Agusan."

25. Kersson, *Grabbing Power*, 29–30.

26. See León, "Rebellion," 202–4; Jung, "African Palm."

27. Edelman and León, "Cycles of Land Grabbing," 1714; Kersson, *Grabbing Power*, 90–99, 106–12.

28. Pietilainen and Otero, "Power and Dispossession"; Guereña and Zepeda, "Power of Oil Palm"; Hervas, "Land, Development and Contract Farming on the Guatemalan Oil Palm Frontier."

29. Taussig, *Palma Africana*, 23.

30. Quoted in Reyes, Vargas, and Kaffure, "Estado, Poder y Dominio," 195, my translation; also see propaganda in Ospina Bozzi and Ochoa Jaramillo, *La palma africana en Colombia*, 160.

31. "Indupalma, 50 Años Generando Desarrollo," 22.

32. Valencia, "Agroindustria y conflicto armado. El caso de la palma de aceite," 182; Rettberg, "Business-Led Peacebuilding."

33. Marin-Burgos, "Power, Access and Justice," 88; Oslender, *Geographies*, 212; Potter, *Managing Oil Palm Landscapes*, 49.

34. Grajales, "The Rifle and the Title," 771; Hurtado, Pereira-Villa, and Villa, "Oil Palm Development"; Potter, "Colombia's Oil Palm Development in Times of War and 'Peace.'"

35. Mingorance, Minelli, and Le Du, *El cultivo de la palma africana en el Chocó*, 55; Chomsky, "Logic of Displacement," 177.

36. Oslender, *Geographies*, 210; Grajales, "State Involvement, Land Grabbing and Counter-Insurgency in Colombia," 222; Maher, "Rooted in Violence"; Taussig, *Palma Africana*.

37. Ouedraogo, Gnisci, and Hitimana, "Land Reform"; Chimhowu, "The 'New' African Customary Land Tenure. Characteristic, Features and Policy Implications of a New Paradigm."

38. Dewi, "Reconciling"; Malaysiakini et al., "Secret Deal."

39. Brook et al., "Momentum Drives the Crash"; Fitzherbert et al., "How Will Oil Palm Expansion Affect Biodiversity?"

40. World Resources Institute, "By the Numbers."

41. De Koninck, Bernard, and Bissonnette, "Agricultural Expansion: Focusing on Borneo"; Jim, "The Forest Fires in Indonesia 1997–98"; Varkkey, *Haze Problem*.

42. Chamorro, Minnemeyer, and Sargent, "Exploring Indonesia's Long and Complicated History of Forest Fires."

43. Filer, "Why Green Grabs Don't Work in Papua New Guinea," 606; Nelson et al., "Oil Palm and Deforestation in Papua New Guinea."

44. Varkkey, *Haze Problem*, 51–53.

45. Potter, "How Can the People's Sovereignty Be Achieved?," 329–30; Schoneveld et al., "Modeling Peat- and Forestland Conversion."

46. Sayam and Cheyns, "From the Coffee-Cocoa"; Jannot, "Emplois, économie, environnement," 399.

47. Castiblanco, Etter, and Aide, "Oil Palm Plantations in Colombia."

48. Wolff, "Economics of Oil Palm"; Castellanos-Navarrete and Jansen, "Oil Palm Expansion without Enclosure"; Pischke, Rouleau, and Halvorsen, "Public Perceptions towards Oil Palm Cultivation in Tabasco, Mexico."

49. Watkins, "African Oil Palms, Colonial Socioecological Transformation and the Making of an Afro-Brazilian Landscape in Bahia, Brazil"; and see Soluri, "Campesinos."

50. A. Escobar, *Territories of Difference*, 82.

51. Marin-Burgos, Clancy, and Lovett, "Contesting Legitimacy of Voluntary Sustainability Certification Schemes"; Cheyns and Rafflegeau, "Family Agriculture," 115.

52. Watkins, "Afro-Brazilian Landscape," 424, quote on 425–26.

53. León, "Rebellion," 122–23; also see Rhoades, *Pacifico Costarricense*, 102, 132.

54. Marin-Burgos, "Power, Access and Justice," 162; Cheyns, Colin, and Ruf, "Relations Entre Agro-industries," 36.

55. Turner and Gillbanks, *Oil Palm Cultivation*, 527; Piggott, *Growing Oil Palms*, 123; on solutions, see Jiwan, "Political Ecology," 61–65; Rhoades, *Pacifico Costarricense*, 270–73, 290.

56. Varkkey, Tyson, and Choiruzzad, "Palm Oil Intensification and Expansion in Indonesia and Malaysia."

57. García and Benítez, "History of Research"; Torres et al., "Bud Rot," 321, Chinchilla, "Many Faces of Spear Rots"; Rhoades, *Pacifico Costarricense*, 82–84.

58. Torres et al., "Bud Rot."

59. Paterson, "Ganoderma Boninense Disease of Oil Palm to Significantly Reduce Production"; Paterson and Lima, "Climate Change Affecting Oil Palm Agronomy."

60. Corley and Tinker, *Oil Palm*; N. Smith et al., *Tropical Forests*, 237; "INDONESIA: Palm Oil Production Prospects Dampened by El Niño Drought"; and for a critical note on *E. oleifera*, see Taussig, *Palma Africana*.

61. Soh et al., "Commercial-Scale Propagation," 938–39; Ong-Abdullah et al., "Loss of *Karma* Transposon Methylation Underlies the Mantled Somaclonal Variant of Oil Palm"; see also Chao, "Seed Care in the Palm Oil Sector."

62. Byerlee, Falcon, and Naylor, *Tropical Oil Crop Revolution*, 133.

63. Byerlee, Falcon, and Naylor, *Tropical Oil Crop Revolution*, 43, 211–12.

64. Pye, "Biofuel Connection"; Peluso, Afiff, and Rachman, "Claiming the Grounds for Reform."

65. McCarthy, "Certifying in Contested Spaces," 1872.

66. "Palm Oil," https://www.nestle.com/csv/raw-materials/palm-oil.

67. Dauvergne and Lister, *Eco-Business*, 83–84.

68. Pye, "Commodifying Sustainability," 222.

69. RSPO data reported at rspo.org/impacts; Ostfeld et al., "Peeling Back the Label."

70. Guindon and Dodson, "Palm Oil Assessment Summary—November 2018"; see also Levin et al., "Profitability and Sustainability in Palm Oil Production."

71. Nikolakis, John, and Krishnan, "How Blockchain Can Shape Sustainable Global Value Chains."

72. "Minister Allys Oil Palm Smallholders' Fears over Malaysian Sustainable Palm Oil Certification," *Malay Mail*, 5 August 2019.

73. See Bose, "Oil Palm Plantations vs. Shifting Cultivation for Indigenous Peoples."

74. Gatti et al., "Sustainable Palm Oil May Not Be So Sustainable"; Gatti and Velichevskaya, "Certified 'Sustainable' Palm Oil"; Cattau, Marlier, and DeFries, "Effectiveness of RSPO"; but see Morgans et al., "Evaluating the Effectiveness."

75. Darrel Webber, quoted in Askew, "RSPO CEO Talks"; see also Meijaard and Sheil, "The Moral Minefield of Ethical Oil Palm and Sustainable Development."

76. On manufacturers as consumers, see Robins, "Slave Cocoa and Red Rubber."

77. Dauvergne, "The Global Politics of the Business of 'Sustainable' Palm Oil," 45; McCarthy, "Certifying in Contested Spaces," 1881.

78. HC Deb. 13 December 1950, vol. 482 cc1127-8.

79. Ojo et al., *Oil Palm Plantations*; Schoneveld, "The Politics of the Forest Frontier."

80. Fallon, "When People Power Meets Litigation."

81. Badgley, "When Wall Street Went to Africa"; Byerlee, Falcon, and Naylor, *Tropical Oil Crop Revolution*, 52.

82. Wedel, "When We Lost the Forest"; Jong, "Indonesian Court"; and see Colchester and Jiwan, *Ghosts*, and other reports from Sawit Watch and Forest People's Programme.

83. See database at http://www.askrspo.force.com/Complaint; McCarthy, "Certifying in Contested Spaces."

84. Serrano, "Can Small-Scale Farming Save Oil Palm?"

85. Marin-Burgos, Clancy, and Lovett, "Contesting Legitimacy of Voluntary Sustainability Certification Schemes"; García Arboleda, *El exterminio*; Taussig, *Palma Africana*.

86. Li, "After the Land Grab"; Li, "The Price of Un/Freedom."

87. "Indonesia: Transmigration—World Bank and Survival International," World Bank memo, 15 May 1986, FCO 15/4636, TNA; see also FCO 15/5373.

88. Prasetijo, "Living without the Forest."

89. Byerlee, Falcon, and Naylor, *Tropical Oil Crop Revolution*, 195–201; Li, "Social Impacts"; Li, "After the Land Grab."

90. Ayodele, "African Case Study"; Ayodele et al., "Nigeria."

91. Bishop, "Ex Post Evaluation of Technology Diffusion in the African Palm Oil Sector," 238; Adjei, "Making of Quality."

92. "Freedom Mill 2 Liberia Palm Oil Machine holiday promotion 2009," https://www.youtube.com/watch?v=vLtoFTXg2KM; Bishop, "Ex Post Evaluation of Technology Diffusion in the African Palm Oil Sector"; Poku, *Small-Scale Palm Oil Processing in Africa*; Hyman, "An Economic Analysis of Small-Scale Technologies for Palm Oil Extraction in Central and West Africa."

93. Author interview with Gero Leson, 27 August 2019.

94. Dallinger, "Oil Palm Development," 28, 40; Byerlee, Falcon, and Naylor, *Tropical Oil Crop Revolution*, 198.

95. Abdullah, *Planter's Tales*, 164.

96. Corley, "How Much Palm Oil Do We Need?"; Byerlee, Falcon, and Naylor, *Tropical Oil Crop Revolution*, 114, estimate between 217 to 325 million tons.

97. Bissonnette and Bernard, "Oil Palm Plantations in Sabah," 130; on smallholders as active agents, see Byerlee, "Learning"; Byerlee, "Fall and Rise."

98. See Potter, "Alternative Pathways," 176.

99. Mukpo, "Industrial Palm Oil." Mongabay maintains a grim news feed on murders of activists: https://news.mongabay.com/list/murdered-activists/.

Bibliography

Archival Sources

Accra, Ghana
- Ghana Public Records and Archives Administration Department (PRAAD-Accra)
  - Administrative records (ADM)
  - Colonial secretary's office (CSO)

Ithaca, N.Y, United States
- Cornell University Library, Division of Manuscripts and Special Collections
  - American Home Economics Association records
  - N.Y. State College of Home Economics records
  - Pamphlet collections

Kuala Lumpur, Malaysia
- Arkib Negara Malaysia (ANM)
  - Colonial administrative records
  - FELDA, RISDA, and other agency minutes and reports
  - PORIM/MPOB reports

London, United Kingdom
- London Metropolitan Archives (LMA)
  - Harrisons & Crosfield and related companies papers
- National Archives of the United Kingdom (TNA)
  - Board of Trade (BT)
  - Colonial Office (CO)
  - Dominions Office and Commonwealth Relations Office (DO)
  - Foreign and Commonwealth Office (FCO)
  - Ministry of Food (MAF)
  - Ministry of Overseas Development (OD)
  - Research institutes (AY)
- Library, Art and Archives, Royal Botanic Gardens, Kew (RBGK)
  - Economic Botany collection (ECB)
  - Correspondence (MR series)
  - Oil Palm and Oil Producing Nuts and Seeds collection (PRO)
- School of Oriental and African Studies Archives & Special Collections (SOAS)
  - Guthrie papers (G)

Port Sunlight, United Kingdom
- Unilever Collections at Unilever Art, Archives and Records Management (UARM)
  - United Africa Company records (UAC)

Wilmington, D.E., United States
Hagley Musuem and Library
Carter Litchfield collection on the history of fatty materials
Pamphlet collection

Published Sources

Abbott, J. S. "Accomplishments of the Margarine Industry." *American Food Journal* 22, no. 4 (April 1927): 113–15.

Abdullah, Mahbob. *Planter's Tales: A Plantation Manager's Stories*. Kuala Lumpur: IPC Services, 2003.

Abood, Sinan A., Janice Ser Huay Lee, Zuzana Burivalova, John Garcia-Ulloa, and Lian Pin Koh. "Relative Contributions of the Logging, Fiber, Oil Palm, and Mining Industries to Forest Loss in Indonesia." *Conservation Letters* 8, no. 1 (2015): 58–67.

Acciaioli, Greg, and Oetami Dewi. "Opposition to Oil Palm Plantations in Kalimantan: Divergent Strategeis, Convergent Outcomes." In *Oil Palm Complex*, edited by R. A. Cramb and John F. McCarthy, 327–77. Singapore: NUS Press, 2016.

Adam, Jean. *Le palmier à huile; habitat, variétés, conditions de végétation, culture, produits, commerce, conservation des palmeraies, amélioration de la production, considérations économiques*. Paris: Challamel, 1910.

Adams, D. W., and Antonio Herron. *Producción y Consumo de Aceites Vegetales en Colombia*. Bogota: Centro Interamericano de Reforma Agraria, 1965.

Adams, John. *Sketches Taken during Ten Voyages to Africa between the Years 1786 and 1800*. London: Hurst, Robinson and Co., 1822.

Addison, W. "The Palm Nut Tree (*Elaeis guineensis*) and Its Uses." *Sierra Leone Studies* 1, no. 1 (1918): 7–19.

Adjei, Betty. "The Making of Quality: A Technography Study of Small Scale Palm Oil Processing in Ghana." PhD diss., University of Wageningen, 2014.

Afigbo, A. E. *Ropes of Sand: Studies in Igbo History and Culture*. Ibadan, Nigeria: Ibadan University Press, 1981.

Afolabi, Abiodun. "The Colonial Taxation Policy among Yoruba of Southwestern Nigeria and Its Implications for Socio-Economic Development." *Journal of the Historical Society of Nigeria* 19 (2010): 63–92.

African Institution. *3rd Report of the Committee of the African Institution*. London: Ellerton and Henderson, 1814.

"The African Oil Palm Industry. III. Machinery." *Bulletin of the Imperial Institute* 15 (1917): 57–78.

"The African Trade." *Maryland Colonizational Journal* 11, no. 15 (April 1842): 161–78.

Africanus, Leo. *A Geographical Historie of Africa. . . .* Translated by John Pory. London: George Bishop, 1600.

Agbasiere, Joseph Thérèse. *Women in Igbo Life and Thought*. London: Routledge, 2000.

Aghalino, S. O. "British Colonial Policies and the Oil Palm Industry in the Niger Delta Region of Nigeria, 1900–1960." *African Study Monographs* 21, no. 1 (January 2000): 19–33.
Agiri, B. A. "Aspects of Socio-Economic Changes among the Awori Egba and Ijebu Remo Communities during the Nineteenth Century." *Journal of the Historical Society of Nigeria* 7, no. 3 (1974): 465–83.
Aiken, Robert. "Losing Ground: Development, Natural Resources, and the Dispossession of Malaysia's Orang Asli." In *Natural Resources in Asia*, edited by Greg Bankoff and Peter Boomgaard, 163–84. Basingstoke: Palgrave, 2007.
Aiken, S. Robert, and Colin H. Leigh. *Vanishing Rain Forests: The Ecological Transition in Malaysia*. Oxford: Clarendon Press, 1992.
Aikin, Arthur, and Charles Rochemont Aikin. *A Dictionary of Chemistry and Mineralogy*. Vol. 2. London: W. Phillips, 1807.
Aitken, A. P. "Feeding Stuffs—Their Composition and Characteristics." *Transactions of the Highland and Agricultural Society of Scotland* 1 (5th series), appendix B (1889): 36–37.
Akiga, and Rupert East. *Akiga's Story: The Tiv Tribe as Seen by One of Its Members*. London: Oxford University Press, 1939.
Akinjogbin, I. A., ed. *War and Peace in Yorubaland, 1793–1893*. Ibadan, Nigeria: Heinemann, 1998.
Akintoye, S. A. "The British and the 1877–93 War in Yorubaland." In *War and Peace in Yorubaland*, edited by I. A. Akinjogbin, 279–94. Ibadan, Nigeria: Heinemann, 1998.
Akyeampong, Emmanuel. *Drink, Power, and Cultural Change: A Social History of Alcohol in Ghana, c. 1800 to Recent Times*. Portsmouth, NH: Heinemann, 1996.
Alabi, Raphael A. "Environment and Subsistence of the Early Inhabitants of Coastal Southwestern Nigeria." *African Archaeological Review* 19, no. 4 (2002): 183–201.
Alagoa, Ebiegberi Joe. *The Small Brave City-State: A History of Nembe-Brass in the Niger Delta*. Ibadan, Nigeria: Ibadan University Press, 1964.
Alencastro, Luiz Felipe de. *O trato dos viventes: formação do Brasil no Atlântico Sul, séculos XVI e XVII*. São Paulo: Companhia das Letras, 2000.
Allen, G. C., and Audrey Donnithorne. *Western Enterprise in Indonesia and Malaysia*. London: George Allen & Unwin, 1957.
Allen, R. R. "Hydrogenation." In *Bailey's Industrial Oil and Fat Products*, edited by Daniel Swern, 4th ed., vol. 2, 1–96. New York: John Wiley & Sons, 1982.
Alonso-Fradejas, Alberto, Juan Liu, Tania Salerno, and Yunan Xu. "Inquiring into the Political Economy of Oil Palm as a Global Flex Crop." *Journal of Peasant Studies* 43, no. 1 (2016): 141–65.
Alpern, Stanley B. *Amazons of Black Sparta: The Women Warriors of Dahomey*. New York: New York University Press, 2011.
——— "Dahomey's Royal Road." *History in Africa* 26 (1999): 11–24.
Alston, R. A. "Fruit-Rot or Bunch-Rot of the Oil Palm." *Malayan Agricultural Journal* 22, no. 10 (August 1934): 360–66.

Amin, Samir. *Neo-Colonialism in West Africa*. Translated by Francis McDonagh. New York: Monthly Review Press, 1973.

Anderson, John Wilson. *The Barilla Question Discussed. . . .* Edinburgh: B. Auchie, 1831.

Anderson, Patrick. "Free, Prior, and Informed Consent? Indigenous Peoples and the Palm Oil Boom in Indonesia." In *The Palm Oil Controversy in Southeast Asia: A Transnational Perspective*, edited by Oliver Pye and Jayati Bhattacharya, 244–58. Singapore: ISEAS Publishing, 2015.

Anene, J. C. *Southern Nigeria in Transition 1885–1906*. Cambridge: Cambridge University Press, 1966.

Anonymous. "The Oil Palm: Future of the West African Industry." *The Planter* 6, no. 3 (October 1925): 76.

Antheaume, Benoît. "La palmeraie du Mono: approche géographique." *Cahiers d'Études africaines* 12, no. 47 (1972): 458–84.

Apeh, Apex A., and Christian C. Opata. "The Oil Palm Wine Economy of Rural Farmers in Nigeria: Evidence from Enugu Ezike, South-Eastern Nigeria." *Rural History* 30, no. 2 (2019): 111–28.

Askew, Katy. "RSPO CEO Talks Sustainable Palm Oil's Big Challenges: Smallholders and Africa." *FoodNavigator*, April 5, 2019. https://www.foodnavigator.com/Article/2019/04/05/RSPO-CEO-talks-sustainable-palm-oil-s-big-challenges-Smallholders-and-Africa.

Aso, Michitake. *Rubber and the Making of Vietnam: An Ecological History, 1897–1975*. Chapel Hill: University of North Carolina Press, 2018.

Association of London and Country Soap Manufacturers. *Case of the Soap Duties*. London: William Clowes & Sons, 1846.

"At a Meeting of the Directors of Price's Patent Candle Company." 1852. https://www.jstor.org/stable/60219055.

Atkins, John. *A Treatise on the Following Chirurgical Subjects: Chap. I. On Ruptures. II. On Fractures of the Skull. III. On Fractures Simple and Compound. IV. On Amputations. V. On Some African Distempers. VI. Of Luxations. VII. On the Venereal Disease*. London: T. Warner, 1724.

———. *A Voyage to Guinea, Brasil and the West Indies. . . .* London: Caesar Ward and Richard Chandler, 1735.

Atkinson, Philippa. "Palm Oil in Liberia: Missed Opportunities and Second Chances." Consulting Study. High Carbon Stock Science Study, 2015. http://www.simedarbyplantation.com/sites/default/files/sustainability/high-carbon-stock/consulting-reports/socio-economic/hcs-consulting-report-17-palm-oil-in-liberia-missed-opportunities-and-second-chances.pdf.

Aubrey, T. *The Sea-Surgeon, or the Guinea Man's Vade Mecum*. London: John Clarke, 1729.

Auchinleck, G. G. *The Plantation Oil-Palm Industry in the East*. Gold Coast Department of Agriculture Bulletin 8. Accra: Gov. Printer, 1926.

———. "The Abolition of the Overseas Slave Trade: A Distorted Theme in West African History." *Journal of the Historical Society of Nigeria* 5, no. 2 (1970): 257–74.

Austen, Ralph A. *African Economic History: Internal Development and External Dependency.* London and Portsmouth, NH: James Currey and Heinemann, 1987.

Austen, Ralph A., and Jonathan A. Derrick. *Middlemen of the Cameroons Rivers: The Duala and Their Hinterland, c.1600–c.1960.* Cambridge: Cambridge University Press, 1999.

Avity, Pierre d'. *The Estates, Empires, & Principallities of the World. . . .* Translated by Edward Grimstone. London: Adam Islip, 1615.

Awe, Bolanle. "Militarism and Economic Development in Nineteenth Century Yoruba Country: The Ibadan Example." *Journal of African History* 14, no. 1 (1973): 65–77.

Ayandele, E. A. *The Ijebu of Yorubaland, 1850–1950: Politics, Economy and Society.* Ibadan, Nigeria: Heinemann, 1992.

Ayodele, Thompson. "African Case Study: Palm Oil and Economic Development in Nigeria and Ghana; Recommendations for the World Bank's 2010 Palm Oil Strategy." Lagos: Initiative for Public Policy Analysis, 2010. http://www.ippanigeria.org/articles/African%20Case%20Study%20Palm%20Oil%20and%20Economic%20Development.pdf.

Ayodele, Thompson, Olusegun Sotola, Olajide Damilola, Khalil Hegarty, and Illemobola Obe. "Nigeria: A Smallholder Case Study." High Carbon Stock Case Study, 2015. http://www.simedarbyplantation.com/sites/default/files/sustainability/high-carbon-stock/consulting-reports/socio-economic/hcs-consulting-report-13-nigeria-a-smallholder-case-study.pdf.

Bacon, L. B., Friedrich C. Schloemer, and Henry C. Taylor. *World Trade in Agricultural Products: Its Growth; Its Crisis; and the New Trade Policies.* Rome: International Institute of Agriculture, 1940.

Badgley, Christiane. "When Wall Street Went to Africa." *Foreign Policy* (blog). Accessed July 26, 2020. https://foreignpolicy.com/2014/07/11/when-wall-street-went-to-africa/.

Badrun, Muhammad. *Milestone of Change: Developing a Nation through Oil Palm "PIR."* Jakarta: Directorate General of Estate Crops, Ministry of Agriculture of the Republic of Indonesia in cooperation with Citra Widya Edukasi Palm Oil Polytechnic Institute, 2011.

Baeke, Viviane. "Wuli Witchcraft." *JASO* 26, no. 1 (1995): 21–41.

Baglioni, Elena, and Peter Gibbon. "Land Grabbing, Large- and Small-Scale Farming: What Can Evidence and Policy from 20th Century Africa Contribute to the Debate?" In *Global Land Grabs: History, Theory and Method*, edited by Marc Edelman, Carlos Oya, and Saturnino M. Borras, 42–65. Abingdon, UK: Routledge, 2015.

Baikie, William Balfour. *Narrative of an Exploring Voyage up the Rivers Kwóra and Bínue (Commonly Known as the Niger and Tsádda) in 1854. . . .* London: J. Murray, 1856.

Bailey, Alton E. *Industrial Oil and Fat Products.* New York: Interscience Publishers, 1945.

Bakulu, Jacques. "History of Oil Palm Use in the Bas-Fleuve District, DRC." *GRAIN* (blog), September 22, 2014. https://www.grain.org/article/entries/5034-oil-palm-production-in-west-and-central-africa.

Bancroft, Edward. *An Essay on the Natural History of Guiana, in South America.* . . . London: Thomas Becket et al., 1769.

Banks, A., and T. P. Hilditch. "A Note on the Composition of Some Fatty Materials Found in Ancient Egyptian Tombs." *Analyst* 58, no. 686 (1933): 265–69.

Baptista, Robert J., and Anthony S. Travis. "I. G. Farben in America: The Technologies of General Aniline & Film." *History and Technology* 22, no. 2 (June 2006): 187–224.

Barbot, Jean. *A Description of the Coasts of North and South-Guinea.* . . . London: Churchill, 1732.

Barham, Henry. *Hortus Americanus: Containing an Account of the Trees, Shrubs, and Other Vegetable Productions of South-America and the West India Islands, and Particularly of the Island of Jamaica.* . . . Kingston: Alexander Aikman, 1794.

Baring, Elevyn [Lord Howick]. "Some Aspects of Commonwealth Development." *African Affairs* 63, no. 252 (1964): 221–32.

Barlow, Colin. *The Natural Rubber Industry: Its Development, Technology and Economy in Malaysia.* Kuala Lumpur: Oxford University Press, 1978.

Barnard, Timothy P. *Nature's Colony: Empire, Nation and Environment in the Singapore Botanic Gardens.* Singapore: National University of Singapore Press, 2016.

Barnes, A. C. *Mechanical Processes for the Extraction of Oil.* Special Bulletin of the Agricultural Department 2. Lagos: Government Printer, 1925.

Barral, Stephanie. "Paternalistic Supervision of Labour in Indonesian Plantations: Between Dependence and Autonomy." *Journal of Agrarian Change* 14, no. 2 (April 2014): 240–59.

Basden, George Thomas. *Among the Ibos of Nigeria: An Account of the Curious & Interesting Habits, Customs, & Beliefs of a Little Known African People.* . . . Philadelphia: J. B. Lippincott, 1921.

———. *Niger Ibos: A Description of the Primitive Life, Customs and Animistic Beliefs, Etc.* London: Cass, 1938.

Bauer, P. T. *West African Trade; a Study of Competition, Oligopoly and Monopoly in a Changing Economy.* Cambridge: Cambridge University Press, 1958.

Baxendale, Cyril. "The Plantation Rubber Industry." *Federated Malay States Agricultural Bulletin* 1, no. 3 (December 1912): 180–91.

Bay, Edna G. *Wives of the Leopard: Gender, Politics, and Culture in the Kingdom of Dahomey.* Charlottesville: University of Virginia Press, 1998.

Bay, Edna G., and Kristin Mann, eds. *Rethinking the African Diaspora: The Making of a Black Atlantic World in the Bight of Benin and Brazil.* London: Frank Cass, 2001.

Beccari, Edoardo. *Contributo Alla Conoscenza Della Palma a Olio.* Florence: Instituto Agricolo Coloniale Italiano, 1914.

Beckman, Bjorn. "Ghana, 1951–78: The Agrarian Basis of the Post-Colonial State." In *Rural Development in Tropical Africa*, edited by Judith Heyes, Pepe Roberts, and Gavin Williams, 143–167. Basingstoke: Palgrave, 1981.

Bede, Cuthbert. "The Literature of the Shop: A Country Rector's Commentary." *Once a Week* 5, no. 111 (August 10, 1861): 189–90.

Bederman, S. H. "Plantation Agriculture in Victoria Division, West Cameroon: An Historical Introduction." *Geography* 51, no. 4 (1966): 349–60.

Beggs, Emily, and Ellen Moore. *The Social Landscape of African Oil Palm Production in the Osa and Golfito Region, Costa Rica*. San Jose: INOGO, Stanford Woods Institute for the Environment, 2013.

Béhi, Yavo Eric Norbert, Matthias Mollet, Olivier Girardin, Jean-Pierre Sorg, and Felix Herzog. "Le Vin de Palme, Aliment et Source de Revenu Pour Les Populations Rurales En Côte d'Ivoire." *Schweizerische Zeitschrift für Forstwesen* 153, no. 4 (April 2002): 123–29.

Behrendt, Stephen D., A. J. H. Latham, and David Northrup, eds. *The Diary of Antera Duke, an Eighteenth-Century African Slave Trader*. Oxford: Oxford University Press, 2010.

Beirnaert, A. "Que Pouvons-Nous Attendre Des Palmeraies Ameliorees Du Congo Belge?" In *Conférence Du Palmier à Huile (20 et 21 Octobre 1937)*, 47–62. Marseille: Institut Colonial, 1937.

Beirnaert, A., and R. Vanderweyen. *Contribution à l'étude genetique et biometrique des variétés d'Elaeis Guineensis Jacquin*. Nairobi: East African Standard, 1941.

Belfort, Roland, and Alfred Johannes Hoyer. *All about Coconuts*. London: St. Catherine Press, 1914.

Belloc, Hilaire, and Basil Temple Blackwood. *The Modern Traveller*. London: Edward Arnold, 1898.

Belt, Deb. "Recipe Ruined? Berger Cookies Could Disappear with Proposed Ban on Trans Fat." *Lutherville-Timonium, MD Patch*, November 21, 2013.

Benedikt, R. *Chemical Analysis of Oils Fats Waxes and of the Commercial Products Derived Therefrom*. Translated by Julius Lewkowitsch. London: Macmillan, 1895.

Berger, Kurt, and S. M. Martin. "Palm Oil." In *Cambridge World History of Food*, edited by Kenneth F. Kiple and Kriemhild Coneè Ornelas, vol. 1, 397–410. Cambridge: Cambridge University Press, 2000.

Berlin, Ira, and Philip D. Morgan. "Introduction." In *The Slaves' Economy: Independent Production by Slaves in the Americas*, edited by Ira Berlin and Philip D. Morgan, 1–27. London: Frank Cass, 1991.

Bersselaar, Dmitri van den. *The King of Drinks: Schnapps Gin from Modernity to Tradition*. Leiden: Brill, 2007.

Billows, Harry Clyde, and Harold Beckwith. *Palm Oil and Kernels. "The Consols of the West Coast." An Exposition of the Palm Oil Industry, &c.* Liverpool: Charles Birchall, 1913.

Biobaku, Saburi O. *The Egba and Their Neighbours, 1842–1872*. Oxford: Clarendon Press, 1957.

———. *Sources of Yoruba History*. Oxford: Clarendon Press, 1973.

Biron, Claude. *Curiositez de la nature et de l'art, aportées dans deux voyages des Indes. . . .* Paris: Jean Moreau, 1703.

Bishop, Catherine P. "Ex post Evaluation of Technology Diffusion in the African Palm Oil Sector: The Caltech Expeller in Cameroon, Benin, and Liberia." *World Development* 112 (2018): 233–43.

Bissonnette, Jean-François, and Stephane Bernard. "Oil Palm Plantations in Sabah: Agricultural Expansion for Whom?" In *Borneo Transformed*, edited by Rodolphe De Koninck, Stephane Bernard, and Jean-François Bissonnette, 120–51. Singapore: NUS Press, 2011.

Bissonnette, Jean-François, and Rodolphe De Koninck. "The Return of the Plantation? Historical and Contemporary Trends in the Relation between Plantations and Smallholdings in Southeast Asia." *Journal of Peasant Studies* 44, no. 4 (July 2017): 918–38.

Blackwell, Elizabeth. *A Curious Herbal. Containing Five Hundred Cuts, of the Most Useful Plants, Which Are Now Used in the Practice of Physick*. Vol. 2 (2 vols). London, 1739.

Bleichmar, Daniela. "Books, Bodies, and Fields: Sixteenth-Century Transatlantic Encounters with New World Materia Medica." In *Colonial Botany*, edited by Londa Schiebinger and Claudia Swan. Philadelphia: University of Pennsylvania Press, 2005.

Blommendaal, H. N. "The Manufacture of Palm Oil." *Malayan Agricultural Journal* 16, no. 6 (1928): 235–54.

Blume, Hans. *Organisational Aspects of Agro-Industrial Development Agencies*. Munich: Weltforum Verlag, 1971.

Bojer, Wenzel. *Hortus Mauritianus, Ou, Énumération Des Plantes, Exotiques et Indigènes, Qui Croissent a l'Ile Maurice. . . .* Mauritius: Aimé Mamarot, 1837.

Bold, Edward. *Merchant's and Mariner's African Guide: Containing an Accurate Description of the Coast, Bays, Harbours and Adjacent Islands of West Africa*. London: Cushing & Appleton, 1822.

Boldingh, J. "Research." In *Margarine: An Economic, Social and Scientific History 1896–1969*, edited by J. H. van Stuyvenberg, 167–225. Liverpool: Liverpool University Press, 1969.

Bonneuil, Christophe. "«Mise En Valeur» de l'empire Colonial et Naissance de l'agronomie Tropicale." In *Du Jardin d'essais Colonial à La Station Expérimentale 1880–1930: Eléments Pour Une Histoire Du CIRAD*, 13–66. Montpellier: CIRAD, 1993.

Boomgaard, Peter. "Land Rights and the Environment in the Indonesian Archipelago, 800–1950." *Journal of the Economic and Social History of the Orient* 54 (2011): 478–96.

Booth, Anne. *Colonial Legacies: Economic and Social Development in East and Southeast Asia*. Honolulu: University of Hawai'i Press, 2007.

Booth, Henry. "Patent Axle-Grease and Lubricating Fluid, No. 6814." *Repertory of Patent Inventions (New Series)* 4 (December 1835): 144–47.

Borras, Saturnino M., Jennifer C. Franco, S. Ryan Isakson, Les Levidow, and Pietje Vervest. "The Rise of Flex Crops and Commodities: Implications for Research." *Journal of Peasant Studies* 43, no. 1 (2016): 93–115.

Bose, Purabi. "Oil Palm Plantations vs. Shifting Cultivation for Indigenous Peoples: Analyzing Mizoram's New Land Use Policy." *Land Use Policy* 81 (February, 2019): 115–23.

Bosma, Ulbe. *The Making of a Periphery: How Island Southeast Asia Became a Mass Exporter of Labor.* New York: Columbia University Press, 2019.

Bosman, Willem. *Nauwkeurige beschryving van de Guinese Goud- Tand- en Slave-kust. . . .* Utrecht: Anthony Schouten, 1704.

———. *A New and Accurate Description of the Coast of Guinea, Divided into the Gold, the Slave, and the Ivory Coasts*. London: J. Knapton et al., 1705.

Boss, Eelco. "The Scramble for Palms: A Comparative Historical Investigation: Two Companies Sourcing Two Commodities in Two Colonies." MA thesis, University of Utrecht, 2014.

Bostoen, Koen, Bernard Clist, Charles Doumenge, Rebecca Grollemund, Jean-Marie Hombert, Joseph Koni Muluwa, Jean Maley, Roger Blench, Pierpaolo Di Carlo, and Jeff Good. "Middle to Late Holocene Paleoclimatic Change and the Early Bantu Expansion in the Rain Forests of Western Central Africa." *Current Anthropology* 56, no. 3 (2015): 367–68.

Bostoen, Koen, Rebecca Grollemund, and Joseph Koni Muluwa. "Climate-Induced Vegetation Dynamics and the Bantu Expansion: Evidence from Bantu Names for Pioneer Trees (*Elaeis guineensis*, *Canarium schweinfurthii*, and *Musanga cecropioides*)." *Comptes Rendus Geoscience*, 345, nos. 7–8 (July 2013): 336–49.

Boutwell, Wayne, Harry O. Doty, Duane Donald Hacklander, and Alan Walter. *Analysis of the Fats and Oils Industry to 1980: With Implications for Palm Oil Imports*. Washington, DC: USDA, 1976.

Bowers, Brian. *Lengthening the Day: A History of Lighting Technology.* Oxford: Oxford University Press, 1998.

Boyle, Robert. *General Heads for the Natural History of a Country Great or Small Drawn out for the Use of Travellers and Navigators*. London: John Taylor and S. Hedford, 1692.

Bradbury, R. E. *The Benin Kingdom and the Edo-Speaking Peoples of South-Western Nigeria: Western Africa*. Oxford: International African Institute, 1957.

Brandão, Frederico, and George Schoneveld. *The State of Oil Palm Development in the Brazilian Amazon: Trends, Value Chain Dynamics, and Business Models.* Bogor: CIFOR, 2015.

Brannt, William T. *A Practical Treatise on Animal and Vegetable Fats and Oils.* Philadelphia: Henry Carey Baird & Co., 1888.

Brass, Tom, and Henry Bernstein. "Introduction: Proletarianisation and Deproletarianisation on the Colonial Plantation." In *Plantations, Proletarians and Peasants in Colonial Asia*, edited by E. Valentine Daniel, Henry Bernstein, and Tom Brass, 1–40. London: Frank Cass, 1992.

Breman, Jan. *Taming the Coolie Beast: Plantation Society and the Colonial Order in Southeast Asia*. Oxford: Oxford University Press, 1989.

Breyer, Melissa. "25 Sneaky Names for Palm Oil." *Treehugger* (blog). May 3, 2020. https://www.treehugger.com/sneaky-names-palm-oil-4858743.

Bristowe, Lindsay W. "Gold Coast Palms: A Neglected Source of Wealth." *Engineering Review* 11, no. 12 (1904): 409–14.

Brocard, Paul. "Afrique Occidentale Francaise." In *Congrès d'agriculture Coloniale 21–25 Mai 1918. 2, Section Des Oléagineux: Compte Rendu Des Travaux*. Paris: Challamel, 1920.

Brook, Barry W., Corey J. A. Bradshaw, Lian Pin Koh, and Navjot S. Sodhi. "Momentum Drives the Crash: Mass Extinction in the Tropics." *Biotropica* 38, no. 3 (2006): 302–5.

Brookfield, H. C., Yvonne Byron, and Lesley M. Potter. *In Place of the Forest: Environmental and Socio-Economic Transformation in Borneo and the Eastern Malay Peninsula*. Tokyo: United Nations University Press, 1995.

Brooks, George E. "Peanuts and Colonialism: Consequences of the Commercialization of Peanuts in West Africa, 1830–71." *Journal of African History* 16, no. 1 (January 1975): 29–54.

Brouwers, J. H. a. M. "Rural People's Response to Soil Fertility Decline: The Adja Case (Benin)." PhD diss., Wageningen Agricultural University, 1993.

Brown, Christopher Leslie. *Moral Capital: Foundations of British Abolitionism*. Chapel Hill: University of North Carolina Press, 2006.

———. "The Origins of 'Legitimate Commerce.'" In *Commercial Agriculture, the Slave Trade and Slavery in Atlantic Africa*, edited by Robin Law, Suzanne Schwarz, and Silke Strickrodt, 138–57. Woodbridge, UK: James Currey, 2013.

Browne, Walter. "On Car Axle-Boxes and Lubricants in Europe." *Railway Gazette* 10 (August 1878): 389.

"The Bubble of the Age." *Quarterly Review* (London) 84, no. 167 (December 1848): 1–65.

Bücher, Herman, and Ernst Fickendey. *Die Ölpalme (Elaeis guineensis)*. Berlin: Auswärtiges Amt, 1919.

Buckley, Anthony D. *Yoruba Medicine*. Oxford: Clarendon Press, 1985.

Buckley, Stephen A., and Richard P. Evershed. "Organic Chemistry of Embalming Agents in Pharaonic and Graeco-Roman Mummies." *Nature* 413, no. 6858 (October 2001): 837–41.

Buelens, Frans, and Ewout Frankema. "Colonial Adventures in Tropical Agriculture: New Estimates of Returns to Investment in the Netherlands Indies, 1919–1938." *Cliometrica* 10, no. 2 (May 2016): 197–224.

Buell, Raymond Leslie. *The Native Problem in Africa*. Vol. 2. New York: Macmillan, 1928.

Bühnen, Stephan. "In Quest of Susu." *History in Africa* 21 (1994): 1–47.

Bujang, Abdul Hakim. "Doppa Disagrees with Kok's Statement to Stop Expansion of Oil Palm Plantations." *Borneo Post Online*, September 9, 2018. https://www.theborneopost.com/2018/09/09/doppa-disagrees-with-koks-statement-to-stop-expansion-of-oil-palm-plantations/.

Bunting, B., B. J. Eaton, and C. D. V. Georgi. "The Oil Palm in Malaya." *Malayan Agricultural Journal* 15, no. 9–10 (October 1927): 303–37.

Burke, John French. *British Husbandry: Exhibiting the Farming Practice in Various Parts of the United Kingdom*. Vol. 2. London: Robert Baldwin, 1848.

Burke, Timothy. *Lifebuoy Men, Lux Women: Commodification, Consumption, and Cleanliness in Modern Zimbabwe*. Chapel Hill, NC: Duke University Press, 1996.

Burkill, H., and W. Birstwistle. *Dictionary of the Economic Products of the Malay Peninsula*. London: J. B. Srivenor, 1935.

Burton, Richard F. *Abeokuta and the Camaroons Mountains: An Exploration. . . .* London: Tinsley Brothers, 1863.

———. *A Mission to Gelele, King of Dahome. . . .* London: Tinsley Brothers, 1864.

Butt, John. "James Young, Scottish Industrialist and Philanthropist." PhD diss., University of Glasgow, 1963.

Buxton, Sir Thomas Fowell. *The African Slave Trade*. London: John Murray, 1839.

Bye-Gones: Relating to Wales and the Border Counties. Owestry: Caxton, 1892.

Byerlee, Derek. "Are We Learning from History?" In *The Global Farms Race*, edited by Michael Kugelman and Susan L. Levenstein, 21–44. Washington, DC: Island Press, 2013.

———. "The Fall and Rise Again of Plantations in Tropical Asia: History Repeated?" *Land* 3, no. 3 (2014): 574–97.

Byerlee, Derek, Walter P. Falcon, and Rosamond L. Naylor. *The Tropical Oil Crop Revolution: Food, Feed, Fuel, and Forests*. Oxford: Oxford University Press, 2016.

Cà da Mosto, Alvise. *The Voyages of Cadamosto and Other Documents on Western Africa in the Second Half of the Fifteenth Century*. Translated by G. R. Crone. London: Hakluyt Society, 1937.

Cain, P. J., and A. G. Hopkins. *British Imperialism, 1688–2000*. 2nd ed. London: Longman, 2001.

Calliauw, Gijs. "Dry Fractionation." *AOCS Lipid Library*, n.d. https://lipidlibrary.aocs.org/edible-oil-processing/dry-fractionation.

Calvert, Albert Frederick. *Togoland*. London: T. W. Laurie, 1918.

Cameron, Charles, Geo. Blake, J. Blake, Richd. Vaughan, and Wm Lund. "Improved Method of Bleaching Palm Oil." *Transactions of the Society, Instituted at London, for the Encouragement of Arts, Manufactures, and Commerce* 54 (1843): 14–20.

Camp, James M. "Relation of the Iron and Steel Industries to the Chemical Industries." *Blast Furnace and Steel Plant* 9 (December 1921): 701–6.

Campbell, Robert. *A Few Facts, Relating to Lagos, Abbeokuta, and Other Sections of Central Africa*. Philadelphia: King & Baird, 1860.

———. *A Pilgrimage to My Motherland: An Account of a Journey among the Egbas and Yorubas. . . .* London: T. Hamilton, 1861.

"Candles, Slavery and War." *Chambers's Journal* 1, no. 18 (May 1854).

Canot, Theodore. *Captain Canot, or, Twenty Years of an African Slaver: Being an Account of His Career and Adventures on the Coast, in the Interior, on Shipboard, and in the West Indies*. New York: D. Appleton, 1854.

Carland, John M. *The Colonial Office and Nigeria, 1898–1914*. Stanford: Hoover Institution Press, 1985.

Carmichael, John. *A Treatise on Soap-Making: Containing an Account of the Alkaline Materials, Test for Discovering the Presence of an Alkali. . . .* Edinburgh: Oliver & Co., 1807.

Carnes, Joshua A. *Journal of a Voyage from Boston to the West Coast of Africa with a Full Description of the Manner of Trading with the Natives on the Coast.* Boston: J. P. Jewett & Co., 1852.

Carney, Judith, and Richard Nicholas Rosomoff. *In the Shadow of Slavery: Africa's Botanical Legacy in the Atlantic World.* Berkeley: University of California Press, 2011.

Carrere, Ricardo. "Oil Palm in Africa: Past, Present and Future Scenarios." Montevideo, Uruguay: World Rainforest Movement, 2013. https://wrm.org.uy/wp-content/uploads/2014/08/Oil_Palm_in_Africa_2013.pdf.

Carrière De Belgarric, René. "Les huileries de palme du plan: équipement et exploitation rationnelle des palmeraies naturelles de l'Afrique française." *Oléagineux* 10 (1955): 381–86.

———. "Souhaits de Bienvenue à Pobé, Le Lundi 16 Janvier 1956." In *Conférence Franco-Brittanique sur le palmier à huile*, 9–10. 14. Paris: Centre Technique d'Agriculture Tropicale and IRHO, 1957.

Casson, Anne. *The Hesitant Boom: Indonesia's Oil Palm Sub-Sector in an Era of Economic Crisis and Political Change.* Bogor, Indonesia: CIFOR, 1999.

Castaneda, Jaime, and Mark Giodano. *Palm Oil Prospects for 2005.* ERS Staff Paper. Washington, DC: USDA, 1995. http://hdl.handle.net/2027/umn.31951d01234239w.

Castellanos-Navarrete, Antonio, and Kees Jansen. "Oil Palm Expansion without Enclosure: Smallholders and Environmental Narratives." *Journal of Peasant Studies* 42, no. 3–4 (July 4, 2015): 791–816.

Castiblanco, Carmenza, Andrés Etter, and T. Mitchell Aide. "Oil Palm Plantations in Colombia: A Model of Future Expansion." *Environmental Science & Policy* 27 (March 2013): 172–83.

A Catalogue of the Exotic Plants Cultivated in the Mauritius: At the Botanic Garden, Monplaisir, Reduit, and Other Places. Mauritius: Baron & Souvignec, 1822.

Cattau, Megan E., Miriam E. Marlier, and Ruth DeFries. "Effectiveness of Roundtable on Sustainable Palm Oil (RSPO) for Reducing Fires on Oil Palm Concessions in Indonesia from 2012 to 2015." *Environmental Research Letters* 11, no. 10 (October 2016): 105007.

CDC Group. "A Joint Statement from CDC Group, BIO, DEG and FMO on the Long-Term Role of Development Finance Institutions in Poverty-Stricken Regions of the World," November 25, 2019. https://www.cdcgroup.com/en/news-insight/news/a-joint-statement-from-cdc-group-bio-deg-and-fmo-on-the-long-term-role-of-development-finance-institutions-in-poverty-stricken-regions-of-the-world/.

Chakrabarti, Pratik. *Materials and Medicine: Trade, Conquest and Therapeutics in the Eighteenth Century.* Manchester: Manchester University Press, 2010.

Chambers, William, ed. “The Light Question.” *Chambers's Journal of Popular Literature, Science and Arts, Jan. 1854–Nov. 1897; London*, no. 227 (8 May 1858): 294–96.

Chamorro, Andres, Susan Minnemeyer, and Sarah Sargent. “Exploring Indonesia's Long and Complicated History of Forest Fires.” World Resources Institute, February 16, 2017. https://www.wri.org/blog/2017/02/exploring-indonesias-long-and-complicated-history-forest-fires.

Chanderbali, David Sinjeet. “Indian Indenture in the Straits Settlements, 1872–1910: Policy and Practice in Province Wellesley.” PhD diss., Australian National University, Canberra, 1983.

Chao, Sophie. “Seed Care in the Palm Oil Sector.” *Environmental Humanities* 10, no. 2 (November 2018): 421–46.

“Cheerily, Cheerily!” *Household Words* 6, no. 131 (25 September 1852): 25–31.

Chevalier, Auguste. *Documents sur le palmier à huile*. Paris: A. Challamel et Laboratoire d'agronomie coloniale, 1910.

———. “Le Palmier à huile à la Côte d'Ivoire.” *Journal d'agriculture traditionnelle et de botanique appliquée* 11, no. 116 (1931): 213–29.

Cheyns, Emmanuelle. “La consommation urbaine de l'huile de palme rouge en Côte d'Ivoire: quels marchés?” *OCL. Oléagineux, Corps gras, Lipides* 8, no. 6 (2001): 641–45

Cheyns, Emmanuelle, J.-Ph. Colin, and François Ruf. “Relations Entre Agro-industries et Dynamiques d'évolution Des Exploitations Familiales et Du Milieu Rural: Agriculture Contractuelle et Industrie Du Palmier à Huile En Côte d'Ivoire.” Grands Investissements Agricoles et Inclusion Des Petits Producteurs. Montpellier and Rome: CIRAD and FAO, 2012.

Cheyns, Emmanuelle, and S. Rafflegeau. “Family Agriculture and the Sustainable Development Issue: Possible Approaches from the African Oil Palm Sector. The Example of Ivory Coast and Cameroon.” *OCL. Oléagineux, Corps Gras, Lipides* 12, no. 2 (2005): 111–20.

Chimhowu, Admos. “The ‘New’ African Customary Land Tenure. Characteristic, Features and Policy Implications of a New Paradigm.” *Land Use Policy* 81 (February 2019): 897–903.

Chinchilla, Carlos M. “The Many Faces of Spear Rots in Oil Palm and the Need for an Integrated Management Approach.” *ASD Oil Palm Papers*, no. 32 (2008): 1–10.

Chinery, David. *West African Slavery, the African Slave-Trade and the Real and Practicable Means for Its Suppression*. London: M. Lownds, 1864.

Chomsky, Aviva. “The Logic of Displacement: Afro-Colombians and the War in Colombia.” In *Beyond Slavery: The Multilayered Legacy of Africans in Latin America and the Caribbean*, edited by Darién J. Davis, 171–98. Lanham, MD: Rowman & Littlefield, 2007.

Choudhury, Ishani. “A Palatable Journey through the Pages: Bengali Cookbooks and the ‘Ideal’ Kitchen in the Late Nineteenth and Early Twentieth Century.” *Global Food History* 3, no. 1 (January 2017): 24–39.

Chubb, L. T. *Ibo Land Tenure*. Ibadan, Nigeria: Ibadan University Press, 1961.

Chuku, Gloria. "'Crack Kernels, Crack Hitler': Export Production Drive and Igbo Women during the Second World War." In *Gendering the African Diaspora: Women, Culture, and Historical Change in the Caribbean and Nigerian Hinterland*, edited by Judith Byfield, LaRay Denzer, and Anthea Morrison, 219–44. Bloomington: Indiana University Press, 2010.

——. *Igbo Women and Economic Transformation in Southeastern Nigeria, 1900–1960*. London: Routledge, 2005.

Churchill, Awnsham, ed. *A Collection of Voyages and Travels, Some Now First Printed from Original Manuscripts, Others Now First Published in English. . . .* Vol. 6. London: John Walthoe et al., 1732.

Ciment, James. *Another America: The Story of Liberia and the Former Slaves Who Ruled It*. New York: Hill and Wang, 2013.

A Circular Addressed by the Soap Manufacturers of London to the Manufacturers in the Country. London: W. Clowes and Sons, 1835.

Clapp, B. W. *An Environmental History of Britain since the Industrial Revolution*. Abingdon, UK: Routledge, 2014.

Clarence-Smith, William G. *Cocoa and Chocolate, 1765–1914*. London: Routledge, 2000.

——. "The Rivaud-Hallet Plantation Group in the Economic Crises of the Inter-War Years." In *Private Enterprises during Economic Crises: Tactics and Strategies*, edited by P. Lanthier, 117–32. Ottawa: Legas, 1997.

——. "Rubber Cultivation in Indonesia and the Congo from the 1910s to the 1950s: Divergent Paths." In *Colonial Exploitation and Economic Development: The Belgian Congo and the Netherlands Indies Compared*, edited by Ewout Frankema and Frans Buelens, 41–59. London: Routledge, 2013.

Clark, J. Desmond. "Prehistoric Populations and Pressures Favoring Plant Domestication." In *Origins of African Plant Domestication*, edited by Jack R. Harlan, Jan J. M. de Wet, and Ann B. Stemler, 67–106. The Hague: Mouton Publishers, 1976.

Clayton, William. *Margarine*. London, Longmans, 1920.

Cleary, M. C. "Plantation Agriculture and the Formulation of Native Land Rights in British North Borneo c. 1880–1930." *Geographical Journal* 158, no. 2 (1992): 170–81.

Clement, Piet. "The Land Tenure System in the Congo, 1885–1960." In *Colonial Exploitation and Economic Development: The Belgian Congo and the Netherlands Indies Compared*, edited by Ewout Frankema and Frans Buelens, 88–108. London: Routledge, 2013.

Clifford, Jim. "London's Soap Industry and the Development of Global Ghost Acres in the Nineteenth Century." *Environment and History* 26, no. 3 (2019).

Clozel, F. J. "Land Tenure among the Natives of the Ivory Coast." *Journal of the Royal African Society* 1, no. 4 (1902): 399–415.

Clusius, Carolus. *Aromatum et simplicium aliquot medicamentorum apud Indios nascentium historia*. Antwerp, Belgium: Plantin, 1567.

——. *Curae posteriores*. Leiden, Netherlands: Officina Plantiniana Raphelengii, 1611.

Cochard, Benoît, Benjamin Adon, Samah Rekima, Norbert Billotte, Roch Desmier de Chenon, Anatole Koutou, Bruno Nouy, et al. "Geographic and Genetic Structure of African Oil Palm Diversity Suggests New Approaches to Breeding." *Tree Genetics & Genomes* 5, no. 3 (July 2009): 493–504.

Cognard, A. "La Palmeraie Du Dahomey." In *Conférence Franco-Britannique*, 91–98. 14. Paris: Centre Technique d'Agriculture Tropicale and IRHO, 1957.

Cohen, Benjamin R. "Analysis as Border Patrol: Chemists along the Boundary between Pure Food and Real Adulteration." *Endeavour* 35, no. 2 (June, 2011): 66–73.

——. *Pure Adulteration: Cheating on Nature in the Age of Manufactured Food*. Chicago: University of Chicago Press, 2019.

Colchester, Marcus. "The Struggle for Land: Tribal Peoples in the Face of the Transmigration Programme." *The Ecologist* 16, nos. 2–3 (1986): 77–88.

——. "Unity and Diversity: Indonesia's Policy Towards Tribal Peoples." *The Ecologist* 16, nos. 2–3 (1986): 89–98.

Colchester, Marcus, and Norman Jiwan. *Ghosts on Our Own Land: Indonesian Oil Palm Smallholders and the Roundtable on Sustainable Palm Oil*. Moreton-in-Marsh, UK, and Bogor: Forest Peoples' Programme and Sawit Watch, 2006.

Cole, William. *Life in the Niger, or, The Journal of an African Trader*. London: Saunders, Otley, 1862.

Coleman, Deirdre. *Romantic Colonization and British Anti-Slavery*. Cambridge: Cambridge University Press, 2004.

Coleman, J. *The Cattle of Great Britain: Being a Series of Articles on the Various Breeds of Cattle of the United Kingdom, the History, Management, &c*. London: "The Field" Office, 1875.

Collecção de Noticias Para a Historia e Geografia Das Nações Ultramarinas, Que Vivem Nos Dominios Portuguezes. . . . Vol. 2. Lisbon: Academia Real das Sciencias, 1812.

Colombijn, Freek. "The Ecological Sustainability of Frontier Societies in Eastern Sumatra." In *Paper Landscapes*, edited by Peter Boomgaard, Freek Colombijn, and David Henley, 309–40. Leiden: KITLV Press, 1997.

Commerical Manufacturing Company. *A Brief History of the Mége Discovery*. New York: Commercial Manufacturing Company Consolidated, 1881.

Committee on Edible and Oil-Producing Nuts and Seeds. *Minutes of Evidence*. Cd. 8248. London: HMSO, 1916.

Committee on Improved and Increased Production of Palm Oil and Palm Kernels in West Africa. *Palm Oil and Palm Kernels. Report of a Committee Appointed by the Secretary of State for the Colonies, September, 1923, to Consider the Best Means of Securing Improved and Increased Production*. Colonial No. 10. London: HMSO, 1925.

"Composition and Spermaceti Candles." *Magazine of Domestic Economy* 6 (1840): 105–111.

Conférence Franco-Britannique Sur Le Palmier à Huile. Paris: Centre Technique d'Agriculture Tropicale and IRHO, 1957.

CONGOPALM. *Production, Properties and Uses of Belgian Congo Palm Oil*. Brussels: CONGOPALM, 1958.

Cook, O. F. "A Brazilian Origin for the Commercial Oil Palm." *Scientific Monthly* 54, no. 6 (1942): 577–80.

——. "Oil Palms in Florida, Haiti, and Panama." *National Horticultural Magazine* 19, no. 1 (January 1940): 10–35.

Coombes, Rebecca. "Trans Fats: Chasing a Global Ban." *British Medical Journal* 343, no. 7822 (2011): 506–9.

Coote, Belinda. "Behind the Palm Trees." *New Statesman*, November 8, 1983.

Copley, Mark S., Pamela J. Rose, Alan Clapham, David N. Edwards, Mark C. Horton, and Richard P. Evershed. "Detection of Palm Fruit Lipids in Archaeological Pottery from Qasr Ibrim, Egyptian Nubia." *Proceedings of the Royal Society B* 268 (2000): 593–97.

Coquery-Vidrovitch, Catherine. "French Colonization in Africa to 1920: Administration and Economic Development." In *Colonialism in Africa*, edited by Lewis H. Gann and Peter Duignan, 165–98. Cambridge: Cambridge University Press, 1969.

——. *Le Congo au temps des grandes compagnies concessionnaires 1898–1930*. 2 vols. Paris: Éditions de l'École des hautes études en sciences sociales, 2001.

Corlett, Wilfred John. *The Economic Development of Detergents*. London: Duckworth, 1958.

Corley, R. H. V. "How Much Palm Oil Do We Need?" *Environmental Science & Policy* 12, no. 2 (April 2009): 134–39.

Corley, R. H. V, and P. B Tinker. *The Oil Palm*. 4th ed. Chichester, UK: Wiley Blackwell, 2008.

——. *The Oil Palm*. 5th ed. Chichester, UK: Wiley Blackwell, 2015.

Correspondence with the British Commissioners at Sierra Leone, the Havana, Rio de Janeiro, and Suriname, relating to the Slave Trade. P.P 1841 (Class A) xiv 1. London: HMSO, 1841.

Correspondence with Foreign Powers, Parties to Treaties under which captured Vessels are to be tried by Mixed Commissions. . . . P.P. 1838 (Class B) xvi 355. London: HMSO, 1838.

Counter, Max. "Land on Trial: Law, Forced Displacement and Land Restitution in Colombia." PhD diss., University of Colorado, Boulder, 2019.

Cowan, Alexander M. *Liberia, as I Found It, in 1858*. Frankfort: A. Hodges, 1858.

Cowen, Mike. "Early Years of the Colonial Development Corporation: British State Enterprise Overseas during Late Colonialism." *African Affairs* 83, no. 330 (1984): 63–75.

Cramb, R. A. "Agrarian Transition in Sarawak: Intensification and Expansion Reconsidered." In *Borneo Transformed*, edited by Rodolphe De Koninck, Stephane Bernard, and Jean-François Bissonnette, 44–93. Singapore: NUS Press, 2011.

——. "The Political Economy of Large-Scale Oil Palm Development in Sarawak." In *The Oil Palm Complex*, edited by R. A. Cramb and John F. McCarthy, 189–246. Singapore: NUS Press, 2016.

——. "Re-Inventing Dualism: Policy Narratives and Modes of Oil Palm Expansion in Sarawak, Malaysia." *Journal of Development Studies* 47, no. 2 (February 2011): 274–93.

Cramb, R. A., and G. N. Curry. "Oil Palm and Rural Livelihoods in the Asia-Pacific Region: An Overview: Oil Palm and Rural Livelihoods." *Asia Pacific Viewpoint* 53, no. 3 (December 2012): 223–39.

Cramb, R. A., and John F. McCarthy. "Characterising Oil Palm Production in Indonesia and Malaysia." In *The Oil Palm Complex: Smallholders, Agribusiness and the State in Indonesia and Malaysia*, edited by R. A. Cramb and John F. McCarthy, 27–77. Singapore: NUS Press, 2016.

——. "Conclusion." In *The Oil Palm Complex: Smallholders, Agribusiness and the State in Indonesia and Malaysia*, edited by R. A. Cramb and John F. McCarthy, 442–64. Singapore: NUS Press, 2016.

——, eds. *The Oil Palm Complex: Smallholders, Agribusiness and the State in Indonesia and Malaysia*. Singapore: NUS Press, 2016.

Cramb, R. A., and Patrick S. Sujang. "The Mouse Deer and the Crocodile: Oil Palm Smallholders and Livelihood Strategies in Sarawak, Malaysia." *Journal of Peasant Studies* 40, no. 1 (January 2013): 129–54.

Cramer, P. J. S. "A Comparison between Oilpalms and Coconuts." *The Planter* 26, no. 8 (August 1950): 341–48.

——. "De Oliepalm (*Elaeis guineensis*). Historisch Onderzoek Over Den Oliepalm in Nederlandsch-Indie." *Teysmannia* 28 (1917): 445–49.

Crawford, Douglas. *African Oil Palm in Central America*. Foreign Agriculture Report 52. Washington, DC: USDA, 1950.

Crow, Hugh. *Memoirs of the Late Captain Hugh Crow, of Liverpool. . . .* London: Longman, 1830.

Crowder, Michael. *West Africa under Colonial Rule*. Evanston, IL: Northwestern University Press, 1968.

Crowther, Samuel. *A Charge Delivered on the Banks of the River Niger in West Africa*. London: Seely, Jackson, & Halliday, 1867.

——. *The River Niger: A Paper Read before the Royal Geographical Society, June 11th, 1877: And a Brief Account of Missionary Operations Carried on under the Superintendence of Bishop Crowther in the Niger Territory*. London: Church Missionary House, 1877.

Cruickshank, Brodie. *Eighteen Years on the Gold Coast of Africa: Including an Account of the Native Tribes and Their Intercourse with Europeans*. London: Hurst and Blackett, 1853.

"Cultivation of the African Oil Palm, with Special Reference to the East Indies." *Bulletin of the Imperial Institute* 18 (1920): 209–52.

Cunyngham-Brown, Sjovald. *The Traders: A Story of Britain's South-East Asian Commercial Adventure*. London: Newman-Neame, 1971.

Curry-Machado, Jonathan. “In Cane’s Shadow: Commodity Plantations and the Local Agrarian Economy on Cuba’s Mid-Nineteenth-Century Sugar Frontier.” In *Global Histories, Imperial Commodities, Local Interactions*, edited by Jonathan Curry-Machado, 143–67. Cambridge Imperial and Post-Colonial Studies Series. Basingstoke: Palgrave Macmillan, 2013.

Curtin, Philip D. *The Rise and Fall of the Plantation Complex: Essays in Atlantic History*. Cambridge: Cambridge University Press, 1990.

Curtin, Timothy. “Privatization Policy in Papua New Guinea.” In *Policy Making and Implementation: Studies from Papua New Guinea*, edited by Ronald James May, 345–68. Canberra: ANU Press, 2009.

Cushman, Gregory. *Guano and the Opening of the Pacific World*. New York: Cambridge University Press, 2013.

Customs Tariffs of the United Kingdom from 1800 to 1897: With Some Notes upon the History of the More Important Branches of Receipt from the Year 1660. London: HMSO, 1897.

Dada, Ladé A. “The African Export Industry: What Happened and How Can It Be Revived? Case Study on the Nigerian Oil Palm Industry.” Agricultural Management, Marketing and Finance Working Document. Rome: FAO, 2007. http://www.fao.org/3/ap303e/ap303e.pdf.

Daddieh, Cyril Kofie. “Contract Farming and Palm Oil Production in Cote d’Ivoire and Ghana.” In *Living under Contract: Contract Farming and Agrarian Transformation in Sub-Saharan Africa*, 188–215. Madison: University of Wisconsin Press, 1994.

“Dahomey.” In *Conférence Du Palmier à Huile (20 et 21 Octobre 1937)*, 81. Marseille: Institut Colonial, 1937.

“The Dahomey Oil Palm.” *Paint, Oil and Drug Review* 49, no. 4 (26 January 1910): 28.

Dallinger, Jonas. “Oil Palm Development in Thailand: Economic, Social and Environmental Considerations.” In *Oil Palm Expansion in Southeast Asia: Trends and Experiences of Local Communities and Indigenous Peoples*, edited by Sophie M. H. Chao and Marcus Colchester, 24–51. Moreton-in-Marsh, UK, and Bogor, Indonesia: Forest Peoples’ Programme and Sawit Watch, 2011.

Dalrymple-Smith, Angus. “A Comparative History of Commercial Transition in Three West African Slave Trading Economies, 1630 to 1860.” PhD diss., University of Wageningen, 2017.

Dalrymple-Smith, Angus, and Ewout Frankema. “Slave Ship Provisioning in the Long 18th Century. A Boost to West African Commercial Agriculture?” *European Review of Economic History* 21, no. 2 (May, 2017): 185–235.

Dalzel, Archibald. *The History of Dahomy, an Inland Kingdom of Africa; Compiled from Authentic Memoirs; with an Introduction and Notes*. London: Spilsbury and Son, 1793.

Dampier, William. *A Voyage to New Holland &c.* Vol. 3. London: James Knapton, 1703.

D’Andrea, A. Catherine, Amanda L. Logan, and Derek J. Watson. “Oil Palm and Prehistoric Subsistence in Tropical West Africa.” *Journal of African Archaeology* 4, no. 2 (2006): 195–222.

Daniel, Jean. *Le palmier à huile au Dahomey.* Paris: Challamel, 1902.
Dauvergne, Peter. *Environmentalism of the Rich.* Cambridge, MA: MIT Press, 2016.
——. "The Global Politics of the Business of 'Sustainable' Palm Oil." *Global Environmental Politics* 18, no. 2 (May 2018): 34–52.
——. *Shadows in the Forest: Japan and the Politics of Timber in Southeast Asia.* Cambridge, MA: MIT Press, 1997.
Dauvergne, Peter, and Jane Lister. *Eco-Business: A Big-Brand Takeover of Sustainability.* Cambridge, MA: MIT Press, 2014.
Dauvergne, Peter, and Kate J. Neville. "The Changing North-South and South-South Political Economy of Biofuels." *Third World Quarterly* 30, no. 6 (September 2009): 1087–1102.
Davids, Mila. "Technology as the New Frontier: Unilever and the Rise of Becel Margarine." *Journal of Modern European History* 14, no. 1 (February 2016): 101–18.
Davies, Peter N. *The Trade Makers; Elder Dempster in West Africa, 1852–1972.* London: Allen & Unwin, 1973.
Dawson, Kevin. *Undercurrents of Power: Aquatic Culture in the African Diaspora.* Philadelphia: University of Pennsylvania Press, 2018.
Dean, Warren. *Brazil and the Struggle for Rubber: A Study in Environmental History.* Cambridge: Cambridge University Press, 2002.
De Bruijn Kops, G. F. "Notulen van de een en twintigste vergadering der Nederlandsch-Indische Maatschappij van Nijverheid." *Tijdschrift voor nijverheid in Nederlandsch Indië* 3 (1856): 384–90.
DeCicco, John M., Danielle Yuqiao Liu, Joonghyeok Heo, Rashmi Krishnan, Angelika Kurthen, and Louise Wang. "Carbon Balance Effects of U.S. Biofuel Production and Use." *Climatic Change* 138, no. 3 (October, 2016): 667–80.
De Koninck, Rodolphe. "La Paysannerie Comme Fer de Lance Territorial de L'Etat: Le Cas de La Malaysia." *Cahiers Des Sciences Humaines* 22, no. 3–4 (1986): 355–70.
De Koninck, Rodolphe, Stéphane Bernard, and Jean-François Bissonnette. "Agricultural Expansion: Focusing on Borneo." In *Borneo Transformed*, 10–43. Singapore: National University of Singapore Press, 2011.
Delaney, Martin. *Official Report of the Niger Valley Exploring Party.* New York: T. Hamilton, 1861.
Demeur, John. "To John Demeur . . . [Patent No. 6256, 13 April 1832]." *London Journal and Repertory of Arts, Sciences, and Manufactuers*, no. 61 (1837): 30–32.
Dentan, Robert Knox. *The Semai: A Nonviolent People of Malaya.* New York: Holt, Rinehart and Winston, 1968.
"Design and Construction of Railway Carriages and Wagons." *The Railway Engineer* 14 (September 1893): 283–88.
Despatches Relating to the Sierra Leone Oil Palm Industry and the Establishment of Oil Palm Plantations. Freetown, Sierra Leone: Government Printer, 1925.
Dewi, Oetami. "Reconciling Development, Conservation, and Social Justice in West Kalimantan." In *The Palm Oil Controversy in Southeast Asia: A Transnational Perspective*, edited by Oliver Pye and Jayati Bhattacharya, 164–78. Singapore: ISEAS Publishing, 2015.

De Wildeman, É. *Remarques a Propos de La Foret Équatoriale Congolaise*. Brussels: Librarie Falk fils, 1934.

De Witte, Ludo. "The Suppression of the Congo Rebellions and the Rise of Mobutu, 1963–65." *International History Review* 39, no. 1 (January 2017): 107–25.

Díaz, María. "Palma Africana En La Costa Caribe: Un Semillero de Empresas Solidarias." Centro de Estudios Economicos Regionales. Cartagena de Indias: Banco de la República de Colombia, 2002.

Dickens, Charles, ed. "Our Oil-Flasks." *All the Year Round* (London) 9, no. 211 (May 9, 1863): 260–64.

Diet, Nutrition and the Prevention of Chronic Disease. WHO Technical Report Series 916. Geneva: WHO, 2003.

Dijk, Han van, Neree Awana Onguene, and Thomas W. Kuyper. "Knowledge and Utilization of Edible Mushrooms by Local Populations of the Rain Forest of South Cameroon." *Ambio* 32, no. 1 (2003): 19–23.

Dijkstra, Albert. "How Chevreul (1786–1889) Based His Conclusions on His Analytical Results." *OCL Oléagineux, Corps Gras, Lipides* 16 (2009): 8–13.

"'Dika Nuts' from Southern Nigeria." *Bulletin of the Imperial Institute* 4 (1906): 19–22.

Dike, K. Onwuka. *Trade and Politics in the Niger Delta: 1830–1885*. Oxford: Oxford University Press, 1956.

Dike, K. Onwuka, and Felicia Ifeoma Ekejiuba. *The Aro of South-Eastern Nigeria, 1650–1980: A Study of Socio-Economic Formation and Transformation in Nigeria*. Ibadan, Nigeria: Ibadan University Press, 1990.

Dillon, Maureen. *Artificial Sunshine: A Social History of Domestic Lighting*. London: National Trust, 2002.

Dimashqi, Shams al-Din Muhammad ibn Abi Talib. *Manuel de la cosmographie du moyen âge*. Copenhagen: C. A. Reitzel, 1874.

Diodorus Siculus. *Library of History*. Edited by Bill Thayer. http://penelope.uchicago.edu/Thayer/E/Roman/Texts/Diodorus_Siculus/home.html.

"Diplomacy and Commerce No. IV." Glasgow, 1840. https://www.jstor.org/stable/60209095.

"Diseases of the Oil Palm in West Africa." *Federated Malay States Agricultural Bulletin* 8, no. 4 (December 1920): 244–46.

Dissou, Machioudi. "Développement et mise en valeur des plantations de palmier à huile au Dahomey." *Cahiers d'Études africaines* 12, no. 47 (1972): 485–99.

———. *Économie de la culture du palmier à huile au Bénin et en Côte d'Ivoire: approche comparative des politiques agricoles en Afrique*. Lomé: les Nouvelles éd. africaines, 1988.

Dolmat, Mohd Tayeb. *Technologies for Planting Oil Palm on Peat*. Kuala Lumpur: Malaysian Palm Oil Board, 2005.

Donelha, André. *An Account of Sierra Leone and the Rivers of Guinea of Cape Verde (1625)*. Lisbon: Junta de Investigacoes Cientificas do Ultramar, 1977.

Dop, Henk, and Phillip Robinson. *Travel Sketches from Liberia: Johann Büttikofer's 19th Century Rainforest Explorations in West Africa*. Leiden: Brill, 2012.

Dove, Michael. *The Banana Tree at the Gate: A History of Marginal Peoples and Global Markets in Borneo*. New Haven, CT: Yale University Press, 2011.
Downs, Ian. *The Australian Trusteeship: Papua New Guinea 1945–75*. Canberra: Australian Government Publishing Service, 1980.
Drabble, John. *An Economic History of Malaysia, c. 1800–1990: The Transition to Modern Economic Growth*. Basingstoke, UK: Palgrave Macmillan, 2000.
———. *Rubber in Malaya 1876–1922: The Genesis of the Industry*. Kuala Lumpur: Oxford University Press, 1973.
Draschouossoff, V. "Agricultural Change in the Belgian Congo: 1945–1960." *Food Research Institute Studies* 5, no. 2 (1965): 137–200.
Drescher, Seymour. *Econocide: British Slavery in the Era of Abolition*. 2nd ed. Chapel Hill, NC: University of North Carolina Press, 2010.
Duchesne, C. "Du Droit Des Indigènes Sur Les Palmeraies Naturelles." In *Institut Colonial de Marseille. Mémoires et Rapports Sur Les Matières Grasses. T. III. Le Palmier à Huile*. Marseille: Institut colonial, 1928.
Dudgeon, Gerald Cecil. *The Agricultural and Forest Products of British West Africa*. London: J. Murray, 1911.
Due, Jean M. "Agricultural Development in the Ivory Coast and Ghana." *Journal of Modern African Studies* 7, no. 4 (1969): 637–60.
Dumont, René. *Afrique Noire Développement Agricole. Reconversion de l'économie Agricole: Guinee, Côte d'Ivoire, Mali*. Paris: Presses Universitaires des France, 1961.
Dunbar, Donald Earl. *The Tin-Plate Industry: A Comparative Study of Its Growth in the United States and in Wales*. New York: Houghton Mifflin, 1915.
Dyke, Frederick Montague. *Report on the Oil Palm Industry in British West Africa*. Lagos: Government Printer, 1927.
Eaton, B. J. "Annual Report of the Chemical Division for 1924." *Malayan Agricultural Journal* 13, no. 7 (1925): 179–88.
———. "A New Fuel." *Federated Malay States Agricultural Bulletin* 9, no. 2 (1921): 105–6.
Eckey, E. W. *Vegetable Fats and Oils*. New York: Reinhold Publishing, 1954.
Eddy, Walter H. "Something New in Foods." *Good Housekeeping* 92, no. 5 (May 1931): 1902.
Edelman, Marc, and Andrés León. "Cycles of Land Grabbing in Central America: An Argument for History and a Case Study in the Bajo Aguán, Honduras." *Third World Quarterly* 34, no. 9 (2013): 1697–1722.
Ekeh, Peter Palmer. *History of the Urhobo People of Niger Delta*. Buffalo, NY: Urhobo Historical Society, 2007.
Ekundare, R. Olufemi. *An Economic History of Nigeria, 1860–1960*. London: Methuen, 1973.
Energy Post. "Biofuels Turn Out to Be a Climate Mistake," November 15, 2016. https://energypost.eu/biofuels-turn-climate-mistake/.
Engerman, Stanley L. "Servants to Slaves to Servants: Contact Labour and European Expansion." In *Colonialism and Migration*, edited by P. C. Emmer, 263–94. Dordrecht: Martinus Nijhoff Publishers, 1986.

Equiano, Olaudah. *The Interesting Narrative of the Life of Olaudah Equiano; or Gustavus Vassa, the African*. London: Penryn, 1815.
E. R. *The Experienced Farrier, or, Farring Compleated in Two Books Physical and Chyrurgical. . . .* London: Richard Northcott, 1681.
Ergo, André-Bernard. "Histoire de l'Elaéiculture au Congo Belge." unpublished manuscript, n.d. http://abergo1.e-monsite.com/medias/files/palmhisco-5.pdf.
"Escape From Tropical Fats." *Soybean Digest* (June/July 1987): 2.
Escobar, Arturo. *Territories of Difference: Place, Movements, Life, Redes*. Durham, NC: Duke University Press, 2008.
Escobar, R., and F. Peralta. "The Oil Palm Industry in Costa Rica: Country Report." *ASD Oil Palm Papers (Costa Rica)* 31 (2007): 17–20.
Evans, Sterling. *Bound in Twine: The History and Ecology of the Henequen-Wheat Complex for Mexico and the American and Canadian Plains, 1880–1950*. College Station: Texas A&M University Press, 2007.
"Extracts from the Records of Surrey County." *William and Mary Quarterly* 11, no. 2 (October 1902): 79.
Fairhead, James, Tim Geysbeek, Svend E. Holsoe, and Melissa Leach, eds. *African-American Exploration in West Africa: Four Nineteenth-Century Diaries*. Bloomington: Indiana University Press, 2003.
Fairhead, James, and Melissa Leach. *Misreading the African Landscape: Society and Ecology in a Forest-Savanna Mosaic*. Cambridge: Cambridge University Press, 1996.
——. *Reframing Deforestation: Global Analyses and Local Realities: Studies in West Africa*. London: Routledge, 1998.
Fallon, Amy. "When People Power Meets Litigation: A History-Making Success Story in Sierra Leone." MobLab, May 9, 2019. https://mobilisationlab.org/stories/how-namati-equipped-sierra-leone-community-to-win-land-back/.
Falola, Toyin. *Colonialism and Violence in Nigeria*. Bloomington: Indiana University Press, 2009.
——. *Economic Reforms and Modernization in Nigeria, 1945–1965*. Kent, OH: Kent State University Press, 2004.
——. *The Political Economy of a Pre-Colonial African State: Ibadan, 1830–1900*. Ile-Ife, Nigeria: University of Ife Press, 1984.
——. "The Yoruba Toll System: Its Operation and Abolition." *Journal of African History* 30, no. 1 (January 1989): 69–88.
FAO. "First African Small-Scale Palm Oil Processing Workshop." Rome: FAO, 1981.
Farquhar, J. H. J. *The Oil Palm and Its Varieties*. Edited by H. N. Thompson. London: Waterlow and Sons, 1912.
Fauconnier, Henri. "Essais de Culture Du Palmier à Huile En Extrême-Orient." *Bulletin des matières grasses* 1 (1920): 20–24.
——. *Malaisie*. London: Macmillan, 1931.
Faulkner, O. T., and J. R. Mackie. *West African Agriculture*. Cambridge: Cambridge University Press, 1933.
Fearnside, Philip M. "Transmigration in Indonesia: Lessons from Its Environmental and Social Impacts." *Environmental Management* 21, no. 4 (July, 1997): 553–70.

Feintrenie, Laurène. "Transfer of the Asian Model of Oil Palm Development: From Indonesia to Cameroon." Annual World Bank Conference on Land and Poverty: Land Governance in a Rapidly Changing Environment, Washington, DC, April 23–26, 2012. http://agritrop.cirad.fr/565370/.

Fenske, James. "Imachi Nkwu: Trade and the Commons." *Journal of Economic History* 74, no. 1 (March 2014): 39–68.

——. "Land Abundance and Economic Institutions: Egba Land and Slavery, 1830–1914." *Economic History Review* 65, no. 2 (May, 2012): 527–55.

Fentiman, Alicia. "The Anthropology of Oil: The Impact of the Oil Industry on a Fishing Community in the Niger Delta." *Social Justice* 23, no. 4 (1996): 87–99.

Ferme, Mariane C. *The Underneath of Things: Violence, History, and the Everyday in Sierra Leone*. Berkeley: University of California Press, 2001.

Ferrand, M. "L'Avenir Du Palmier à Huile." In *Mémoires et rapports sur les matières grasses. T. III. Le Palmier à huile*, 217–27. Marseille: Institut Colonial, 1928.

Ferwerda, J. D. "Questions Relevant to Replanting in Oil Palm Cultivation." PhD diss., Wageningen Agricultural University, 1955.

Fickendey, Ernst. *Die Ölpalme an der Ostküste von Sumatra*. Berlin: Kolonialwirtschaftliches Komitee, 1922.

——. "Über Die Verwendungsfähigkeit Des Palmöls Als Speisefett." *Der Tropenpflanzer* 10 (1910): 566–70.

Field, M. J. "The Agricultural System of the Manya-Krobo of the Gold Coast." *Africa: Journal of the International African Institute* 14, no. 2 (1943): 54–65.

Fieldhouse, David Kenneth. *Merchant Capital and Economic Decolonization: The United Africa Company, 1929–1987*. Wotten-under-Edge, UK: Clarendon Press, 1994.

——. *Unilever Overseas: The Anatomy of a Multinational, 1895–1965*. Stanford: Hoover Institution Press, 1978.

Fife, Bruce. *The Shocking Truth about Palm Oil: How It Affects Your Health and the Environment*. Colorado Springs: Piccadilly Books, 2017.

Filer, Colin. "Asian Investment in the Rural Industries of Papua New Guinea: What's New and What's Not?" *Pacific Affairs* 86, no. 2 (2013): 305–25.

——. "Why Green Grabs Don't Work in Papua New Guinea." *Journal of Peasant Studies* 39, no. 2 (April 2012): 599–617.

Fitzherbert, Emily B., Matthew J. Struebig, Alexandra Morel, Finn Danielsen, Carsten A. Brühl, Paul F. Donald, and Ben Phalan. "How Will Oil Palm Expansion Affect Biodiversity?" *Trends in Ecology & Evolution* 23, no. 10 (October, 2008): 538–45.

Fleming, T. "The Company View." In *The Hoskins Development*, 7–20. New Guinea Research Bulletin, no. 49. Port Moresby: Australian National University, 1972.

Flenley, John, and Mark Bush. *Tropical Rainforest Responses to Climatic Change*. Berlin: Springer, 2007.

Flint, John. *Sir George Goldie and the Making of Nigeria*. London: Oxford University Press, 1960.

Flower, Philip William. *A History of the Trade in Tin: A Short Description of Tin Mining and Metallurgy*. . . . London: G. Bell and Sons, 1880.

Floyd, Barry. *Eastern Nigeria: A Geographical Review*. New York: Praeger, 1969.
Fochtman, E. G. L. C. Kinney, G. G. Ference, R. W. Riemenschneider, S. G. Morris, and W. C. Ault. *Animal Fats in Hot Dip Tinning*. Philadelphia: USDA Bureau of Agricultural and Industrial Chemistry, 1953.
Fold, Niels. "Oiling the Palms: Restructuring of Settlement Schemes in Malaysia and the New International Trade Regulations." *World Development* 28, no. 3 (March 2000): 473–86.
Fomento del Cultivo de la Palma Africana en el Valle del Aguan. Tegucigalpa, Honduras: BID-IICA-CONSUPLAN, 1972.
Fontenay, Catherine C. de. "The Dual Role of Market Power in the Big Push: From Evidence to Theory." *Journal of Development Economics* 75, no. 1 (2004): 221–38.
Forbes, Frederick Edwyn. *Dahomey and the Dahomans, Being the Journals of Two Missions to the King of Dahomey, and Residence at His Capital in the Years 1849 and 1850*. London: A. and W. Galignani, 1851.
Forbes, Robert James. *More Studies in Early Petroleum History*. Leiden: Brill, 1959.
Forde, C. Daryll. "Land and Labour in a Cross River Village, Southern Nigeria." *Geographical Journal* 90, no. 1 (1937): 24–47.
——. "The Rural Economies." In *The Native Economies of Nigeria*, edited by Margery Perham, 29–216. London: Faber & Faber, 1946.
Forero Rueda, Sebastián. "La Lucha de Los Trabajadores de Indupalma, Un Conflicto de Vieja Data." *El Espectador*, October 14, 2018. https://www.elespectador.com/noticias/nacional/la-lucha-de-los-trabajadores-de-indupalma-un-conflicto-de-vieja-data-articulo-817928.
Forty-Fourth Annual Report on Alkali, &c. Works. P.P. 1908 (182) xi 247. London: HMSO, 1908.
Fouqet, Roger, and Peter J. G. Pearson. "Seven Centuries of Energy Services: The Price and Use of Light in the United Kingdom (1300–2000)." *Energy Journal* 27, no. 1 (2006): 139–77.
Fouque, C. *Histoire raissonée du commerce de Marseille: appliquée aux développements des prospérités modernes*. Paris: Chez Roret, 1843.
Fox, Renee C., Willy de Craemer, and Jean-Marie Ribeaucourt. "'The Second Independence': A Case Study of the Kwilu Rebellion in the Congo." *Comparative Studies in Society and History* 8, no. 1 (1965): 78–109.
Frankema, Ewout. "Colonial Taxation and Government Spending in British Africa, 1880–1940: Maximizing Revenue or Minimizing Effort?" *Explorations in Economic History* 48, no. 1 (2011): 136–49.
——. "Raising Revenue in the British Empire, 1870–1940: How 'Extractive' Were Colonial Taxes?" *Journal of Global History* 5, no. 3 (2010): 447–77.
Frankema, Ewout, and M. F. M. van Waijenburg. "Metropolitan Blueprints of Colonial Taxation? Lessons from Fiscal Capacity Building in British and French Africa, c. 1880–1940." *Journal of African History* 55, no. 3 (2014): 371–400.
Frankema, Ewout, Jeffrey Williamson, and Pieter Woltjer. "An Economic Rationale for the West African Scramble? The Commercial Transition and the

Commodity Price Boom of 1835–1885." *Journal of Economic History* 78, no.1 (2018): 231–67.
Fraser, A. C., ed. "Candlemaking and Christianity." *North British Review* 20, no. 39 (November 1853): 161–78.
Freas, Alan D'Yarmett, Albert Wade Brant, Alice Louise Robert, Arthur Charles Dillman, Coit Alfred Suneson, George Arceneaux, Harold Morrison, et al. *The Demand and Price Structure for Food Fats and Oils: With Emphasis on Analyses Designed to Measure the Effects of Increased Yields of Cottonseed Oil on Prices and Total Returns.* Washington, DC: USDA, 1953.
Friedel, C. "Sur des matières grasses trouvées dans des tombes égyptiennes d'Abydos." *Comptes rendus hebdomadaires des séances de l'Académie des sciences* 124 (1897): 648–53.
Frobenius, Leo. *Volksdichtungen aus Oberguinea.* Jena: Diederichs, 1924.
Froger, François. *A Relation of a Voyage Made in the Years 1695, 1696, 1697, on the Coasts of Africa, Streights of Magellan, Brasil, Cayenna, and the Antilles. . . .* London: M. Gillyflower et al., 1698.
FUNDESVIC. *. . . Y Empezó Nuestro Sueño.* Buscaramanga, Colombia: FUNDESVIC, Asociacion Minga, and Sintraproaceites, 2011. https://issuu.com/asominga/docs/y_empezo_nuestro_sue__o.
——. *De Siervos a Obreros.* Buscaramanga: FUNDESVIC, Asociacion Minga, and Sintraproaceites, 2013. https://issuu.com/asominga/docs/de_siervos_a_obreros.
Further Papers Relating to the Slave Trade; Viz. Communications to the Admiralty, and Instructions to Naval Officers; Since the 6th of February 1821. P.P. 1822 (223) iv 7. London: HMSO, 1822.
Fyfe, Christopher. "Charles Heddle: An African 'Merchant Prince.'" In *Enterprises et Entrepreneurs En Afrique, 19–20 Siècles*, 235–47. Paris: L'Harmattan, 1983.
——. *A History of Sierra Leone.* Oxford: Oxford University Press, 1962.
Gallegos, Danielle. "Palm Oil Tensions." *New Formations* 74, no. 74 (2011): 18–32.
Gandhi, M. K. *Diet and Diet Reform.* Edited by Bharatan Kumarappa. Ahemadabad, India: Navajivan Publishing House, 1949.
García Arboleda, Juan Felipe. *El exterminio de la Isla de Papayal: etnografías sobre el Estado y la construcción de paz en Colombia.* Bogotá, Colombia: Editorial Pontificia Universidad Javeriana, 2019.
García, Celsa Quinteiro, and Édgar Benítez. "The History of Research on Oil Palm Bud Rot (*Elaeis guineensis* Jacq.) in Colombia." *Agronomía Colombiana* 32, no. 3 (2014): 390–98.
Gardner, Leigh A. *Taxing Colonial Africa: The Political Economy of British Imperialism.* Oxford: Oxford University Press, 2012.
Gascon, J. P., J. M. Noiret, and J. Meunier. "Oil Palm." In *Oil Crops of the World*, edited by Gerhard Röbbelen, R. Keith Downey, and Amram Ashri, 475–93. New York: McGraw-Hill, 1989.
Gaskell, Joanne Catherine. "The Palm Oil Revolution in Asia." PhD diss., Stanford University, 2012.
Gater, B. A. R. "Insects on African Oil-Palms." *Malayan Agricultural Journal* 13, no. 8 (August 1925): 250–56.

———. “Les Insectes et le Palmier à Huile.” In *Mémoires et rapports sur les matières grasses. T. III. Le Palmier à huile*, 193–98. Marseilles: Institut colonial de Marseille, 1928.

Gatti, Roberto Cazzolla, and Alena Velichevskaya. “Certified ‘Sustainable’ Palm Oil Took the Place of Endangered Bornean and Sumatran Large Mammals Habitat and Tropical Forests in the Last 30 Years.” *Science of the Total Environment* 742 (2020): 140712.

Gatti, Roberto Cazzolla, Jingjing Liang, Alena Velichevskaya, and Mo Zhou. “Sustainable Palm Oil May Not Be So Sustainable.” *Science of the Total Environment* 652 (2019): 48–51.

Gatto, Marcel, Meike Wollni, and Matin Qaim. “Oil Palm Boom and Land-Use Dynamics in Indonesia: The Role of Policies and Socioeconomic Factors.” *Land Use Policy* 46 (July, 2015): 292–303.

Gellert, Paul. “Palm Oil Expansion in Indonesia: Land Grabbing as Accumulation by Dispossession.” *Current Perspectives in Social Theory* 34 (2015): 65–99.

Georgi, C. D. V. “A Comparison of the Press and Centrifugal Methods for Treatment of Oil Palm Fruit.” *Malayan Agricultural Journal* 21, no. 3 (1933): 103–18.

———. “Malayan Palm Kernels.” *Malayan Agricultural Journal* 20, no. 2 (1932): 53–60.

———. “The Removal of Plant Nutrients in Oil Palm Cultivation.” *Malayan Agricultural Journal* 19, no. 10 (October 1931): 484–89.

Georgia Writers’ Project—Savannah Unit. *Drums and Shadows: Survival Studies among the Georgia Coastal Negroes*. Edited by Charles Joyner. Athens, GA: University of Georgia Press, 1986.

Gertzel, Cherry J. “Commercial Organization on the Niger Coast, 1852–1891.” In *Historians in Tropical Africa*, 289–304. Salisbury: University College of Rhodesia and Nyasaland, 1962.

Getz, Trevor R. *Slavery and Reform in West Africa: Toward Emancipation in Nineteenth-Century Senegal and the Gold Coast*. Athens, OH: Ohio University Press, 2004.

Ghesquière, J. “L’Elaeis guineensis Jacq. est-il africain ou américain? (suite à la Note de M. le Pr Aug. Chevalier).” *Journal d’agriculture traditionnelle et de botanique appliquée* 14, no. 153 (1934): 340–43.

Giacomin, Valeria. “The Emergence of an Export Cluster: Traders and Palm Oil in Early Twentieth-Century Southeast Asia.” *Enterprise & Society* 19, no. 2 (June 2018): 272–308.

———. “The Transformation of the Global Palm Oil Cluster: Dynamics of Cluster Competition between Africa and Southeast Asia (c. 1900–1970).” *Journal of Global History* 13, no. 3 (November 2018): 374–98.

Gibbins, E. A. *Britain’s Railways: The Reality*. Stoke on Trent: Leisure Products, 2003.

Giles-Vernick, Tamara. *Cutting the Vines of the Past: Environmental Histories of the Central African Rain Forest*. Charlottesville: University of Virginia Press, 2002.

Girshick, Paula Ben-Amos, and John Thornton. "Civil War in the Kingdom of Benin, 1689–1721: Continuity or Political Change?" *Journal of African History* 42, no. 3 (2001): 353–76.
Gittins, L. "Soapmaking in Britain, 1824–1851: A Study in Industrial Location." *Journal of Historical Geography* 8, no. 1 (January 1982): 29–40.
Glickman, Lawrence. *Buying Power: A History of Consumer Activism in America.* Chicago: University of Chicago Press, 2009.
Gold Coast Agricultural Development Corporation. *Mensah the Oil Palm Farmer.* Accra, Ghana: Government Printer, 1956.
Gomes, Alberto. *Looking for Money: Capitalism and Modernity in an Orang Asli Village*. Melbourne: Trans-Pacific Press, 2004.
Gordon, Alec. "Contract Labour in Rubber Plantations: Impact of Smallholders in Colonial South-East Asia." *Economic and Political Weekly* 36, no. 10 (2001): 847–60.
———. "Towards a Model of Asian Plantation Systems." *Journal of Contemporary Asia* 31, no. 3 (2001): 306–30.
Goss, Andrew. *The Floracrats: State-Sponsored Science and the Failure of the Enlightenment in Indonesia*. Madison: University of Wisconsin Press, 2011.
Goucher, Candice. *Congotay! Congotay! A Global History of Caribbean Food.* Armonk, NY: M. E. Sharpe, 2014.
Grace, John. *Domestic Slavery in West Africa: With Particular Reference to the Sierra Leone Protectorate, 1896–1927.* New York: Barnes & Noble Books, 1975.
Graham, Edgar, Ingrid Floering, and David Fieldhouse. *The Modern Plantation in the Third World*. London: Croom Helm, 1984.
Grajales, Jacobo. "The Rifle and the Title: Paramilitary Violence, Land Grab and Land Control in Colombia." *Journal of Peasant Studies* 38, no. 4 (October 2011): 771–92.
———. "State Involvement, Land Grabbing and Counter-Insurgency in Colombia." *Development and Change* 44, no. 2 (March 2013): 211–32.
Grandin, Greg. *Fordlandia: The Rise and Fall of Henry Ford's Forgotten Jungle City.* London: Icon, 2010.
Green, M. M. *Ibo Village Affairs*. London: A. Brown and Sons, 1947.
Greeney, Spring. "When Laundry Detergent Was Edible." *Edge Effects*, 2018. https://edgeeffects.net/laundry-soap-history-edible-tide-pods/.
Grieve, Robert B. "The Oil Palm Industry of Papua New Guinea." *Australian Geographer* 17, no. 1 (1986): 72–76.
Grist, D. H. "Agricultural Education." *Federated Malay States Agricultural Bulletin* 9, no. 2 (June 1921): 122–30.
Grivot, R. *Réactions Dahoméennes*. Paris: Editions Berger-Levrault, 1954.
Grossman, Daniel. "Deep in the Jungle, Scientists Explore the Links between the Congo and Climate Change." *Ensia* (blog), November 3, 2017. https://ensia.com/features/congo-rainforest/.
Gruca, Marta, Tinde R. van Andel, and Henrik Balslev. "Ritual Uses of Palms in Traditional Medicine in Sub-Saharan Africa: A Review." *Journal of Ethnobiology and Ethnomedicine* 10, no. 60 (July 2014).

Guereña, Arantxa, and Ricardo Zepeda. "The Power of Oil Palm: Land Grabbing and Impacts Associated with the Expansion of Oil Palm Crops in Guatemala: The Case of the Palmas Del Ixcán Company." Oxfam America, 2013. https://www.oxfamamerica.org/explore/research-publications/power-of-oil-palm-guatemala/.

Gueye, M'baye, and A. Adu Boahen. "African Initiatives and Resistance in West Africa, 1880–1914." In *UNESCO General History of Africa*, edited by A. Adu Boahen, vol. 7, 114–48. Berkeley: University of California Press, 1985.

Guézo, Anselme. "The Impact of British Abolition on the States in the Hinterland of the Slave Coast." *Journal of the Historical Society of Nigeria* 19 (2010): 132–56.

Guindon, Michael, and Annabelle Dodson. "Palm Oil Assessment Summary—November 2018." SPOTT.org, 2018. https://www.spott.org/palm-oil-summary/.

"Guinée Française." In *Conférence Du Palmier à Huile (20 et 21 Octobre 1937)*, 69. Marseille: Institut Colonial, 1937.

Gunggut, Haijon, Dg Siti Noor Saufidah Ag Mohd Saufi, Zuraidah Zaaba, and May Siaw-Mei Liu. "Where Have All the Forests Gone? Deforestation in Land below the Wind." *Procedia—Social and Behavioral Sciences*, AMER International Conference on Quality of Life, 153 (2014): 363–69.

Gyasi, Edwin A. "Emergence of a New Oil Palm Belt in Ghana." *Tijdschrift Voor Economische En Sociale Geografie* 83, no. 1 (1992): 39–49.

———. "The RISONPALM Nucleus Estate of Nigeria: An Agricultural Plantation and Its Effect on the Surrounding Area." *Malaysian Journal of Tropical Geography* 15 (1987): 26–38.

Hailey, William M. *An African Survey: A Study of Problems Arising in Africa South of the Sahara*. London: Oxford University Press, 1938.

Hall, Derek, Philip Hirsch, and Tania Murray Li. *Powers of Exclusion: Land Dilemmas in Southeast Asia*. Singapore: National University of Singapore Press, 2011.

Hall, Gwendolyn Midlo. *Slavery and African Ethnicities in the Americas: Restoring the Links*. Chapel Hill: University of North Carolina Press, 2005.

Hall, James. "To Emigrants." *Maryland Colonization Journal*, new series, vol. 1, no. 3 (August 1841): 40–46

Hallet, Adrien. "Les Plantations d'Elaeis En Malaisie." *Revue de Botanique Appliquée*, no. 29 (January 1924): 46–47.

———. "Note Sur Le Palmier à Huile." *Bulletin de l'Association Des Planteurs de Caoutchouc* 5, no. 11 (November 1913): 279.

Hallett, Robin. *The Niger Journal of Richard and John Lander*. London: Routledge, 1965.

Hambloch, Caroline. "Governance and Land Reform in the Palm Oil Value Chain in the Philippines." *SOAS Department of Economics Working Paper Series* 209, no. 209 (2018).

Hamilton, James. *Excelsior: Helps to Progress in Religion, Science, and Literature*. London: James Nisbet and Co., 1856.

Hammond, George B. "The Manufacture of Tin-Plates." *Journal of the Iron and Steel Institute* 52, no. 2 (1897): 24–39.

Hancock, Sir Keith. *Survey of British Commonwealth Affairs*. Vol. 2, pt. 1. London: Oxford University Press, 1940.

Hand, Jane. "Marketing Health Education: Advertising Margarine and Visualising Health in Britain from 1964–c. 2000." *Contemporary British History* 31, no. 4 (October 2, 2017): 477–500.

Handler, Jerome S. "Slave Medicine and Obeah in Barbados, circa 1650 to 1834." *New West Indian Guide* 74, no. 1–2 (2000): 57–90.

Handler, Jerome, and Diane Wallman. "Production Activities in the Household Economies of Plantation Slaves: Barbados and Martinique, Mid-1600s to Mid-1800s." *International Journal of Historical Archaeology* 18, no. 3 (2014): 441–66.

Handover, Will P. "The Eradication of Lalang." *The Planter* 1, no. 1 (August 1920): 31–32.

Haraway, Donna. "Anthropocene, Capitalocene, Plantationocene, Chthulucene: Making Kin." *Environmental Humanities* 6 (2015): 159–65.

Harding, L. "Hamburg's West African Trade in the 19th Century." In *Figuring the African Trade*, edited by G. Liesegang, H. Pasch, and A. Jones, 363–91. Berlin: Dietrich Reimer Verlag, 1986.

Hardgrove, Anne. "The Politics of Ghee Adulteration and Its Public Resolution in Calcutta, c. 1917." In *Commodifying Everything: Relationships of the Market*, edited by Susan Strasser, 191–213. New York: Routledge, 2003.

Hardjono, J. M. *Transmigration in Indonesia*. Kuala Lumpur: Oxford University Press, 1977.

Hardon, J., and R. L. Thomas. "Breeding and Selection of Oil Palm in Malaya." *Oléagineux* 23, no. 2 (1968): 85–90.

Hargreaves, Susan M. "The Political Economy of Nineteenth Century Bonny: A Study of Power, Authority, Legitimacy and Ideology in a Delta Trading Community from 1790 to 1914." PhD diss., University of Birmingham, 1987.

Harlan, Jack R., Jan J. M. de Wet, and Ann B. Stemler. "Plant Domestication and Indigenous African Agriculture." In *Origins of African Plant Domestication*, 3–22. The Hague: Mouton Publishers, 1976.

Harms, Robert W. *River of Wealth, River of Sorry: The Central Zaire Basin in the Era of the Slave and Ivory Trade, 1500–1891*. New Haven, CT: Yale University Press, 1981.

Harper, T. N. *The End of Empire and the Making of Malaya*. Cambridge: Cambridge University Press, 1999.

———. "The Politics of the Forest in Colonial Malaya." *Modern Asian Studies* 31, no. 1 (1997): 1–29.

Harris, David R. "Traditional Systems of Plant Food Production and the Origins of Agriculture in West Africa." In *Origins of African Plant Domestication*, edited by Jack R. Harlan, Jan J. M. de Wet, and Ann B. Stemler, 311–56. The Hague: Mouton Publishers, 1976.

Hartley, C. W. S. "The Improvement of the West African Natural Palm Groves." In *Conférence Franco-Britannique*, 67–75. Paris: Centre Technique d'Agriculture Tropicale et IRHO, 1957.

——. *The Oil Palm (Elaeis guineensis Jacq.).* London: Longmans, 1967.
——. *The Oil Palm (Elaeis guineensis Jacq.).* 2nd ed. London: Longmans, 1977.
——. *The Oil Palm.* 3rd ed. New York: Longman, 1988.
Harvey, F. "Chopping Oil in West Africa." *Littell's Living Age,* August 8, 1896.
Harvey, Gideon. *The Family-Physician, and the House-Apothecary. . . .* London: T. Rookes, 1678.
Hatton, Joshua, ed. "Price's Patent Candle Company." *New Monthly Magazine* 120, no. 718 (1881): 402–14.
Havik, Philip J. "Hybridising Medicine: Illness, Healing and the Dynamics of Reciprocal Exchange on the Upper Guinea Coast (West Africa)." *Medical History* 60, no. 2 (April 2016): 181–205.
Hayati, A., R. Wickneswari, I. Maizura, and N. Rajanaidu. "Genetic Diversity of Oil Palm (*Elaeis guineensis* Jacq.) Germplasm Collections from Africa: Implications for Improvement and Conservation of Genetic Resources." *Theoretical and Applied Genetics* 108, no. 7 (May 2004): 1274–84.
Hayes, K. C., and Andrzej Pronczuk. "Replacing Trans Fat: The Argument for Palm Oil with a Cautionary Note on Interesterification." *Journal of the American College of Nutrition* 29, no. 3 suppl (June 2010): 253S–84S.
Headrick, Daniel R. "Botany, Chemistry, and Tropical Development." *Journal of World History* 7, no. 1 (1996): 1–20.
Heap, Simon. "Those That Are Cooking the Gins: The Business of Ogogoro in Nigeria during the 1930s." *Contemporary Drug Problems* 35 (2008): 573.
"Hearing on Tropical Fats." *Journal of the American Oil Chemists' Society* 64, no. 11 (1987): 1506.
Helleiner, Gerald K. "The Fiscal Role of Marketing Boards in Nigerian Economic Development, 1947–1961." In *Growth and Development of the Nigerian Economy,* edited by Carl Eicher and Carl Liedholm, 119–55. Lansing: Michigan State University Press, 1970.
——. *Peasant Agriculture, Government, and Economic Growth in Nigeria.* Homewood, IL: Richard D. Irwin, 1966.
Hellermann, Pauline von. "The Chief, the Youth and the Plantation: Communal Politics in Southern Nigeria." *Journal of Modern African Studies* 48, no. 2 (June 2010): 259–83.
——. "Reading Farm and Forest: Colonial Forest Science and Policy in Southern Nigeria." In *Engaging Colonial Knowledge: Reading European Archives in World History,* edited by Ricardo Roque and Kim A. Wagner, 89–114. London: Palgrave Macmillan, 2012.
Henderson, L. "The African Oil Palm in Malaya: An Attractive Alternative to Rubber and Coconuts." *The Planter* 1, no. 9–12 (July 1921).
Henderson, Richard N. *The King in Every Man: Evolutionary Trends in Onitsha Ibo Society and Culture.* New Haven, CT: Yale University Press, 1972.
Henriet, Benoît. "'Elusive Natives': Escaping Colonial Control in the Leverville Oil Palm Concession, Belgian Congo, 1923–1941." *Canadian Journal of African Studies* 49, no. 2 (May 2015): 339–61.

Henriques, Sofia Teives, and Paul Warde. "Fuelling the English Breakfast: Hidden Energy Flows in the Anglo-Danish Trade 1870–1913." *Regional Environmental Change* 18, no. 4 (April, 2018): 965–77.
Henry, Yves. "Documents Sur Le Palmier à Huile à Sumatra." In *Mémoires et rapports sur les matières grasses. T. III. Le Palmier à huile*, 199–215. Marseille: Institut Colonial, 1928.
———. "L'exploitation Du Palmier á Huile En Afrique Tropicale." In *Matières Premières Africaines*, edited by Yves Henry, 265–371. Paris: Emile Larose, 1918.
Herodotus. *Histories*. Translated by George Rawlinson. New York: D. Appleton & Co., 1860.
———. *Histories*. Translated by G. C. Macaulay. London: MacMillan and Co., 1890.
———. *Histories*. Translated by A. D. Godley. London: William Heinemann, 1920.
Hervas, Anastasia. "Land, Development and Contract Farming on the Guatemalan Oil Palm Frontier." *Journal of Peasant Studies* 46, no. 1 (January 2019): 115–41.
Heywood, Linda M., and John Thornton. *Central Africans, Atlantic Creoles, and the Foundation of the Americas, 1585–1660*. Cambridge: Cambridge University Press, 2007.
Higman, B. W. *Jamaican Food: History, Biology, Culture*. Kingston: University of the West Indies Press, 2008.
Hinds, Allister. "Chiefs and the Making of Industrial Policy in Nigeria." In *Agency and Action in Colonial Africa: Essays for John E. Flint*, edited by Chris Youé and Stapleton, 141–55. Basingstoke: Palgrave, 2001.
———. "Government Policy and the Nigerian Palm Oil Export Industry, 1939–49." *Journal of African History* 38, no. 3 (1997): 459–78.
"A History of Price's Candles." Accessed April 4, 2018. http://www.prices-candles.co.uk/history/historydetail.asp.
Hochschild, Adam. *King Leopold's Ghost: A Story of Greed, Terror, and Heroism in Colonial Africa*. Boston: Houghton Mifflin, 1999.
Hodge, Joseph Morgan. *Triumph of the Expert: Agrarian Doctrines of Development and the Legacies of British Colonialism*. Athens: Ohio University Press, 2007.
Hofstra, Sjoerd. "The Social Significance of Oilpalm in the Life of the Mendi." *Internationales Archiv fur Ethnographie* 34, no. 5–6 (1937): 105–18.
Hogarth, Rana A. *Medicalizing Blackness: Making Racial Difference in the Atlantic World, 1780–1840*. Chapel Hill: University of North Carolina Press, 2017.
Hogendorn, Jan, and Marion Johnson. *The Shell Money of the Slave Trade*. Cambridge: Cambridge University Press, 2003.
Holman, James. *Travels in Madeira, Sierra Leone, Teneriffe, St. Jago, Cape Coast, Fernando Po, Princes Island, Etc. Etc.*, 2d ed. London: G. Routledge, 1840.
Hoover, Charles L. "The Popular Cooking Fat in Austria." In *Daily Consular and Trade Reports*. Washington, DC: GPO, 1913.
Hopkins, A. G. "An Economic History of Lagos, 1880–1914." PhD diss., University of London (SOAS), 1964.
———. *An Economic History of West Africa*. New York: Columbia University Press, 1973.

———. "Economic Imperialism in West Africa: Lagos, 1880–92." *Economic History Review* 21, no. 3 (1968): 580–606.

Horowitz, Roger. *Kosher USA: How Coke Became Kosher and Other Tales of Modern Food*. New York: Columbia University Press, 2016.

Houard, A., L. Castelli, and J. Lavergne. *Contribution à l'étude Du Palmier à Huile En Afrique Occidentale Française. . . .* Paris: Librairie Émile Larose, 1927.

Houben, Vincent, and Julia Seibert. "(Un)Freedom: Colonial Labor Relations in Belgian Congo and the Netherlands Indies Compared." In *Colonial Exploitation and Economic Development*, edited by Ewout Frankema and Frans Buelens, 178–92. London: Routledge, 2013.

Houtart, François. *Agrofuels: Big Profits, Ruined Lives and Ecological Destruction*. Translated by Victoria Bawtree. London: Pluto Press, 2010.

Howard, Martha C. "The Margarine Industry in the United States: Its Development under Legislative Control." PhD diss., Columbia University, 1951.

Hubbard, John Waddington. *The Sobo of the Niger Delta: A Work Dealing with the History and Languages of the People Inhabiting the Sobo (Urhobo) Division, Warri Province, Southern Nigeria, and the Geography of Their Land*. Zaria, Nigeria: Gaskiya Corporation, 1948.

Hubert, Paul. *Le palmier à huile*. Paris: H. Dunod et E. Pinat, 1911.

Huff, W. G. "Entitlements, Destitution, and Emigration in the 1930s Singapore Great Depression." *Economic History Review* 54, no. 2 (2001): 290–323.

Hughes, Griffith. *The Natural History of Barbados*. London, 1750.

Hulme, David. "An Economic Appraisal of the Hoskins Oil Palm Scheme." *Australian Geographer* 15, no. 5 (1983): 330–34.

Human Rights Watch. "A Dirty Investment." Human Rights Watch, 2019. https://www.hrw.org/report/2019/11/25/dirty-investment/european-development-banks-link-abuses-democratic-republic.

Hunger, Friedrich Wilhelm Tobias. *De Oliepalm (Elaeis guineensis): Historisch Onderzoek over Den Oliepalm in Nederlandsch-Indië*. 2nd ed. Leiden: Brill, 1924.

Hunt, K. E. "Raw Materials." In *Margarine: An Economic, Social and Scientific History 1896–1969*, edited by J. H. van Stuyvenberg, 37–82. Liverpool: Liverpool University Press, 1969.

Hupfeld, Fr. "Palm Oil in the German Colonies." *Journal of the Royal Society of Arts* 62, no. 3217 (1914): 762–63.

Hurtado, Mónica, Catherine Pereira-Villa, and Edgar Villa. "Oil Palm Development and Forced Displacement in Colombia: Causal or Spurious?" *Cuadernos de Economía* 36, no. 71 (2017): 441–68.

Hyam, Ronald. *Understanding the British Empire*. Cambridge: Cambridge University Press, 2010.

Hyman, Eric L. "An Economic Analysis of Small-Scale Technologies for Palm Oil Extraction in Central and West Africa." *World Development* 18, no. 3 (March 1990): 455–76.

Ibrahim, Haitham, and William J. Baker. "Medemia Argun—Past, Present, and Future." *Palms* 53, no. 1 (2009): 9–19.

IBRD. "Current Economic Position and Prospects of Nigeria." Washington, DC: IBRD, 1964.
——. "Economic Growth of Nigeria, Vol. IV." Washington, DC: IBRD, 1965.
——. *The Economic Development of Malaya*. Baltimore: Johns Hopkins University Press, 1955.
——. *The Economic Development of Nigeria*. Washington, DC: IBRD, 1955.
Ickowitz, Amy. "Shifting Cultivation and Deforestation in Tropical Africa: Critical Reflections." *Development and Change* 37, no. 3 (2006): 599–626.
Imbua, David Lishilinimle. "Colonial Rule and the Anti-Pawning Campaign in Obudu Division, 1900–1938." *Journal of the Historical Society of Nigeria* 24 (2015): 73–94.
"An Imperial Industry." *Journal of the Royal African Society* 15, no. 60 (1916): 320–34.
"INDONESIA: Palm Oil Production Prospects Dampened by El Niño Drought." Commodity Intelligence Report. Washington, DC: USDA Foreign Agricultural Service, December 15, 2015. https://ipad.fas.usda.gov/highlights/2015/12/Indonesia/Index.htm.
"Indupalma, 50 Años Generando Desarrollo En El Campo Colombiano." *El Palmicultor* 477 (November 2011): 21–22.
Indupalma. "Nuestra historia." http://emindupalma.blogspot.com/p/h.html.
Inikori, Joseph E. "Africa and the Globalization Process: Western Africa, 1450–1850." *Journal of Global History* 2, no. 1 (2007): 63–86.
——. *Africans and the Industrial Revolution in England: A Study in International Trade and Development*. Cambridge: Cambridge University Press, 2002.
——. "West Africa's Seaborne Trade, 1750–1850: Volume, Structure and Implications." In *Figuring African Trade*, edited by G. Liesegang, Adam Jones, and H. Pasch, 49–88. Berlin: D. Reimer, 1986.
International Exhibition of Australia. *Official Record*. Melbourne: Blundell & Co., 1867.
"Investigations in Connection with the African Palm Oil Industry." *Bulletin of the Imperial Institute* 8, no. 4 (1909): 357–94.
Irwin, Emily. "The Spermaceti Candle and the American Whaling Industry." *Historia* 21 (2012): 45–53.
Isichei, Elizabeth Allo. *Igbo Worlds: An Anthology of Oral Histories and Historical Descriptions*. London: Macmillan, 1977.
J. V. *Golgotha, or, A Looking-Glass for London, and the Suburbs Thereof Shewing the Causes, Nature and Efficacy of the Present Plagues. . . .* London, 1665.
Jacks, David, Kevin H. O'Rourke, and Jeffrey G. Williamson. "Commodity Price Volatility and World Market Integration since 1700." *Review of Economics and Statistics* 93, no. 3 (2011): 800–13.
Jackson, James C. *Planters and Speculators: Chinese and European Agricultural Enterprise in Malaya, 1786–1921*. Kuala Lumpur: University of Malaya Press, 1968.
Jacquin, Nikolaus Joseph. *Selectarum stirpium Americanarum historia. . . .* Vienna: Officina Krausiana, 1763.

Jagoe, R. B. "Deli Oil Palms and Early Introductions of *Elaeis guineensis* to Malaya." *Malayan Agricultural Journal* 35 (1952): 3–10.
——. "Observations and Experiments in Connection with Pollination of Oil Palms." *Malayan Agricultural Journal* 22 (1934): 598–606.
Jain, Ravindra K. "Tamilian Labour and Malayan Plantations, 1840–1938." *Economic and Political Weekly* 28, no. 43 (1993): 2363–70.
Jamieson, Robert. *Western Central Africa: A Review of the Measures Adopted for the Suppression of the Slave Trade from Thence, and of Their Effects on the Interests of British Commerece in Africa. . . .* London: Effingham Wilson, 1859.
Jannot, Claude. "Emplois, économie, environnement: le développement de la filière palmier à huile en Côte-d'Ivoire." *OCL. Oléagineux Corps gras Lipides* 17, no. 6 (2010): 393–99.
Janssens, P. "Le Palmier à Huile Au District Du Kasai." *Bulletin Agricole Du Congo Belge* 8 (1917): 218–44.
Jelsma, Idsert, G. C. Schoneveld, Annelies Zoomers, and A. C. M. van Westen. "Unpacking Indonesia's Independent Oil Palm Smallholders: An Actor-Disaggregated Approach to Identifying Environmental and Social Performance Challenges." *Land Use Policy* 69 (2017): 281–97.
Jewsiewicki, Bogumil. "African Peasants in the Totalitarian Colonial Society of the Belgian Congo." In *Peasants in Africa*, edited by Martin Klein, 45–76. Beverly Hills: Sage Publications, 1980.
Jim, C. Y. "The Forest Fires in Indonesia 1997–98: Possible Causes and Pervasive Consequences." *Geography* 84, no. 3 (1999): 251–60.
Jiwan, Norman. "The Political Ecology of the Indonesian Palm Oil Industry." In *The Palm Oil Controversy in Southeast Asia: A Transnational Perspective*, edited by Oliver Pye and Jayati Bhattacharya, 48–75. Singapore: ISEAS Publishing, 2015.
Johnson, Samuel, and O. (Obadiah) Johnson. *The History of the Yorubas: From the Earliest Times to the Beginning of the British Protectorate*. Lagos, Nigeria: C.M.S. Bookshops, 1921.
Johnston, Harry H. "British West Africa and the Trade of the Interior." *Proceedings of the Royal Colonial Institute* 20 (1888): 90–129.
——. *George Grenfell and the Congo*. London: D. Appleton & Company, 1910.
——. *Liberia*. Vol. 1. London: Hutchinson & Co., 1906.
Jomo Kwame Sundaram. *A Question of Class: Capital, the State, and Uneven Development in Malaya*. Singapore: Oxford University Press, 1986.
Jomo Kwame Sundaram, Y. T. Chang, and K. J. Khoo. *Deforesting Malaysia: The Political Economy and Social Ecology of Agricultural Expansion and Commercial Logging*. London: Zed Books, 2004.
Jones, Adam. *Brandenburg Sources for West African History, 1680–1700*. Stuttgart: F. Steiner Verlag Wiesbaden, 1985.
——. *German Sources for West African History, 1599–1669*. Wiesbaden: Steiner, 1983.

Jones, Adam, and Peter Sebald. *An African Family Archive: The Lawsons of Little Popo/Aneho (Togo) 1841–1938*. Oxford: Oxford University Press, 2006.

Jones, Christopher F. *Routes of Power: Energy and Modern America*. Cambridge, MA: Harvard University Press, 2014.

Jones, G. Howard. *The Earth Goddess: A Study of Native Farming on the West African Coast*. London: Longmans, Green and Co., 1936.

Jones, G. I. *From Slaves to Palm Oil: Slave Trade and Palm Oil Trade in the Bight of Biafra*. Cambridge: African Studies Centre, 1989.

——. *The Trading States of the Oil Rivers*. Oxford: Oxford University Press, 1963.

Jones, Geoffrey. *Renewing Unilever: Transformation and Tradition*. Oxford: Oxford University Press, 2005.

Jones, J. H. *The Tinplate Industry*. London: P. S. King & Son, 1914.

Jong, Hans Nicholas. "Indonesian Court Jails Indigenous Farmers for 'Stealing' from Land They Claim." *Mongabay Environmental News*, June 23, 2020. https://news.mongabay.com/2020/06/indonesia-indigenous-farmers-palm-oil-jail-sentence-conflict-hmbp-best-group/.

Jorgensen, Dolly, Finn Arne Jorgensen, and Sara B. Pritchard. *New Natures: Joining Environmental History with Science and Technology Studies*. Pittsburgh: University of Pittsburgh Press, 2013.

Juhé-Beaulaton, Dominique. "La palmeraie du Sud Bénin avant la colonisation: essai d'analyse historique." In *Plantes et paysages d'Afrique, une histoire à explorer*, edited by Monique Chastanet, 327–52. Paris: Karthala, 1998.

Jung, Laura S. "African Palm and Afro-Indigenous Resistance: Race and Dispossession of Garifuna Lands on Honduras' Northern Coast." MA thesis, DePaul University, 2011.

Junker, Wilhelm. *Travels in Africa during the Years 1875–1886*. Translated by A. H. Keane. London: Chapman and Hall, 1890.

Kadandale, Sowmya, Robert Marten, and Richard Smith. "The Palm Oil Industry and Noncommunicable Diseases." *Bulletin of the World Health Organization* 97, no. 2 (2019): 73–168.

Kajisa, Kei, Mywish K. Maredia, and Duncan Boughton. "Transformation versus Stagnation in the Oil Palm Industry: A Comparison between Malaysia and Nigeria." Staff Paper. East Lansing: Department of Agricultural Economics, Michigan State University, 1997. https://ageconsearch.umn.edu/record/11483/.

Kaniki, Martin. "Economical Technology against Technical Efficiency in the Oil Palm Industry of West Africa." *Development and Change* 11, no. 2 (1980): 273–84.

Kantor, Helene J. "Plant Ornament: Its Origin and Development in the Ancient Near East." PhD diss., University of Chicago, 1945.

Kathirithamby-Wells, Jeyamalar. *Nature and Nation: Forests and Development in Peninsular Malaysia. Man and Nature in Asia*. Copenhagen: NIAS Press, 2005.

Kaur, Amarjit. *Economic Change in East Malaysia: Sabah and Sarawak since 1850*. Basingstoke: Palgrave, 1998.

——. "Plantation Systems, Labour Regimes and the State in Malaysia, 1900–2012." *Journal of Agrarian Change* 14, no. 2 (April 2014): 190–213.

———. “Tappers and Weeders: South Indian Plantation Workers in Peninsular Malaysia, 1880–1970.” *South Asia: Journal of South Asian Studies* 21, no. 1 (January, 1998): 73–102.

Kay, Cristóbal. “Reflections on Latin American Rural Studies in the Neoliberal Globalization Period: A New Rurality?” *Development and Change* 39, no. 6 (2008): 915–43.

Kennedy, Kathleen. “Gripping It by the Husk: The Medieval English Coconut.” *Medieval Globe* 3, no. 1 (2017).

Kerr, Robert. *A General History and Collection of Voyages and Travels, Arranged in Systematic Order. . . .* Edinburgh: James Ballantyne and Co., 1824.

Kersson, Tanya K. *Grabbing Power: The New Struggles for Land, Food and Democracy in Northern Honduras*. Oakland: FoodFirst Books, 2013.

Khor, Yuleng. “The Felda Case Study.” High Carbon Stock Case Study. Oxford: LMC International, 2015.

Kilby, Peter. *Industrialization in an Open Economy: Nigeria, 1945–1966*. Cambridge: Cambridge University Press, 1969.

———. “Reply to Malcolm Purvis’ Oil Palm Report.” *Food Research Institute Studies* 8, no. 2 (1968): 199–202.

King, J. B. “Details of Explorations of the Old Calabar River, in 1841 and 1842.” *Journal of the Royal Geographical Society* 14 (1844): 260–83.

King, Victor T. “Models and Realities: Malaysian National Planning and East Malaysian Development Problems.” *Modern Asian Studies* 22, no. 2 (1988): 263–98.

Kiple, Kenneth F. *A Movable Feast: Ten Millennia of Food Globalization*. Cambridge: Cambridge University Press, 2007.

Klaveren, Marieke Van. “Death among Coolies: Mortality of Chinese and Javanese Labourers on Sumatra in the Early Years of Recruitment, 1882–1909.” *Itinerario* 21, no. 1 (March 1997): 111–24.

Klein, Herbert S. *The Atlantic Slave Trade*. Cambridge: Cambridge University Press, 1999.

Klonoff, David C. “Replacements for Trans Fats—Will There Be an Oil Shortage?” *Journal of Diabetes Science and Technology* 1, no. 3 (May 2007): 415–22.

Knapp, Friedrich Ludwig. *Chemical Technology: Or, Chemistry, Applied to the Arts and to Manufactures*. London: H. Baillière, 1848.

Knoll, Arthur J. *Togo under Imperial Germany, 1884–1914: A Case Study in Colonial Rule*. Stanford: Hoover Institution Press, 1978.

Knothe, Gerhard. “George Chavanne and the First Biodiesel.” *Inform* 28, no. 7 (2017): 21–24.

———. “Historical Perspectives on Vegetable Oil-Based Diesel Fuels.” *Inform* 12 (2001): 103–7.

Knox, Andrew M. *Coming Clean: A Postscript after Retirement from Unilever*. London: Heinemann, 1976.

Koczberski, Gina. “Loose Fruit Mamas: Creating Incentives for Smallholder Women in Oil Palm Production in Papua New Guinea.” *World Development* 35, no. 7 (July 1, 2007): 1172–85.

Koczberski, Gina, Katherine Gibson, and George Nicolas Curry. *Improving Productivity of the Smallholder Oil Palm Sector in Papua New Guinea: A Socio-Economic Study of the Hoskins and Popondetta Schemes*. Canberra: Australian National University, 2001.

Köler, Hermann. *Einige Notizen Über Bonny an Der Küste von Guinea, Seine Sprache Und Seine Bewohner*. . . . Göttingen: Dieterichschen Univ. Buchdruckerei, 1848.

Kone, Kadidja. "Le Palmier à Huile En Côte d'Ivoire." *GRAIN* (blog), September 22, 2014. https://www.grain.org/article/entries/5034-oil-palm-production-in-west-and-central-africa.

Konings, Piet. "Unilever, Contract Farmers and Co-operatives in Cameroon: Crisis and Response." *Journal of Peasant Studies* 26, no. 1 (October 1998): 112–38.

———. *Unilever Estates in Crisis and the Power of Organizations in Cameroon*. Leiden: African Studies Centre, 1998.

Korieh, Chima J. "Alcohol and Empire: 'Illicit' Gin Prohibition and Control in Colonial Eastern Nigeria." *African Economic History*, no. 31 (2003): 111–34.

———. *The Land Has Changed: History, Society and Gender in Colonial Eastern Nigeria*. Calgary: University of Calgary Press, 2010.

———. "The Nineteenth Century Commercial Transition in West Africa: The Case of the Biafra Hinterland." *Canadian Journal of African Studies* 34, no. 3 (2000): 588–615.

Kozlik, Adolf, and Peter Boomgaard. *The Demand for Fats and Oils in the Soap Industry*. Research Bulletin 311. Ames, Iowa: Agricultural Experiment Station, Iowa State College of Agriculture and Mechanic Arts, 1943.

Kratoska, Paul H. *The Japanese Occupation of Malaya: A Social and Economic History*. Honolulu: University of Hawai'i Press, 1997.

Kris-Etherton, P. M., and F. H. Seligson. "The Heart of the Tropical Oils Issue." *Journal of the American Dietetic Association* 89, no. 5 (1989): 627–28.

Kroef, Justus M. van der. "Private Armies and Extrajudicial Violence in the Philippines." *Asian Affairs* 13, no. 4 (1986): 1–21.

Kurniawati, Dewi. "Where Losing the Forest Means Losing Your Identity." *Jakarta Globe*, December 23, 2009.

Kurten, Philip. *The Art of Manufacturing Soaps, Including the Most Recent Discoveries*. . . . Philadelphia: Lindsay & Blakiston, 1854.

Kwa, B. H. "Environmental Change, Development and Vectorborne Disease: Malaysia's Experience with Filariasis, Scrub Typhus and Dengue." *Environment, Development and Sustainability* 10, no. 2 (2008): 209–17.

Kwang, King Chuan. "Interchangeability of Important Vegetable Oils and Fats and Its Implications to Malaysian Oil Palm Industry." Occassional Paper of the Faculty of Economics. Kuala Lumpur: National University of Malaysia, September 1981.

La Fleur, James D. *Fusion Foodways of Africa's Gold Coast in the Atlantic Era*. Leiden: Brill, 2012.

"Lagos Palm Oil. (*Elæis guineensis*, Jacq.)." *Bulletin of Miscellaneous Information (Royal Botanic Gardens, Kew)* 1892, no. 69 (1892): 200–208.

Lai, Bryan Paul. *A Glance of Tawau in the Sixties: Land below the Wind.* Singapore: Partridge Publishing, 2016.

Laird, MacGregor. *Narrative of an Expedition into the Interior of Africa, by the River Niger.* London: Bentley, 1837.

Lam, Man Kee, Kok Tat Tan, Keat Teong Lee, and Abdul Rahman Mohamed. "Malaysian Palm Oil: Surviving the Food versus Fuel Dispute for a Sustainable Future." *Renewable and Sustainable Energy Reviews* 13, no. 6–7 (August 2009): 1456–64.

Lamb, Nicole. "A Time of Normalcy: Javanese 'Coolies' Remember the Colonial Estate." *Journal of the Humanities and Social Sciences of Southeast Asia* 170, no. 4 (2014): 530–56.

Lamborn, Leebert Lloyd. *Modern Soaps, Candles and Glycerin: A Practical Manual of Modern Methods of Utilization of Fats and Oils in the Manufacture of Soap and Candles, and of the Recovery of Glycerin.* New York: Van Nostrand, 1906.

Lampe, Markus, and Paul Sharp. *A Land of Milk and Butter: How Elites Created the Modern Danish Dairy Industry.* Chicago: University of Chicago Press, 2019.

Lanagan, Kevin J. *The Palm Oil Industry in West Malaysia.* Washington, DC: GPO, 1977.

Langdon, John. "Three Voyages to the West Coast of Africa, 1881–1884." In *Travel, Trade and Power in the Atlantic 1765–1884*, edited by Martin Lynn, 165–283. Cambridge: Cambridge University Press, 2002.

Larson, Donald F. "Indonesia's Palm Oil Subsector." WPS Policy Research Working Paper. Washington, DC: World Bank, 1996.

Latham, A. J. H. *Old Calabar, 1600–1891; the Impact of the International Economy upon a Traditional Society.* Oxford: Clarendon Press, 1973.

——. "Palm Oil Exports from Calabar 1812–1887 (with a Note on Price Formation)." In *Figuring the African Trade*, edited by G. Liesegang, H. Pasch, and Adam Jones, 265–96. Berlin: Dietrich Reimer Verlag, 1986.

Laucks, Irving Fink. *Commercial Oils, Vegetable and Animal, with Special Reference to Oriental Oils.* New York: John Wiley & Sons, 1919.

Laurent, C. K. *Investment in Nigerian Tree Crops: Smallholder Production.* Consortium for the Study of Nigerian Rural Development, CSNRD-18. Ibadan, Nigeria, and East Lansing, MI: NISER/CSNRD, 1968.

Law, Robin. "The 'Crisis of Adaptation' Revisited: The Yoruba Wars of 1877–1893." In *Africa, Empire and Globalization: Essays in Honor of AG Hopkins*, edited by Toyin Falola and Emily Brownell, 125–43. Durham: Carolina Academic Press, 2011.

——. *The English in West Africa, 1691–1699.* Fontes Historiae Africanae, New Ser. 8. Oxford and New York: Oxford University Press for The British Academy, 2006.

——. *From Slave Trade to "Legitimate" Commerce: The Commercial Transition in Nineteenth-Century West Africa.* Cambridge: Cambridge University Press, 1995.

——. "Introduction." In *From Slave Trade to "Legitimate" Commerce*, 1–31. Cambridge: Cambridge University Press, 1995.

——. "'There's Nothing Grows in the West Indies but Will Grow Here': Dutch and English Projects of Plantation Agriculture on the Gold Coast, 1650s–1780s." In *Commercial Agriculture, the Slave Trade and Slavery in Atlantic Africa*, 116–37. Woodbridge, UK: Boydell and Brewer, 2013.

——. "Trade and Gender in Yorubaland and Dahomey." In *From Slave Trade to "Legitimate" Commerce: The Commercial Transition in Nineteenth-Century West Africa*, 195–214. Cambridge: Cambridge University Press, 1995.

——. "Wheeled Transport in Pre-Colonial West Africa." *Africa: Journal of the International African Institute* 50, no. 3 (1980): 249–62.

Law, Robin, Suzanne Schwarz, and Silke Strickrodt, eds. *Commercial Agriculture, the Slave Trade and Slavery in Atlantic Africa*. Woodbridge, UK: James Currey, 2013.

Lawrance, Benjamin N. "'En Proie à La Fièvre Du Cacao': Land and Resource Conflict on an Ewe Frontier, 1922–1939." *African Economic History*, no. 31 (2003): 135–81.

——. *Locality, Mobility, and "Nation": Periurban Colonialism in Togo's Eweland, 1900–1960*. Rochester, NY: University of Rochester Press, 2007.

Leach, Melissa. *Rainforest Relations: Gender and Resource Use among the Mende of Gola, Sierra Leone*. Edinburgh: Edinburgh University Press, 1994.

League of Nations. *Statistical Yearbook*. Geneva: League of Nations, 1926–1944.

Leal, Claudia. *Landscapes of Freedom: Building a Postemancipation Society in the Rainforests of Western Colombia*. Tuscon: University of Arizona Press, 2018.

Lees, Lynn Hollen. *Planting Empire, Cultivating Subjects: British Malaya, 1786–1941*. Cambridge: Cambridge University Press, 2017.

Leicester, Henry M. *The Historical Background of Chemistry*. New York: John Wiley & Sons, 1956.

Leis, Philip E. "Palm Oil, Illicit Gin, and the Moral Order of the Ijaw." *American Anthropologist*, New Series, 66, no. 4 (August 1964): 828–38.

Leith-Ross, Sylvia. *African Women: A Study of the Ibo of Nigeria*. London: Faber and Faber, 1939.

Lemery, Nicolás. *Traité Universel Des Drogues Simples, Mises an Ordre Alphabetique. . . .* Paris: Laurent d'Houry, 1714.

Lennartson, Anders. *The Chemical Works of Carl Wilhelm Scheele*. Berlin: Springer, 2017.

León, Andrés. "Rebellion under the Palm Trees: Memory, Agrarian Reform and Labor in the Aguán, Honduras." PhD diss., City University of New York, 2015.

LePlae, Edmond. "Agriculture in the Belgian Congo." *Bulletin of the Imperial Institute* 12 (1914): 60–75.

——. "Oil Palm Groves in the Belgian Congo." In *Oil Palms and Their Fruit*, edited by A. A. L. Rutgers, 8–17. London: Griffith & Co, 1922.

——. *Le Palmier à Huile En Afrique: Son Exploitation Au Congo Belge et En Extrême-Orient*. Brussels: Librairie Falk fils, 1939.

Leroy, Jean-François. "Rapport sommaire sur une mission en A.O.F. (10 janvier–10 mars 1957)." *Journal d'agriculture traditionnelle et de botanique appliquée* 4, no. 3 (1957): 170–75.

Leubuscher, Charlotte. "Marketing Schemes for Native-Grown Produce in African Territories." *Africa: Journal of the International African Institute* 12, no. 2 (1939): 163–88.
——. *The Processing of Colonial Raw Materials: A Study in Location*. London: HMSO, 1951.
Leverhulme, William Hulme Lever. *Viscount Leverhulme*. Boston: Houghton Mifflin, 1927.
Levin, Joshua, Ginny Ng, Desmond Fortes, Sarah Garcia, Samantha Lacey, and Donald Grubba. "Profitability and Sustainability in Palm Oil Production." Washington, DC: WWF, FMO, and CDC, 2012.
Levitt, Benjamin. *Oil, Fat, and Soap*. New York: Chemical Publishing, 1951.
Lewicki, Tadeusz, and Marion Johnson. *West African Food in the Middle Ages: According to Arabic Sources*. Cambridge: Cambridge University Press, 1974.
Lewis, Brian. *So Clean: Lord Leverhulme, Soap and Civilisation*. Manchester: Manchester University Press, 2008.
Lewis, William. *An Experimental History of the Materia Medica, or of the Natural and Artificial Substances Made Use of in Medicine*. 3rd ed. London: J. Johnson and R. Baldwin, 1784.
——. *The New Dispensatory. . . .* , 4th edition. Dublin: James Potts, 1778.
Lewkowitsch, Julius. *Cantor Lectures on Oils and Fats: Their Uses and Applications*. London: W. Trounce, 1904.
——. *Chemical Technology and Analysis of Oils, Fats and Waxes*. London: Macmillan, 1914.
Leys, Colin. *The Rise & Fall of Development Theory*. Bloomington: Indiana University Press, 1996.
Li, Tania Murray. "After the Land Grab: Infrastructural Violence and the 'Mafia System' in Indonesia's Oil Palm Plantation Zones." *Geoforum* 96 (2018): 328–37.
——. "Intergenerational Displacement in Indonesia's Oil Palm Plantation Zone." *Journal of Peasant Studies* 44, no. 6 (2017): 1158–76.
——. "The Price of Un/Freedom: Indonesia's Colonial and Contemporary Plantation Labor Regimes." *Comparative Studies in Society and History; Cambridge* 59, no. 2 (April 2017): 245–76.
——. "Situating Transmigration in Indonesia's Oil Palm Labour Regime." In *Oil Palm Complex*, edited by John F. McCarthy and R. A. Cramb, 354–77. Singapore: NUS Press, 2016.
——. "Social Impacts of Oil Palm in Indonesia: A Gendered Perspective from West Kalimantan." Occassional Paper. Bogor, Indonesia: CIFOR, 2015.
Lian Pin Koh, and David S. Wilcove. "Oil Palm: Disinformation Enables Deforestation." *Trends in Ecology & Evolution* 24, no. 2 (February 2009): 67–68.
Lief, Alfred. *It Floats: The Story of Procter & Gamble*. New York: Rinehart, 1958.
Ligon, Richard. *A True & Exact History of the Island of Barbados Illustrated with a Mapp of the Island. . . .* London: Humphrey Moseley, 1657.
Lim Teck Ghee. *Peasants and Their Agricultural Economy in Colonial Malaya, 1874–1941*. Kuala Lumpur: Oxford University Press, 1977.

Lim Teck Ghee, and Richard Dorall. "Contract Farming in Malaysia: With Special Reference to FELDA Land Schemes." In *Contract Farming in Southeast Asia*, edited by Lim Teck Ghee and David Glover, 71–118. Kuala Lumpur: Vinlin Press, 1992.

Lindblad, J. Thomas. *Bridges to New Business: The Economic Decolonization of Indonesia*. Leiden: KITLV Press, 2008.

List, Gary R., and Michael A. Jackson. "The Battle over Hydrogenation (1903–1920)." *AOCS Lipid Library* (blog), 2014. http://lipidlibrary.aocs.org/history/hydrogenation/index.htm.

Little, I. M. D. *A Social Cost Benefit Analysis of the Kulai Oil Palm Estate: West Malaysia*. Paris: Development Centre of the OECD, 1972.

Lloyd, Peter Cutt. "Some Problems of Tenancy in Yoruba Land Tenure." *African Studies* 12 (1955): 93–103.

Loh, Francis Kok Wah. "Where Has All the Violence Gone? Mapping Old and New Conflict." In *Building Bridges, Crossing Boundaries: Everyday Forms of Inter-Ethnic Peace Building in Malaysia*, edited by Francis Kok Wah Loh, 15–51. Jakarta and Kajang: Ford Foundation and Malaysian Social Science Association, 2010.

Loh, Willie. "Trans Fat Reformulation Is Not a Technical Objective!" In *Trans Fats Alternatives*, edited by Dharma R. Kodali and Gary R. List, 96–105. Urbana, IL: AOCS Publishing, 2005.

Long, Edward. *The History of Jamaica. Or, General Survey of the Antient and Modern State of the Island. . . .* London: T. Lowndes, 1774.

Longayroux, J. P. "Hoskins Oil Palm Project: An Introduction." In *The Hoskins Development*, 1–6. Port Moresby: Australian National University, 1972.

Lorenzini, Sara. *Global Development: A Cold War History*. Princeton, NJ: Princeton University Press, 2019.

Lovejoy, Paul E. *Transformations in Slavery: A History of Slavery in Africa*. 2nd ed. Cambridge: Cambridge University Press, 2000.

Lovejoy, Paul E., and David Richardson. "From Slaves to Palm Oil: Afro-European Commercial Relations in the Bight of Biafra, 1741–1841." In *Maritime Empires*, edited by David Killingray, Margarette Lincoln, and Nigel Rigby, 13–29. Woodbridge, UK: Boydell and Brewer, 2004.

———. "'This Horrid Hole': Royal Authority, Commerce and Credit at Bonny, 1690–1840." *Journal of African History* 45, no. 3 (2004): 363–92.

Lucas, A. *Ancient Egyptian Materials and Industries*. 2nd ed. London: Edward Arnold, 1934.

Lund, Charles E. "Fats and Oils Industry Report–September 1947." Bureau of Foreign and Domestic Commerce. Washington, DC: U.S. Department of Commerce, 1947.

Lundberg, Donald E. *Adventure in Costa Rica*. Tallahassee: Dixie Publishers, 1960.

Luytjes, M. A. "La Situation de La Culture Du Palmier à Huile Sur La Côte Orientale de Sumatra et Dans La Province d'Atjeh." In *Mémoires et Rapports Sur Les Matières Grasses. T. III. Le Palmier à Huile*, 229–51. Marseille: Institut Colonial, 1928.

Lynn, Martin. *Commerce and Economic Change in West Africa: The Palm Oil Trade in the Nineteenth Century*. Cambridge: Cambridge University Press, 1997.

——. "Factionalism, Imperialism and the Making and Breaking of Bonny Kingship c. 1830–1885." *Outre-Mers. Revue d'histoire* 82, no. 307 (1995): 169–92.

——. "Liverpool and Africa in the Nineteenth Century: The Continuing Connection." *Transactions of the Historic Society of Lancashire and Cheshire* 147 (1998): 27–54.

Maas, J. A. "Het Planten van Oliepalmen." *Teysmannia* 32 (1921): 495–521.

——. "La Culture et la Sélection du Palmier Elaeis aux Indes Néerlandaises." In *Mémoires et rapports sur les matières grasses. T. III. Le Palmier à huile*, 187–92. Marseilles: Institut colonial de Marseille, 1928.

Maat, H. *Science Cultivating Practice: A History of Agricultural Science in the Netherlands and Its Colonies, 1863–1986*. Dordrecht: Springer Netherlands, 2001.

MacCormack, Carol P. "Control of Land, Labor, and Capital in Rural Southern Sierra Leone." In *Women and Work in Africa*, edited by Edna G. Bay, 35–54. Boulder: Westview Press, 1982.

MacFadyen, Joshua. *Flax Americana: A History of the Fibre and Oil That Covered a Continent*. Montreal: McGill-Queen's University Press, 2018.

MacKendrick, Paul L., and Herbert M. Howe. *Classics in Translation, Volume I: Greek Literature*. Madison: University of Wisconsin Press, 1959.

Mackeown, Robert L. *Twenty-Five Years in Qua Iboe: The Story of a Missionary Effort in Nigeria*. London: Morgan & Scott, 1912.

Mackie, J. A. C. *Konfrontasi: The Indonesia-Malaysia Dispute 1963–1966*. Kuala Lumpur: Oxford University Press, 1974.

Maclean, J. L. *The British Railway System: A Description of the Work Performed in the Principal Departments*. London: McCorquodale & Company, 1883.

MacQueen, James. *The Colonial Controversy: Containing a Refutation of the Calumnies of the Anticolonists. . . .* Glasgow: Khull, Blackie, 1825.

Madge, Bruce. "Elizabeth Blackwell—the Forgotten Herbalist?" *Health Information & Libraries Journal* 18, no. 3 (2001): 144–52.

Maher, David. "Rooted in Violence: Civil War, International Trade and the Expansion of Palm Oil in Colombia." *New Political Economy* 20, no. 2 (2015): 299–330.

Maier, Donna J. E. "Precolonial Palm Oil Production and Gender Division of Labor in Nineteenth-Century Gold Coast and Togoland." *African Economic History* 37, no. 1 (2009): 1–32.

Majid Cooke, Fadzilah. "In the Name of Poverty Alleviation: Experiments with Oil Palm Smallholders and Customary Land in Sabah, Malaysia" *Asia Pacific Viewpoint* 53, no. 3 (December 2012): 240–53.

——. "Maps and Counter-Maps: Globalised Imaginings and Local Realities of Sarawak's Plantation Agriculture." *Journal of Southeast Asian Studies* 34, no. 2 (2003): 265–84.

——. "Vulnerability, Control and Oil Palm in Sarawak: Globalization and a New Era?" *Development and Change* 33, no. 2 (2002): 189–211.

Majid Cooke, Fadzilah, and Dayan Suria Mulia. "Migration and Moral Panic: The Case of Oil Palm in Sabah, East Malaysia." In *The Palm Oil Controversy in Southeast Asia: A Transnational Perspective*, edited by Oliver Pye and Jayati Bhattacharya, 140–63. Singapore: ISEAS Publishing, 2015.

Majid Cooke, Fadzilah, Su Mei Toh, and Justine Vaz. "Making an Informed Choice: A Review of Oil Palm Partnerships in Sabah and Sarawak, East Malaysia." London: International Institute of Environment and Development, 2011. http://www.fao.org/fileadmin/templates/tci/pdf/InternationalInvestment/IIED/MALAYSIA_OIL_PALM_STUDY_IIED_FINAL_MAY_2011__v2__-_LC_-_Clean.pdf.

Malaysiakini, Mongabay, Tempo, and The Gecko Project. "The Secret Deal to Destroy Paradise." *Mongabay Environmental News*, November 2018. https://news.mongabay.com/2018/11/the-secret-deal-to-destroy-paradise/.

Malaysian Palm Oil Board. *Going for Liquid Gold*. Kuala Lumpur: MPOB, 2010.

Malden, Henry Elliot. *The Victoria History of the County of Surrey*. London: A. Constable, 1905.

Malet, C. C. "The Palm Oil Planting Industry in Malaya." *The Planter* 8, no. 11 (June 1928): 323–24.

———. Letter to editor. *The Planter* 11, no. 2 (September 1930): 37–38.

Maley, Jean. "The Impact of Arid Phases on the African Rain Forest through Geological History." In *African Rain Forest Ecology and Conservation*, edited by W. Weber, L. White, A. Vedder, and L. Naughton-Treves, 68–87. New Haven, CT: Yale University Press, 2001.

Maley, Jean, and Alex Chepstow-Lusty. "*Elaeis guineensis* Jacq. (Oil Palm) Fluctuations in Central Africa during the Late Holocene: Climate or Human Driving Forces for This Pioneering Species?" *Vegetation History and Archaeobotany* 10, no. 2 (2001): 117–20.

Mangga Barani, Achmad. *Palm Oil, a Golden Gift from Indonesia to the World*. Jakarta: Directorate General of Estate Crops and Sinar Mas, 2009.

Manicler, Nicolas Hegesippe. "Patent 6121, 2 June 1831." *London Journal and Repertory of Arts, Sciences, and Manufactuers* 1, no. 1 (1832): 240–43.

Mann, Kristin. "Owners, Slaves, and the Struggle for Labour in the Commercial Transition at Lagos." In *From Slave Trade to "Legitimate" Commerce: The Commercial Transition in Nineteenth-Century West Africa*, edited by Robin Law, 144–72. Cambridge: Cambridge University Press, 1995.

———. *Slavery and the Birth of an African City: Lagos, 1760–1900*. Bloomington: Indiana University Press, 2007.

Manning, Patrick. "An Economic History of Southern Dahomey, 1880–1914." PhD diss., University of Wisconsin–Madison, 1969.

———. "Slave Trade, 'Legitimate' Trade, and Imperialism Revisited: The Control of Wealth in the Bights of Benin and Biafra." In *Africans in Bondage*, edited by Paul E. Lovejoy, 203–34. Madison: University of Wisconsin Press, 1986.

———. *Slavery and African Life: Occidental, Oriental, and African Slave Trades*. Cambridge: Cambridge University Press, 1990.

——. *Slavery, Colonialism and Economic Growth in Dahomey, 1640–1960*. Cambridge: Cambridge University Press, 1982.

Manshard, Walther, and William B. Morgan, eds. *Agricultural Expansion and Pioneer Settlements in the Humid Tropics*. Hong Kong: United Nations University Press, 1988.

"The Manufactures of Price's Patent Candle Company. A Day at the Sherwood Works, Battersea." *Chemist and Druggist* 3, no. 31 (March 15, 1862): 65–69.

Marchal, Jules. *Lord Leverhulme's Ghosts: Colonial Exploitation in the Congo*. Translated by Martin Thom. London: Verso, 2008.

Marin-Burgos, Victoria. "Power, Access and Justice in Commodity Frontiers. The Political Ecology of Access to Land and Palm Oil Expansion in Colombia." PhD diss., University of Twente, 2014.

Marin-Burgos, Victoria, Joy S. Clancy, and Jonathan Cranidge Lovett. "Contesting Legitimacy of Voluntary Sustainability Certification Schemes: Valuation Languages and Power Asymmetries in the Roundtable on Sustainable Palm Oil in Colombia." *Ecological Economics* 117 (2015): 303–13.

Markley, Klare S. *Vegetable Oils: Some Observations on Recent Developments in El Salvador, Honduras, Venezuela, and Brazil*. Washington, DC: GPO, 1953.

Marquardt, Steve. "Pesticides, Parakeets, and Unions in the Costa Rican Banana Industry, 1938–1962." *Latin American Research Review* 37, no. 2 (2002): 3–36.

Marsh, T. D. "Oil Palm." *The Planter* 5, no. 11–12 (1925): 313–15, 337–38.

Martin, Anne. *The Oil Palm Economy of the Ibibio Farmer*. Ibadan, Nigeria: Ibadan University Press, 1956.

Martin, Marshall A. "U.S. Palm Oil Imports: An Economic Analysis of Recent Events and Prospects for the Future." *Journal of ASFMRA* 41, no. 2 (1977): 25–31.

Martin, Susan. "An Edible Oil for the World: Malaysian and Indonesian Competition in the Palm Oil Trade, 1945–2000." In *Intra-Asian Trade and the World Market*, edited by A. J. H Latham and Heita Kawakatsu, 209–30. Abingdon, UK: Routledge, 2006.

——. "Gender and Innovation: Farming, Cooking and Palm Processing in the Ngwa Region, South-Eastern Nigeria, 1900–1930." *Journal of African History* 25, no. 4 (1984): 411–27.

——. *Palm Oil and Protest: An Economic History of the Ngwa Region, South-Eastern Nigeria, 1800–1980*. Cambridge: Cambridge University Press, 1988.

——. "Slaves, Igbo Women, and Palm Oil." In *From Slave Trade to "Legitimate" Commerce: The Commercial Transition in Nineteenth-Century West Africa*, edited by Robin Law, 172–94. Cambridge: Cambridge University Press, 1995.

——. *The UP Saga*. Copenhagen: NIAS Press, 2003.

Mason, Margie, and Robin McDowell. "Palm Oil Labor Abuses Linked to World's Top Brands, Banks." *AP News*, September 24, 2020. https://apnews.com/article/virus-outbreak-only-on-ap-indonesia-financial-markets-malaysia-7b634596270cc6aa7578a062a30423bb.

Masson, Paul. *Marseille et la colonisation française: essai d'histoire coloniale*. Marseille: Barlatier, 1906.

Mathe-Shires, Laszlo. "Lagos Colony and the Oil Rivers Protectorate." In *Encyclopedia of African History*, edited by Kevin Shillington, 791–92. London: Routledge, 2004.
Mathur, Murli Manohar. *Vanaspati Industry in India*. New Delhi: Concept Publishing, 1981.
May, Stacy. *Costa Rica: A Study in Economic Development*. New York: Twentieth Century Fund, 1952.
May, Stacy, and Galo Plaza Lasso. *The United Fruit Company in Latin America*. Washington, DC: National Planning Association, 1958.
Mba, Nina Emma. *Nigerian Women Mobilized: Women's Political Activity in Southern Nigeria, 1900–1965*. Berkeley: University of California, Berkeley, Institute of International Studies, 1982.
Mccann, James C. "Climate and Causation in African History." *International Journal of African Historical Studies* 32, no. 2–3 (1999): 261–79.
McCann, Thomas P. *An American Company: The Tragedy of United Fruit*. Edited by Henry Scammell. New York: Crown Publishers, 1976.
McCarthy, John F. "Certifying in Contested Spaces: Private Regulation in Indonesian Forestry and Palm Oil." *Third World Quarterly* 33, no. 10 (November 2012): 1871–88.
McCarthy, John F., and R. A. Cramb. "Policy Narratives, Landholder Engagement, and Oil Palm Expansion on the Malaysian and Indonesian Frontiers." *GEOJ Geographical Journal* 175, no. 2 (2009): 112–23.
McClintock, Anne. *Imperial Leather: Race, Gender, and Sexuality in the Colonial Contest*. New York: Routledge, 1995.
McCook, Stuart. *States of Nature: Science, Agriculture, and Environment in the Spanish Caribbean, 1760–1940*. Austin: University of Texas Press, 2002.
McEwan, Cheryl. "Representing West African Forests in British Imperial Discourse c. 1830–1900." In *Contesting Forestry in West Africa*, edited by Reginald Cline-Cole and Clare Madge, 16–35. Aldershot: Ashgate, 2000.
McGucken, William. *Biodegradable: Detergents and the Environment*. College Station: Texas A&M University Press, 1991.
McIntosh, Marjorie Keniston. *Yoruba Women, Work, and Social Change*. Bloomington: Indiana University Press, 2009.
McKenzie, Peter Rutherford. *Hail Orisha!: A Phenomenology of a West African Religion in the Mid-Nineteenth Century*. Leiden: Brill, 1997.
McNamara, Donald J. "Palm Oil and Health: A Case of Manipulated Perception and Misuse of Science." *Journal of the American College of Nutrition* 29, no. supplement 3 (2010): 240–44.
M'Cormac, Henry. "On the Oil Called Palm Oil." *Glasgow Medical Journal* 5, no. 18 (May 1832): 227–28.
McPhee, Allan. *The Economic Revolution in British West Africa*. New York: Negro Universities Press, 1970.
McWilliam, M. *The Development Business: A History of the Commonwealth Development Corporation*. Basingstoke: Palgrave, 2001.

Meehan, Sarah. "Berger Cookies Needed to Drop Trans Fat. Turns out: They Already Had." *Baltimore Sun*. November 21, 2017, sec. Food & Drink. https://www.baltimoresun.com/food-drink/bs-fo-berger-cookies-trans-fat-20171120-story.html.

Meenakshi, Jonnalagadda Venkata. "Coconuts and Oil Palm in the Indian Oilseeds Economy." MS thesis, Cornell University, 1986.

Meijaard, Erik, and Douglas Sheil. "The Moral Minefield of Ethical Oil Palm and Sustainable Development." *Frontiers in Forests and Global Change* 2, no. 22 (2019).

Menon, Narayana, and Chris Leggett. "The NUPW in the Nineties: Plantation Workers in Malaysia." *Indian Journal of Industrial Relations* 32, no. 1 (1996): 56–68.

Mentelle, Edme. *Cosmographie Élémentaire, Divisée En Parties Astronomique et Géographique. . . .* Paris: E. Mentelle, 1785.

Mercader, Julio, Huw Barton, Jason Gillespie, Jack Harris, Steven Kuhn, Robert Tyler, and Christophe Boesch. "4,300-Year-Old Chimpanzee Sites and the Origins of Percussive Stone Technology." *PNAS* 104, no. 9 (2007): 3043–48.

Meredith, David. "Government and the Decline of the Nigerian Oil-Palm Export Industry, 1919–1939." *Journal of African History* 25, no. 3 (1984): 311–29.

———. "State Controlled Marketing and Economic 'Development': The Case of West African Produce during the Second World War." *Economic History Review* 39, no. 1 (1986): 77–91.

Metropolitan Working Classes' Association for Improving the Public Health. *Bathing and Personal Cleanliness*. London: John Churchill, 1847.

Miller, Brad. "'Killed, Forced, Afraid': Philippine Palm Oil Legacy Incites New Fears." *Mongabay*, May 9, 2017. https://news.mongabay.com/2017/05/killed-forced-afraid-philippine-palm-oil-legacy-incites-new-fears/.

Miller, Joseph. *Botanicum Officinale, or, A Compendious Herbal: Giving an Account of All Such Plants as Are Now Used in the Practice of Physick, with Their Descriptions and Virtues*. London: E. Bell et al., 1722.

Miller, William L. "An Economic Analysis of Oil Palm Fruit Processing in Eastern Nigeria." PhD diss., Michigan State University, 1965.

Milligan, F. W. *The Cultivation of the Oil Palm*. London: Crosby, Lockwood, and Son, 1914.

Milsum, J. N. "The African Oil Palm in Sumatra." *Federated Malay States Agricultural Bulletin* 9, no. 2 (June 1921): 90–104.

Milsum, J. N., and C. D. V. Georgi. "Small Scale Extraction of Palm Oil." *Malayan Agricultural Journal* 26, no. 2 (February 1938): 53.

Minchinton, W. E. *The British Tinplate Industry*. Oxford: Clarendon Press, 1957.

"A Mine of Palm-Oil," *The Colonies and India*, 20 September 1879, 9.

Mingorance, Fidel, Flaminia Minelli, and Hélène Le Du. *El cultivo de la palma africana en el Chocó: legalidad ambiental, territorial y derechos humanos*. Colombia: Editorial Códice Ltda., 2004.

Ministère des colonies. *Le Palmier à Huile (Elaeis guineensis): Huile et Amandes de Palme*. Brussels: Office Colonial, 1935.

Miracle, Marvin P. *Agriculture in the Congo Basin*. Madison: University of Wisconsin Press, 1967.

Mitchell, P. K. "Trade Routes of the Early Sierra Leone Protectorate." *Sierra Leone Studies* 16 (1962): 204–17.

Mitman, Gregg. "Forgotten Paths of Empire: Ecology, Disease, and Commerce in the Making of Liberia's Plantation Economy." *Environmental History* 22, no. 1 (2017): 1–22.

———. "Reflections on the Plantationocene: A Conversation with Donna Haraway and Anna Tsing." *Edge Effects*, October 12, 2019. https://edgeeffects.net/haraway-tsing-plantationocene/.

Mockler-Ferryman, Augustus F. *British Nigeria; a Geographical and Historical Description of the British Possessions Adjacent to the Niger River, West Africa*. London: Cassell, 1902.

Mockler-Ferryman, Augustus F., and Charles Russell Day. *Up the Niger*. London: G. Philip & Son, 1892.

Moll, H. A. J. *The Economics of Oil Palm*. Wageningen: Pudoc, 1987.

Moloney, Cornelius Alfred, Herbert Druce, and G. E. Shelley. *Sketch of the Forestry of West Africa with Particular Reference to Its Present Principal Commercial Products*. London: S. Low, Marston, Searle, & Rivington, 1887.

Mondon, Gaston-Charles-Julien. "Côte d'Ivoire." In *Conférence Du Palmier à Huile (20 et 21 Octobre 1937)*, 81–86. Marseille: Institut Colonial, 1937.

Moody, Sophy. *The Palm Tree*. London: T. Nelson and Sons, 1864.

Moore, Jason W. *Capitalism and the Web of Life: Ecology and the Accumulation of Capital*. London: Verso, 2015.

Moreau de Saint-Méry, Louis-Élie. *Description Topographique, Physique, Civile, Politique et Historique de La Partie Française de l'isle Saint-Domingue*. Vol. 1. Paris: Dupont, 1797.

Morel, E. D. *Affairs of West Africa*. London: W. Heinemann, 1902.

———. *The Black Man's Burden*. Manchester: National Labour Press, 1920.

Morfit, Campbell. *Chemistry Applied to the Manufacture of Soap and Candles. . . .* Philadelphia: Carey and Hart, 1847.

Morgan, W. B. "Farming Practice, Settlement Pattern and Population Density in South-Eastern Nigeria." *Geographical Journal* 121, no. 3 (1955): 320–33.

Morgans, Courtney L., Erik Meijaard, Truly Santika, Elizabeth Law, Sugeng Budiharta, Marc Ancrenaz, and Kerrie A. Wilson. "Evaluating the Effectiveness of Palm Oil Certification in Delivering Multiple Sustainability Objectives." *Environmental Research Letters* 13, no. 6 (June 2018): 064032.

Mosbergen, Dominique. "Palm Oil Is in Everything—And It's Destroying Southeast Asia's Forests." *Huffington Post*, July 23, 2015, sec. WorldPost. http://www.huffingtonpost.com/entry/palm-oil-impacts_us_55a4c391e4b0b8145f737dd5.

Mouser, Bruce, Edwin Nuijten, Florent Okry, and Paul Richards. "Commodity and Anti-Commodity: Linked Histories of Slavery, Emancipation and Red and White Rice at Sierra Leone." *Commodities of Empire Working Paper Series*, no. 19 (2012).

Mukpo, Ashoka. "Industrial Palm Oil Investors Struggle to Gain Foothold in Africa." *GRAIN* (blog), September 25, 2019. https://www.grain.org/en/article/6331-industrial-palm-oil-investors-struggle-to-gain-foothold-in-africa.
Müllendorff, P. "The Development of German West Africa (Kamerun)." *Journal of the Royal African Society* 2, no. 5 (1902): 70–92.
Munier, P. *Le palmier à huile dans la cuvette centrale Congolaise*. Haut-Commissariat de la Republique en A.E.F., Comite de Coordination de la Recherche Agronomique et de la Production Agricole, 1956.
Murfitt, Stephen. "The English Patent System and Early Railway Technology 1800–1852." PhD diss., University of York, 2017.
Murrell, Nathaniel Samuel. *Afro-Caribbean Religions: An Introduction to Their Historical, Cultural, and Sacred Traditions*. Philadelphia: Temple University Press, 2010.
Muspratt, Sheridan. *Chemistry, Theoretical, Practical and Analytical as Applied and Relating to the Arts and Manufactures*. Glasgow: W. Mackenzie, 1853.
Musson, A. E. *Enterprise in Soap and Chemicals: Joseph Crosfield & Sons, Limited 1815–1965*. Manchester: Manchester University Press, 1965.
Naanen, Ben. "'You Are Demanding Tax from the Dead:' The Introduction of Direct Taxation and Its Aftermath in South-Eastern Nigeria, 1928–39." *African Economic History*, no. 34 (2006): 69–102.
Nadel, S. F. *A Black Byzantium: The Kingdom of Nupe in Nigeria*. Oxford: Oxford University Press, 1942.
Nair, Kannan K. *Politics and Society in South Eastern Nigeria 1841–1906*. Evanston, IL: Northwestern University Press, 1972.
——. "Trade in Southern Nigeria from the 1860s to the 1870s: Expansion or Stagnation?" *Journal of the Historical Society of Nigeria* 6, no. 4 (1973): 425–33.
Naismith, John. *Observations on the Different Breeds of Sheep, and the State of Sheep Farming, in the Southern Districts of Scotland*. Edinburgh: W. Smellie, 1795.
Nancy Sullivan & Associates Ltd. "Welpam Woris: The Social Impact of Oil Palm in Oro Province." Social Impact Assessment, 2005. https://www.academia.edu/4492658/Welpam_Woris_The_social_impact_of_oil_palm_in_Oro_Province.
National Nutrition Monitoring Bureau. *Report for the Year 1980*. Hyderabad: Indian Council of Medical Research, 1981.
Nelson, Henry Samuel. *Colonialism in the Congo Basin, 1880–1940*. Athens: Ohio University Center for International Studies, 1994.
Nelson, Paul N., Jennifer Gabriel, Colin Filer, Murom Banabas, Jeffrey A. Sayer, George N. Curry, Gina Koczberski, and Oscar Venter. "Oil Palm and Deforestation in Papua New Guinea." *Conservation Letters* 7, no. 3 (2014): 188–95.
Neslen, Arthur. "Leaked Figures Show Spike in Palm Oil Use for Biodiesel in Europe." *The Guardian*, June 1, 2016. https://www.theguardian.com/environment/2016/jun/01/leaked-figures-show-spike-in-palm-oil-use-for-biodiesel-in-europe.

Nestle, Marion. *Food Politics: How the Food Industry Influences Nutrition and Health*. Berkeley: University of California Press, 2013.

Newbury, C. W. *British Policy towards West Africa 1786–1874*. Oxford: Clarendon Press, 1965.

——. *British Policy towards West Africa: Select Documents 1875–1914*. Oxford: Clarendon Press, 1971.

Newell, Stephanie. "Dirty Whites: 'Ruffian-Writing' in Colonial West Africa." *Research in African Literatures* 39, no. 4 (2008): 1–13.

Newland, H. Osman. *The Planting, Cultivation and Expression of Coconuts, Kernels, Cacao and Edible Vegetable Oils and Seeds of Commerce*. London: Charles Griffin & Co., 1919.

Newman, Jon. *Battersea's Global Reach: The Story of Price's Candles*. Mitcham, UK: History and Social Action Publications, 2009.

Newton, Janice. "Feasting for Oil Palm." *Social Analysis: The International Journal of Social and Cultural Practice*, no. 10 (1982): 63–78.

Ngidang, Dimbab. "Contradictions in Land Development Schemes: The Case of Joint Ventures in Sarawak, Malaysia." *Asia Pacific Viewpoint* 43, no. 2 (August 2002): 157–80.

Nicholls, William H. "Some Economic Aspects of the Margarine Industry." *Journal of Political Economy* 54, no. 3 (1946): 221–42.

Nicolaï, Henri. "Le Congo et l'huile de Palme. Un Siècle. Un Cycle?" *Belgeo: Revue Belge de Géographie* 4 (2013).

Nikolakis, William, Lijo John, and Harish Krishnan. "How Blockchain Can Shape Sustainable Global Value Chains: An Evidence, Verifiability, and Enforceability (EVE) Framework." *Sustainability* 10, no. 11 (2018): 3926.

Nkongho, Raymond N., Thomas E. Ndjogui, and Patrice Levang. "History of Partnership between Agro-Industries and Oil Palm Smallholders in Cameroon." *OCL* 22, no. 3 (May 2015): A301.

Nkongho, Raymond Ndip, Laurène Feintrenie, and Patrice Levang. *The Non-Industrial Palm Oil Sector in Cameroon*. Bogor, Indonesia: CIFOR, 2014.

Noah, M. Efiong. "Political History of the City States of Old Calabar, 1820–60." In *Studies in Southern Nigerian History*, edited by B. I. Obichere, 39–72. London: Frank Cass, 1982.

Nonini, Donald M. *British Colonial Rule and the Resistance of the Malay Peasantry, 1900–1957*. New Haven, CT: Yale University Press, 1992.

Norris, Robert. *Memoirs of the Reign of Bossa Ahádee: King of Dahomy, an Inland Country of Guiney*. . . . London: W. Lowndes, 1789.

Northrup, David. *Beyond the Bend in the River: African Labor in Eastern Zaire, 1865–1940*. Athens: Ohio University Center for International Studies, 1988.

——. "Nineteenth-Century Patterns of Slavery and Economic Growth in Southeastern Nigeria." *International Journal of African Historical Studies* 12, no. 1 (1979): 1–16.

——. "The Compatibility of the Slave and Palm Oil Trades in the Bight of Biafra." *Journal of African History* 17, no. 3 (1976): 353–64.

Nurul, Mohd Ahyan, Samion Syahrullail, and Hai Wen Teng. "Alternative Lubricants: Study on Palm Oil–Based Lubricants in Metal Forming Process." *Journal of Oil Palm Research* 28, no. 1 (2016): 93–103.

Nwabughuogu, Anthony I. "Oil Mill Riots in Eastern Nigeria 1948/51: A Study in Indigenous Reaction to Technological Innovation." *Africa Development* 7, no. 4 (1982): 66–84.

Nwanze, S. C. "The Economics of the Pioneer Oil Mill." *Journal of the West African Institute for Oil Palm Research* 2 (1956): 231–56.

Nwaubani, Ebere. "The Political Economy of Aboh, 1830–1857." *African Economic History*, no. 27 (1999): 93–116.

Nwokeji, G. Ugo. "The Slave Emancipation Problematic: Igbo Society and the Colonial Equation." *Comparative Studies in Society and History* 40, no. 2 (1998): 318–55.

Nworah, K. Dike. "The Politics of Lever's West African Concessions, 1907–1913." *International Journal of African Historical Studies* 5, no. 2 (1972): 248–64.

Nye, David E. *American Technological Sublime*. Cambridge, MA: MIT Press, 1996.

Oas, Sarah E., A. Catherine D'Andrea, and Derek J. Watson. "10,000 Year History of Plant Use at Bosumpra Cave, Ghana." *Vegetation History and Archaeobotany* 24, no. 5 (September 2015): 635–53.

Obeng-Odoom, Franklin. "Understanding Land Reform in Ghana: A Critical Postcolonial Institutional Approach." *Review of Radical Political Economics* 48, no. 4 (2016): 661–80.

Obichere, Boniface. "Women and Slavery in the Kingdom of Dahomey." *Outre-Mers. Revue d'histoire* 65, no. 238 (1978): 5–20.

Ofonagoro, Walter Ibekwe. *Trade and Imperialism in Southern Nigeria, 1881–1929*. Lagos: Nok, 1979.

Ogunremi, Deji. "Economic Development and Warfare in 19th Century Yorubaland." In *War and Peace in Yorubaland*, edited by I. A. Akinjogbin. Ibadan, Nigeria: Heinemann, 1998.

Ohadike, Don C. *Anioma: A Social History of the Western Igbo People*. Athens: Ohio University Press, 1994.

——. *The Ekumeku Movement: Western Igbo Resistance to the British Conquest of Nigeria, 1883–1914*. Athens: Ohio University Press, 1991.

O'Hear, Ann. "The Enslavement of Yoruba." In *The Yoruba Diaspora in the Atlantic World*, edited by Toyin Falola and Matt D. Childs, 58–73. Bloomington: Indiana University Press, 2004.

The Oil Palm. "History and Origin," January 1, 2014. http://theoilpalm.org/history-and-origin/.

"Oil Palm in Labuan: A Success and a Failure (*Elæis guineensis*)." *Bulletin of Miscellaneous Information (Royal Botanic Gardens, Kew)* 1889, no. 35 (1889): 259–67.

Ojo, G. U., R. A. Offiong, S. O. Akhaine, A. Baiyewu-Teru, and F. Allen. *Oil Palm Plantations in Forest Landscapes: Impacts, Aspirations and Ways Forward in Nigeria*. Wageningen: Tropenbos International, 2017.

Okonta, Ike, and Oronto Douglas. *Where Vultures Feast: Shell, Human Rights, and Oil in the Niger Delta*. London: Verso, 2003.

Olabimtan, Kehinde. "Townsend, Henry." In *Dictionary of African Christian Biography*, 2011. https://dacb.org/stories/nigeria/townsend-henry/.

Ollagnier, Michel. "Coopération française en Colombie: la plantation de palmiers à huile de San Alberto." *Oléagineux* 21 (1966): 575–82.

Olomola, Ade. *Financing Oil Palm Production through Contract Farming*. Ibadan, Nigeria: Nigerian Institute of Social and Economic Research, 1991.

Oloruntimehin, B. Olatunji. "The French Estate in West Africa, 1890–1918." *Journal of the Historical Society of Nigeria* 7, no. 3 (1974): 447–63.

Olukoju, Ayodeji. "Confronting the Combines: Producers' and Traders' Militancy in Western Nigeria, 1934–1939." *Nordic Journal of African Studies* 9, no. 1 (2000): 49–69.

———. "Government, the Business Community, and Quality Control Schemes in the Agricultural Export Trade of Nigeria, 1889–1929." *African Economic History*, no. 26 (1998): 99–118.

———. "'Nigeria or Lever-Ia?': Nationalist Reactions to Economic Depression and the 'Menace of Mergers' in Colonial Nigeria." *Journal of Third World Studies* 19, no. 1 (2002): 173–94.

———. *The "Liverpool" of West Africa: The Dynamics and Impact of Maritime Trade in Lagos, 1900–1950*. Trenton: Africa World Press, 2004.

———. "The United Kingdom and the Political Economy of the Global Oils and Fats Business during the 1930s." *Journal of Global History* 4, no. 1 (2009): 105–25.

Ong-Abdullah, Meilina, Jared M. Ordway, Nan Jiang, Siew-Eng Ooi, Sau-Yee Kok, Norashikin Sarpan, Nuraziyan Azimi, et al. "Loss of *Karma* Transposon Methylation Underlies the Mantled Somaclonal Variant of Oil Palm." *Nature* 525, no. 7570 (September 2015): 533–37.

Onwueme, Inno Chukuma. *Like a Lily among Thorns: Colonial African Village Child Transitions to Post-Colonial Modernity, and America*. Bloomington: AuthorHouse, 2014.

Onyioha, K. O. K. *ENDC Oils*. Enugu, Nigeria: ENDC, 1965.

Ooi Jin-Bee. *Peninsular Malaysia*. London: Longman, 1976.

Ooi Keat Gin. "An Economic History of Sarawak during the Period of Brooke Rule, 1841–1946." PhD diss., University of Hull, 1995.

Ooi, S. C., J. C. Nascimento, and Edson B. Da Silva. "The Oil Palm Industry in Brazil—Current Status and Future Potential." In *Oil Palm in the Eighties*, edited by E. Pushparajah and Chew Poh Soon, 369–91. Kuala Lumpur: Incorporated Society of Planters, 1982.

Operations Evaluation Department. "Papua New Guinea: New Britain Smallholder Development Projects." Impact Evaluation Report. Washington, DC: World Bank, 1980.

———. "Project Performance Audit Report: Indonesia First and Second North Sumatra Estates Projects." Washington, DC: World Bank, 1978.

Opsomer, J. E. "Les Premières Descriptions Du Palmier à Huile (*Elaeis guineensis* Jacq.)." *Bulletin Des Seances Academie Royale Des Sciences Coloniales*, new series, 2, no. 2 (1956): 253–72.

Orewa, S. I. "Designing Agricultural Development Projects for the Small Scale Farmers: Some Lessons from the World Bank Assistance Small Holder Oil Palm Development Scheme in Nigeria." *Journal of Applied Sciences* 8, no. 2 (2008): 295–301.

Oriji, J. N. "Igbo Women, Technological Changes and Protests (1946–1953)." *Moebius* 1, no. 1 (2003): 11–21.

———. "A Re-Assessment of the Organisation and Benefits of the Slave and Palm Produce Trade Amongst the Ngwa—Igbo." *Canadian Journal of African Studies* 16, no. 3 (1982): 523–48.

Oroge, E. Adeniyi. "The Institution of Slavery in Yorubaland with Particular Reference to the Nineteenth Century." PhD diss., University of Birmingham, 1971.

Orstom, G. Courade. *Unilever Industrial Plantations in Cameroon*. Yaounde, Cameroon: National Office for Scientific and Technical Research, 1977.

Orta, Garcia da. *Colóquios dos simples e drogas da India* (1563). Edited by Conde de Ficalho. Lisbon: Imprensa Nacional, 1891.

Osborn, Emily Lynn. "From Bauxite to Cooking Pots: Aluminum, Chemistry, and West African Artisanal Production." *History of Science* 54, no. 4 (2016): 425–42.

Osei-Tutu, John Kwadwo, and Victoria Ellen Smith, eds. *Shadows of Empire in West Africa: New Perspectives on European Fortifications*. London: Palgrave Macmillan, 2018.

Oslender, Ulrich. *The Geographies of Social Movements: Afro-Colombian Mobilization and the Aquatic Space*. Durham, NC: Duke University Press, 2016.

Oslisly, Richard, and Lee White. "Human Impact and Environmental Exploitation in Gabon during the Holocene." In *Rethinking Agriculture: Archaeological and Ethnoarchaeological Perspectives*, edited by Timothy P. Denham, José Iriarte, and Luc Vrydaghs, 347–60. Abingdon, UK: Routledge, 2016.

Oslisly, Richard, Lee White, Ilham Bentaleb, Charly Favier, Michel Fontugne, Jean-François Gillet, and David Sebag. "Climatic and Cultural Changes in the West Congo Basin Forests over the Past 5000 Years." *Philosophical Transactions of the Royal Society B: Biological Sciences* 368, no. 1625 (September 2013).

Ospina Bozzi, Martha Luz, and Doris Ochoa Jaramillo. *La palma africana en Colombia: apuntes y memorias*. Santa Fe de Bogotá: FEDEPALMA, 1998.

Ostfeld, Rosemary, David Howarth, David Reiner, and Pawel Krasny. "Peeling Back the Label—Exploring Sustainable Palm Oil Ecolabelling and Consumption in the United Kingdom." *Environmental Research Letters* 14, no. 1 (2019): 014001.

Othman, Jamal B., Jack E. Houston, and Christopher S. McIntosh. "Health Issue Commodity Promotion: Impacts on US Edible Vegetable Oil Demand." *Food Policy* 18, no. 3 (1993): 214–23.

Otite, Onigu. "Rural Migrants as Catalysts in Rural Development: The Urhobo in Ondo State, Nigeria." *Africa: Journal of the International African Institute* 49, no. 3 (1979): 226–34.

Ouedraogo, Hubert, Donata Gnisci, and Leonidas Hitimana. "Land Reform Processes in West Africa: A Review." Sahel and West Africa Club Secretariat, August 2006. https://dlc.dlib.indiana.edu/dlc/bitstream/handle/10535/5486

/06_OECD_LandNetWestAfrica_Land_reform_processes_in_West_Africa.pdf ?sequence=1.

Pacheco Pereira, Duarte. *Esmeraldo de Situ Orbis*. Translated by George H. Kimble. London: Hakluyt Society, 1936.

Pakiam, Geoffrey. "Smallholders and the Making of Malaysia's Oil Palm Industry." Washington, DC, 2019. https://www.iseas.edu.sg/images/pdf /ISEASEWP2019-5Pakiam.pdf.

Pakiam, Geoffrey K. "Smallholder Involvement in Tree Crops in Malaya, with Special Reference to Oil and Coconut Palms in Johor, 1862–1963." PhD diss., University of London (SOAS), 2018.

———. "Why Don't Some Cuisines Travel? Charting Palm Oil's Journey from West African Staple to Malayan Chemical." *Journal of Global History* 15, no. 1 (2020): 39–60.

"Palm Kernel Meal." *Flour and Feed* 17, no. 2 (1916): 27.

"The Palm-Oil Districts of Africa." *Chemist and Druggist* 39 (4 July 1891): 26.

"The Palm Oil Factory at Elaeis Estate, Johore." *Malayan Agricultural Journal* 17, no. 10 (1929): 497–501.

"Palm Oil in Margarine." *The Dairy World* 11 (16 September 1902): 120–21.

Palmer, William. "How Ideology Works: Historians and the Case of British Abolitionism." *Historical Journal* 52, no. 4 (2009): 1039–51.

Panganiban, Chris. "Tension Grips Agusan Plantation as Land Reform Claimants Bicker." *Philippine Daily Inquirer*, December 6, 2017. https://newsinfo.inquirer .net/950133/tension-grips-agusan-plantation-as-land-reform-claimants-bicker.

Pantzaris, T. P. *Pocketbook of Palm Oil Uses*. Kuala Lumpur: MPOB, 2000.

Parés, Luis Nicolau. *The Formation of Candomblé: Vodun History and Ritual in Brazil*. Chapel Hill: University of North Carolina Press, 2013.

Parmer, J. Norman. *Colonial Labor Policy and Administration: A History of Labor in the Rubber Plantation Industry in Malaya, c. 1910–1941*. Locust Valley, NY: Association for Asian Studies, 1960.

Parnell, Edward A. *Applied Chemistry: In Manufactures, Arts and Domestic Economy*. London: Taylor and Walton, 1844.

Parrish, Susan Scott. *American Curiosity: Cultures of Natural History in the Colonial British Atlantic World*. Chapel Hill, NC: University of North Carolina Press, 2006.

Partridge, Charles. *Cross River Natives: Being Some Notes on the Primitive Pagans of Obubura Hill District. . . .* London: Hutchinson and Co., 1905.

Patel, Raj, and Jason W. Moore. *A History of the World in Seven Cheap Things: A Guide to Capitalism, Nature, and the Future of the Planet*. Oakland: University of California Press, 2017.

Patents for Inventions. Abridgements of Specifications Relating to Oils, Animal, Vegetable, and Mineral; Including Lubricants, Candles, and Soap, A.D. 1617–1863. London: Office of the Commissioners of Patents for Inventions, 1865.

Paterson, R. Russell M. "Ganoderma Boninense Disease of Oil Palm to Significantly Reduce Production after 2050 in Sumatra If Projected Climate Change Occurs." *Microorganisms* 7, no. 1 (January 2019): 24.

Paterson, R. Russell M., and Nelson Lima. "Climate Change Affecting Oil Palm Agronomy, and Oil Palm Cultivation Increasing Climate Change, Require Amelioration." *Ecology and Evolution* 8, no. 1 (November 2017): 452–61.

Patiño, Victor Manuel. "Informacion Preliminar Sobre La Palma de Aceite Africana En Colombia." Estacion Agro-Forestal del Pacifico de Calima-Buenaventura, December 1948.

Pavlakis, Dean. *British Humanitarianism and the Congo Reform Movement, 1896–1913*. Farnham, UK: Routledge, 2015.

Péhaut, Yves. "Le cocotier et le palmier à huile en Amérique." *Les Cahiers d'Outre-Mer* 53, no. 209 (2000): 55–94.

Peluso, Nancy Lee. "The Plantation and the Mine: Agrarian Transformation and the Remaking of Land and Smallholders in Indonesia." In *Land and Development in Indonesia: Searching for the People's Sovereignty*, edited by John F. McCarthy and Kathryn Robinson, 35–62. Singapore: ISEAS—Yusof Ishak Institute, 2016.

Peluso, Nancy Lee, Suraya Afiff, and Noer Fauzi Rachman. "Claiming the Grounds for Reform: Agrarian and Environmental Movements in Indonesia." *Journal of Agrarian Change* 8, no. 2–3 (2008): 377–407.

Peluso, Nancy Lee, and Peter Vandergeest. "Genealogies of the Political Forest and Customary Rights in Indonesia, Malaysia, and Thailand." *Journal of Asian Studies* 60, no. 3 (2001): 761–812.

Pelzer, Karl Josef. *Planter and Peasant: Colonial Policy and the Agrarian Struggle in East Sumatra 1863–1947*. Gravenhage: Nijhoff, 1978.

———. *Planters against Peasants: The Agrarian Struggle in East Sumatra 1947–1958*. Gravenhage: Martinus Nijhoff, 1982.

Pepple, King of Grand Bonny. *A Brief Statement of the Case of King Pepple of Grand Bonny, Showing the Wrongs Sustained by Him at the Hands of the British Government*. London: George Coventry, 1857.

Persoon, Gerard. "The Kubu and the Outside World (South Sumatra, Indonesia). The Modification of Hunting and Gathering." *Anthropos* 84, no. 4/6 (1989): 507–19.

Perumal, S. "Large-Scale Development of Oil Palm as a Smallholder Crop." In *Oil Palm in Agriculture in the Eighties*, edited by E. Pushparajah and Chew Poh Soon. Kuala Lumpur: Incorporated Society of Planters, 1981.

Peša, Iva. "Between Waste and Profit: Environmental Values on the Central African Copperbelt." *The Extractive Industries and Society*, August 2020. https://doi.org/10.1016/j.exis.2020.08.004.

Petrick, Gabriella M. "Larding the Larder." *The Senses and Society* 5, no. 3 (November 2010): 382–87.

Phillips, Anne. *The Enigma of Colonialism: British Policy in West Africa*. London: Currey, 1989.

Pietilainen, Emma Pauliina, and Gerardo Otero. "Power and Dispossession in the Neoliberal Food Regime: Oil Palm Expansion in Guatemala." *Journal of Peasant Studies* 46, no. 6 (2019): 1142–66.

Piggott, C. J. *Growing Oil Palms: An Illustrated Guide*. Kuala Lumpur: Incorporated Society of Planters, 1990.
Pim, Alan. *Colonial Agricultural Production: The Contribution Made by Native Peasants and by Foreign Enterprise*. London: Oxford University Press, 1946.
Pinto, Viveat Susan. "40 Years Ago . . . And Now: How Dalda Built, and Lost, Its Monopoly." *Business Standard India*, March 5, 2015. https://www.business-standard.com/article/management/40-years-ago-and-now-how-dalda-built-and-lost-its-monopoly-115030501153_1.html.
Pioch, Daniel, and Gilles Vaitilingom. "Palm Oil and Derivatives: Fuels or Potential Fuels?" *OCL. Oléagineux, Corps Gras, Lipides* 12, no. 2 (March 2005): 161–69.
Pischke, Erin C., Mark D. Rouleau, and Kathleen E. Halvorsen. "Public Perceptions towards Oil Palm Cultivation in Tabasco, Mexico." *Biomass and Bioenergy* 112 (2018): 1–10.
Pletcher, James. "Regulation with Growth: The Political Economy of Palm Oil in Malaysia." *World Development* 19, no. 6 (1991): 623–36.
Ploeg, A. "Sociological Aspects of Kapore Settlement." In *The Hoskins Development*, 23–118. Port Moresby: Australian National University, 1972.
Poku, Kwasi. *Small-Scale Palm Oil Processing in Africa*. FAO Agricultural Services Bulletin 148. Rome: FAO, 2002.
Pomet, Pierre. *A Compleat History of Druggs, Written in French by Monsieur Pomet, Chief Druggist to the Late French King Lewis XIV*. 3rd English edition. London: J. and J. Bonwicke et al., 1737.
Porter, Donna V., and Donna U. Vogt. "Labeling of Tropical Oils: Legislation, Health and Trade Issues." 20 November 1987, Congressional Research Service, Washington, DC.
Potter, Lesley. "Alternative Pathways for Smallholder Oil Palm in Indonesia: International Comparisons." In *The Oil Palm Complex: Smallholders, Agribusiness and the State in Indonesia and Malaysia*, edited by R. A. Cramb and John F. McCarthy, 155–88. Singapore: NUS Press, 2016.
———. "Colombia's Oil Palm Development in Times of War and 'Peace': Myths, Enablers and the Disparate Realities of Land Control." *Journal of Rural Studies* 78 (August 2020): 491–502.
———. "How Can the People's Sovereignty Be Achieved in the Oil Palm Sector? Is the Plantation Model Shifting in Favour of Smallholders?" In *Land and Development in Indonesia*, edited by John F. McCarthy and Kathryn Robinson, 293–320. Singapore: ISEAS—Yusof Ishak Institute, 2016.
———. *Managing Oil Palm Landscapes: A Seven-Country Survey of the Modern Palm Oil Industry in Southeast Asia, Latin America and West Africa*. Bogor, Indonesia: CIFOR, 2015.
Potter, Lesley, and Justin Lee. *Tree Planting in Indonesia: Trends, Impacts and Directions*. Occassional Paper 18. Bogor, Indonesia: CIFOR, 1998.
Prasetijo, Adi. "Living without the Forest: Adaptive Strategy of Orang Rimba." *Senri Ethnological Studies* 95 (2017): 255–78.

Pratten, David. *Man-Leopard Murders*. Edinburgh: Edinburgh University Press, 2007.
Price, Richard. "Subsistence on the Plantation Periphery: Crops, Cooking and Labour among Eighteenth-Century Suriname Maroons." In *The Slaves' Economy: Independent Production by Slaves in the Americas*, edited by Ira Berlin and Philip D. Morgan, 107–30. London: Frank Cass, 1991.
Price's Patent Candle Company. "Letter . . . to the Board of Directors of [the] . . . Company, 4th Nov., 1852. [on welfare work in the Company]." https://www.jstor.org/stable/10.2307/60219055
Prince, Nancy. *The West Indies: Being a Description of the Islands, Progress of Christianity, Education, and Liberty among the Colored Population Generally*. Boston: Dow and Jackson, 1841.
Prodöhl, Ines. "From Dinner to Dynamite: Fats and Oils in Wartime America." *Global Food History* 2, no. 1 (2016): 1–20.
Pugh, Peter. *Great Enterprise: A History of Harrisons & Crosfield*. Edited by Guy Nickalls. London: Harrisons & Crosfield, 1990.
Puplett, P. A. R. *Synthetic Detergents: A Study of the Development and Marketing of a New Product*. London: Sidgwick and Jackson, 1957.
Purchas, Samuel. *Hakluytus Posthumus* (1625). Glasgow: J. MacLehose and sons, 1905.
——. *Purchas His Pilgrimes. Part 1 In Five Bookes*. . . . . London: William Stansby, 1625.
——. *Purchas His Pilgrimes. Part 2 In Five Bookes*. . . . London: William Stansby, 1625.
Purdy, John. *Memoir, Descriptive and Explanatory, to Accompany the New Chart of the Ethiopic or Southern Atlantic Ocean, with the Western Coasts of South-America, from Cape Horn to Panama*. . . . London: R. H. Laurie, 1822.
Purvis, Malcolm. "New Sources of Growth in a Stagnant Smallholder Economy in Nigeria: The Oil Palm Rehabilitation Scheme." In *Growth and Development of the Nigerian Economy*, edited by Carl Eicher and Carl Liedholm, 267–81. Lansing: Michigan State University Press, 1970.
Putzel, James. *A Captive Land: The Politics of Agrarian Reform in the Philippines*. London: Catholic Institute for International Relations, 1992.
Pye, Oliver. "The Biofuel Connection-Transnational Activism and the Palm Oil Boom." *Journal of Peasant Studies* 37, no. 4 (2010): 851–74.
——. "Commodifying Sustainability: Development, Nature and Politics in the Palm Oil Industry." *World Development* 121 (2019): 218–28.
——. "Introduction." In *The Palm Oil Controversy in Southeast Asia: A Transnational Perspective*, edited by Oliver Pye and Jayati Bhattacharya, 1–18. Singapore: ISEAS Publishing, 2015.
"The Quality of Tin Roofs." *Metal Worker* 43 (March 3, 1895): 32.
"La Question Des Palmeraies a La Cote d'Ivoire." *Bulletin des matières grasses* 1, no. 1 (1920): 3–14.
Quincy, John. *Pharmacopoia Officinalis & Extemporanea*. 10th ed. Vol. 1. London: Longman, 1736.

Rabiu, Ademola, Samya Elias, and Oluwaseun Oyekola. "Oleochemicals from Palm Oil for the Petroleum Industry." In *Palm Oil*, edited by Viduranga Yashasvi Waisundara. London: IntechOpen, 2018. https://www.intechopen.com/books/palm-oil.
"Railway and Waggon Grease." *Railway Times* 15, no. 52 (December 26, 1863): 411.
"Railway Grease." *Scientific American*, August 26, 1865.
Raja, Sivachandralingam Sundara. *The Economy of Colonial Malaya: Administrators versus Capitalists*. London: Routledge, 2017.
Ramamurthy, Anandi. *Imperial Persuaders: Images of Africa and Asia in British Advertising*. Manchester: Manchester University Press, 2003.
Ramasamy, P. "Labour Control and Labour Resistance in the Plantations of Colonial Malaya." *Journal of Peasant Studies* 19, no. 3–4 (April 1992): 87–105.
Ramsden, Christopher E., Daisy Zamora, Sharon Majchrzak-Hong, Keturah R. Faurot, Steven K. Broste, Robert P. Frantz, John M. Davis, Amit Ringel, Chirayath M. Suchindran, and Joseph R. Hibbeln. "Re-Evaluation of the Traditional Diet-Heart Hypothesis: Analysis of Recovered Data from Minnesota Coronary Experiment (1968–73)." *BMJ* 353 (April 12, 2016): i1246.
Rao, Gita. "The Changing Position of Igbo- and Ibibio-Speaking Women in the Palm Oil Industry." MA thesis, American University, Washington, DC, 1979.
Rapley, John. *Ivoirien Capitalism*. Boulder: Lynne Rienner Publishers, 1993.
Rasiah, Rajah. "Explaining Malaysia's Export Expansion in Palm Oil and Related Products." In *Technology, Adaptation, and Exports*, edited by Vandana Chandra, 163–92. Washington, DC: World Bank, 2006.
Ray, Utsa. "The Body and Its Purity: Dietary Politics in Colonial Bengal." *Indian Economic & Social History Review* 50, no. 4 (October 2013): 395–421.
Raymond, W. D. "The Oil Palm Industry." *Tropical Science* 3, no. 2 (1961): 69–89.
Read, Hollis. *The Negro Problem Solved*. New York: A. A. Constantine, 1864.
"Recent Progress in Agriculture and the Development of Natural Resources." *Bulletin of the Imperial Institute* 9, no. 3 (1911): 289–313.
"Recent Progress in Agriculture and the Development of Natural Resources." *Bulletin of the Imperial Institute* 12, no. 1 (1914): 126–41.
Reid, John. "Warrior Aristocrats in Crisis: The Political Effects of the Transition from the Slave Trade to Palm Oil Commerce in the Nineteenth Century Kingdom of Dahomey." PhD diss., University of Stirling, 1986.
Rendell, Sir William. *The History of the Commonwealth Development Corporation, 1948–1972*. London: Heinemann Educational Books, 1976.
Report from the Select Committee of the House of Lords, Appointed to Consider The Best Means Which Great Britain Can Adopt for the Final Extinction of the African Slave Trade. . . . P.P. 1849 (53) ix 279. London: HMSO, 1850.
Report from the Select Committee on Africa (Western Coast). P.P. 1865 (412) v 1. London: HMSO, 1865.
Report from the Select Committee on the West Coast of Africa. P.P. 1842 (551, 2 parts) xi–xii. London: HMSO, 1842.
Report of the Departmental Committee on Compensation for Industrial Diseases. Cd. 3495. London: HMSO, 1907.

Report of the FAO Oilseed Mission for Venezuela. Washington, DC: Food and Agriculture Organization, 1949.

Report of the Mission to Enquire into the Production and Transport of Vegetable Oils and Oilseeds in the West African Colonies. Colonial no. 211. London: HMSO, 1947.

Reports made for the Year 1856 to the Secretary of State . . . with a view to exhibit generally the Past and Present State of Her Majesty's Colonial Possessions. P.P. 1857–1858 (2403) xl 1. London: HMSO, 1858.

Rettberg, Angelika. "Business-Led Peacebuilding in Colombia: Fad or Future of a Country in Crisis?" Crisis States Programme working paper. Development Research Centre, London School of Economics, 2004.

Reyes, Paola García, Jennifer Vargas, and Sonia Uribe Kaffure. "Estado, Poder y Dominio. Una Lectura a Propósito de Sus Mecanismos." *Perfiles Latinoamericanos* 26, no. 5 (2018): 191–214.

Reynolds, Edward. "Agricultural Adjustments on the Gold Coast after the End of the Slave Trade, 1807–1874." *Agricultural History* 47, no. 4 (1973): 308–18.

———. *Trade and Economic Change on the Gold Coast, 1807–1874*. London: Longman, 1974.

Rhoades, Patricia Clare. *Los Cambios En La Cadena de Produccion de La Palma Aceitera En El Pacifico Costarricense: Una Historia Economica, Socioambiental y Tecnoscientifica 1950–2007*. San Jose: Socieded Editora Alquimia 2000, 2011.

Richards, Paul. *Indigenous Agricultural Revolution: Ecology and Food Crops in West Africa*. London: Westview Press, 1985.

Richardson, D. L. "La historia del mejoramiento genético de la palma aceitera en la compañía United Fruit en América." *ASD Oil Palm Papers (Costa Rica)* 11 (1995): 1–22.

Richardson, Nathaniel R. *Liberia's Past and Present*. London: Diplomatic Press and Publishing, 1959.

Richardson, Thomas, and Henry Watts. *Chemical Technology; or Chemistry in Its Applications to the Arts and Manufactures*. 2nd ed. London: H. Baillière, 1863.

Ridler, Neil B. "A Case in Import Substitution: Palm Oil in Colombia." *Inter-American Economic Review* 29, no. 4 (1979): 21–34.

Ridley, H. N. "The Oil Palm." *Agricultural Bulletin of the Straits Settlement and Federated Malay States*, no. 2 (1907): 37–40.

Riepma, Siert F. "Margarine in Western Europe." Foreign Agricultural Service. Washington, DC: US Department of Agriculture, May 1960.

Rist, Gilbert. *The History of Development*. Translated by Patrick Camiller. London: Zed Books, 2002.

Rival, Alain, and Patrice Levang. *Palms of Controversies: Oil Palm and Development Challenges*. Bogor, Indonesia: CIFOR, 2014.

Roadnight, Andrew. *United States Policy towards Indonesia in the Truman and Eisenhower Years*. Basingstoke, UK: Palgrave Macmillan, 2002.

Roberts, Jonathan. "Medical Exchange on the Gold Coast during the Seventeenth and Eighteenth Centuries." *Canadian Journal of African Studies / Revue Canadienne Des Études Africaines* 45, no. 3 (2011): 480–523.

Robins, Jonathan E. *Cotton and Race across the Atlantic: Britain, Africa, and America, 1900–1920*. Rochester, NY: University of Rochester Press, 2016.

———. "'Imbibing the Lesson of Defiance': Oil Palms and Alcohol in Colonial Ghana, 1900–40." *Environmental History* 23, no. 2 (2018): 293–317.

———. "Oil Boom: Agriculture, Chemistry, and the Rise of Global Plant Fat Industries, ca. 1850–1920." *Journal of World History* 29, no. 3 (2018): 313–42.

———. "Slave Cocoa and Red Rubber: E. D. Morel and the Problem of Ethical Consumption." *Comparative Studies in Society and History* 54, no. 3 (July 2012): 592–611.

———. "Smallholders and Machines in the West African Palm Oil Industry, 1850–1950." *African Economic History* 46, no. 1 (2018): 69–103.

Rogers, Vance. *Abaca, Cacao, and the African Oil Palm in Costa Rica. A Special Report*. Washington, DC: Institute of Inter-American Affairs, 1947.

Rømer, Ludvig Ferdinand. *A Reliable Account of the Coast of Guinea (1760)*. Translated by Selena Axelrod Winsnes. Oxford: Oxford University Press, 2000.

Rönnbäck, Klas. *Labour and Living Standards in Pre-Colonial West Africa: The Case of the Gold Coast*. Abingdon, UK: Routledge, 2015.

Ross, Corey. *Ecology and Power in the Age of Empire: Europe and the Transformation of the Tropical World*. Oxford: Oxford University Press, 2017.

———. "The Plantation Paradigm: Colonial Agronomy, African Farmers, and the Global Cocoa Boom, 1870s–1940s." *Journal of Global History* 9, no. 1 (March 2014): 49–71.

Ross, David A. "The Career of Domingo Martinez in the Bight of Benin 1833–64." *Journal of African History* 6, no. 1 (1965): 79–90.

Roughley, Thomas. *The Jamaica Planter's Guide; Or, A System for Planting and Managing a Sugar Estate, or Other Plantations in That Island, and Throughout the British West Indies in General*. London: Longman, Hurst, Rees, Orme, and Brown, 1823.

Royal Botanic Gardens, Kew. *The Useful Plants of Nigeria IV*. London: HSMO, 1922.

Rudin, Harry R. *Germans in the Cameroons, 1884–1914: A Case Study in Modern Imperialism*. New Haven, CT: Yale University Press, 1938.

Ruf, Francois O. "The Myth of Complex Cocoa Agroforests: The Case of Ghana." *Human Ecology* 39 (2011): 373–88.

Rupilius, Wolfgang, and Salmiah Ahmad. "The Changing World of Oleochemicals." *Palm Oil Developments* 44 (June 2006): 21–28.

Russell, Colin Archibald, and John A. Hudson. *Early Railway Chemistry and Its Legacy*. Cambridge: Royal Society of Chemistry, 2012.

Rutgers, A. A. L. "African Oil Palm: Developments in Sumatra." *The Planter* 3, no. 11 (1923): 595.

———. "Crop Records of the Oil Palm." *Federated Malay States Agricultural Bulletin* 8, no. 4 (December 1920): 247–55.

———. "The Cultivation of the Oil Palm on the East Coast of Sumatra." In *Oil Palms and Their Fruit*, 1–7. London: Griffith, 1922.

Rutgers, A. A. L., H. N. Blommendaal, F. C. Van Heurn, C. Heusser, J. A. Maas, and C. Yampolsky. *Investigations on Oilpalms Made at the General Experiment Station of the A.V.R.O.S.* Batavia: Ruygrok, 1922.

Rutgers, A. A. L., Edmond LePlae, and Paul Tingey. *Oil Palms and Their Fruit.* London: Griffith, 1922.

Sacerdoti, Guy, and Sheilah Ocampo. "Guthrie and the Angels." *Far Eastern Economic Review* (November 19, 1982): 58–62.

Sagar, H. S. Sathya Chandra, Amani Mabano, Ramya Roopa, Mahmuda Sharmin, Freddie-Jeanne Richard, and Julia Clause. "India in the Oil Palm Era: Describing India's Dependence on Palm Oil, Recommendations for Sustainable Production, and Opportunities to Become an Influential Consumer." *Tropical Conservation Science* 12 (2019): 1940082919838918.

Salleh, Halim. "Peasants, Proletarianisation and the State: FELDA Settlers in Pahang." In *Fragmented Visions,* edited by Joel S. Kahn and Lok Kok Wah, 107–32. North Sydney: Asian Studies Association of Australia, 1992.

Salmon, William. *Medicina Practica, or, Practical Physick Shewing the Method of Curing the Most Usual Diseases Happening to Humane Bodies. . . .* London: T. Howkins et al., 1692.

Salzmann, Ulrich, and Philipp Hoelzmann. "The Dahomey Gap: An Abrupt Climatically Induced Rain Forest Fragmentation in West Africa during the Late Holocene." *The Holocene* 15, no. 2 (2005): 190–99.

Sanders, James. "Palm Oil Production on the Gold Coast in the Aftermath of the Slave Trade: A Case Study of the Fante." *The International Journal of African Historical Studies* 15, no. 1 (1982): 49.

Sanderson, Jean-Paul. "Le Congo belge entre mythe et réalité. Une analyse du discours démographique colonial." *Population* 55, no. 2 (2000): 331–55.

Sandhu, Kernial Singh. *Indians in Malaya: Some Aspects of Their Immigration and Settlement (1786–1957).* Cambridge: Cambridge University Press, 1969.

Santika, Truly, Kerrie A. Wilson, Sugeng Budiharta, Elizabeth A. Law, Tun Min Poh, Marc Ancrenaz, Matthew J. Struebig, and Erik Meijaard. "Does Oil Palm Agriculture Help Alleviate Poverty? A Multidimensional Counterfactual Assessment of Oil Palm Development in Indonesia." *World Development* 120 (2019): 105–17.

Saravanamuttu, Johan. "The Political Economy of Migration and Flexible Labour Regimes: The Case of the Oil Palm Industry in Malaysia." In *The Palm Oil Controversy in Southeast Asia: A Transnational Perspective,* edited by Oliver Pye and Jayati Bhattacharya, 120–39. Singapore: ISEAS Publishing, 2015.

Sarbah, J. "The Oil Palm and Its Uses." *Journal of the Royal African Society* 8 (1909): 232–50.

Sarbah, John Mensah. *Fanti Customary Laws: A Brief Introduction to the Principles of the Native Laws and Customs. . . .* London: W. Clowes and Sons, 1897.

Sato, Hiroaki. "Hunting of the Boyela, Slash-And-Burn Agriculturalists, in the Central Zaire Forest." *African Study Monographs* 4 (1983): 1–54.

Sayam, Vylie Tientcheu, and Emmanuelle Cheyns. "From the Coffee-Cocoa Combination to Oil Palm Cycles: The Case of Dabou and Aboisso in Côte d'Ivoire." In *Economics and Ecological Diversification: The Case of Tropical Tree Crops*, edited by François Ruf and Götz Schroth, 103–20. Dordrecht, Netherlands: Springer, 2015.

Schiebinger, Londa L. *Plants and Empire: Colonial Bioprospecting in the Atlantic World*. Cambridge, MA: Harvard University Press, 2004.

———. *Secret Cures of Slaves: People, Plants, and Medicine in the Eighteenth-Century Atlantic World*. Stanford: Stanford University Press, 2017.

Schisgall, Oscar. *Eyes on Tomorrow: The Evolution of Procter & Gamble*. Chicago: Doubleday, 1981.

Schleifer, David. "Categories Count: Trans Fat Labeling as a Technique of Corporate Governance." *Social Studies of Science* 43, no. 1 (February 2013): 54–77.

———. "Fear of Frying." *N+1* (blog), May 21, 2007. https://nplusonemag.com/online-only/online-only/fear-frying/.

———. "The Perfect Solution: How Trans Fats Became the Healthy Replacement for Saturated Fats." *Technology and Culture* 53, no. 1 (2012): 94–119.

———. "We Spent a Million Bucks and Then We Had To Do Something: The Unexpected Implications of Industry Involvement in Trans Fat Research." *Bulletin of Science, Technology & Society* 31, no. 6 (December, 2011): 460–71.

Schleifer, Philip, and Yixian Sun. "Emerging Markets and Private Governance: The Political Economy of Sustainable Palm Oil in China and India." *Review of International Political Economy* 25, no. 2 (March 2018): 190–214.

Schnapper, Bernard. *La politique et le commerce français dans le golfe de Guinée, de 1838 à 1871*. Paris: Mouton, 1961.

Schoneveld, George Christoffel. "The Politics of the Forest Frontier: Negotiating between Conservation, Development, and Indigenous Rights in Cross River State, Nigeria." *Land Use Policy* 38 (May 2014): 147–62.

Schoneveld, George Christoffel, D. Ekowati, A. Andrianto, and S. van der Haar. "Modeling Peat- and Forestland Conversion by Oil Palm Smallholders in Indonesian Borneo." *Environmental Research Letters* 14, no. 1 (January 2019): 014006.

Schultes, Richard Evans. "Taxonomic, Nomenclatural and Ethnobotanic Notes on Elaeis." *Elaeis* 2, no. 1 (1990): 172–87.

Schwartz, Stuart B. *Slaves, Peasants, and Rebels: Reconsidering Brazilian Slavery*. Urbana: University of Illinois Press, 1992.

Scott, Richenda. "Production for Trade." In *The Native Economies of Nigeria*, edited by Margery Perham, 217–92. London: Faber & Faber, 1946.

Secrett, Charles. "The Environmental Impact of Transmigration." *The Ecologist* 16, no. 2/3 (1986): 77–88.

Seibert, Julia. "More Continuity than Change? New Forms of Unfree Labour in the Belgian Congo, 1908–1930," edited by Marcel van der Linde, 369–86. Leiden: Brill, 2011.

Serrano, Angela. "Can Small-Scale Farming Save Oil Palm?" *Edge Effects*, February 20, 2020. https://edgeeffects.net/small-scale-oil-palm/.

Service agriculture du Côte-d'Ivoire. "Les Travaux d'aménagement et d'extension de La Palmeraie Naturelle de La Région de Dabou (Côte-d'Ivoire)." In

Conférence Franco-Britannique, 111–20. Paris: Centre Technique d'Agriculture Tropicale and IRHO, 1957.

Settié, Louis Edouard. *L'ère de l'économie Des Plantations En Côte d'Ivoire: 1900–1983*. Paris: L'Harmattan, 2014.

Shamsul Bahrin, Tunku, and Lee Boon Thong. *FELDA: 3 Decades of Evolution*. Kuala Lumpur: FELDA, 1998.

Shamsul Bahrin, Tunku, Lee Boon Thong, and Richard F. Dorall. "The Jengka Triangle: A Report on Research in Progress." In *Agricultural Expansion and Pioneer Settlements in the Humid Tropics*, edited by Walther Manshard and William B. Morgan. Hong Kong: United Nations University Press, 1988.

Shankar, Bhavani, and Corinna Hawkes. "India Has a Problem with Palm Oil: A Substantial Tax Could Be Part of the Solution." *BMJ: British Medical Journal* 347, no. 7933 (2013): 9.

Shaw, Thurstan. "Early Crops in Africa: A Review of the Evidence." In *Origins of African Plant Domestication*, edited by Jack R. Harlan, Jan J. M. de Wet, and Ann B. Stemler, 107–54. The Hague: Mouton Publishers, 1976.

Shengo, Michel Lutandula, Meschac-Bill Kime, Matanda Pascal Mambwe, and Trésor Kilwa Nyembo. "A Review of the Beneficiation of Copper-Cobalt-Bearing Minerals in the Democratic Republic of Congo." *Journal of Sustainable Mining* 18, no. 4 (November 2019): 226–46.

Sheridan, Richard B. *Doctors and Slaves: A Medical and Demographic History of Slavery in the British West Indies, 1680–1834*. Cambridge: Cambridge University Press, 1985.

Sherlock, T., and J. Towne. *The Duties, Drawbacks and Bounties of Customs and Excise, Payable in Great Britain*. Liverpool: G. F. Harris's Widow and Brothers, 1819.

Sherwood, Marika. *After Abolition: Britain and the Slave Trade after 1807*. London: I. B. Tauris, 2007.

Shields, Francine. "Palm Oil and Power: Women in an Era of Economic and Social Transition in 19th Century Yorubaland." PhD diss., University of Stirling, 1997.

——. "Those Who Remained Behind: Women Slaves in Nineteenth-Century Yorubaland." In *Identity in the Shadow of Slavery*, edited by Paul E. Lovejoy, 183–201. London: Continuum, 2000.

Sierra Leone Annual Report for 1857–1858. London: HMSO, 1858.

Sills, Morris W., and Harry O. Doty. *Detergents, Emulsifiers, and Emulsion Products as Market Outlets for Fats and Oils*. Marketing Research Report 46. Washington, DC: USDA, 1953.

Simmonds, P. L. *Commercial Products of the Vegetable Kingdom*. London: Petter and Company, 1854.

Simmons, D. "An Ethnographic Sketch of the Efik People." In *Efik Traders of Old Calabar*, edited by Daryll Forde, 1–26. London: Oxford University Press, 1956.

Simmons, W. H, and C. Ainsworth Mitchell. *Edible Fats and Oils: Their Composition, Manufacture and Analysis*. London: Scott, Greenwood & Co., 1911.

Simon, William J. "A Luso-African Formulary of the Late Eighteenth Century: Some Notes on Angolan Contributions to European Knowledge of Materia Medica." *Pharmacy in History* 18, no. 3 (1976): 103–14.

Simpson, Bradley R. *Economists with Guns: Authoritarian Development and U.S.-Indonesian Relations, 1960–1968*. Stanford: Stanford University Press, 2008.

Singh, Rajinder, Meilina Ong-Abdullah, Eng-Ti Leslie Low, Mohamad Arif Abdul Manaf, Rozana Rosli, Rajanaidu Nookiah, Leslie Cheng-Li Ooi, et al. "Oil Palm Genome Sequence Reveals Divergence of Interfertile Species in Old and New Worlds." *Nature* 500, no. 7462 (August 2013): 335–39.

Skertchly, J. Alfred. *Dahomey as It Is: Being a Narrative of Eight Months' Residence in That Country* . . . London: Chapman and Hall, 1874.

Skinner, Snider W. *The Agricultural Economy of the Ivory Coast*. USDA ERS-Foreign 69. Washington, DC: GPO, 1964.

Slater, R. "The Gold Coast: Some Facts and Figures." *Journal of the Royal African Society* 29, no. 116 (July 1930): 343–49.

Sloane, Hans. *A Voyage to the Islands Madera, Barbados, Nieves, S. Christophers and Jamaica*. . . . 2 vols., London: 1707, 1725.

Smiles, Robert. *Memoir of the Late Henry Booth of the Liverpool and Manchester, and Afterwards of the London and North-Western Railway*. London: Wyman & Sons, 1869.

Smith, Harold Hamel, and F. A. G. Pape. *Coco-Nuts: The Consols of the East*. London: Tropical Life, 1912.

Smith, J. *Trade and Travels in the Gulph of Guinea, Western Africa: With an Account of the Manners, Habits, Customs, and Religion of the Inhabitants*. London: Simpkin, Marshall, and Company, 1851.

Smith, Michael Stephen. *The Emergence of Modern Business Enterprise in France, 1800–1930*. Cambridge, MA: Harvard University Press, 2006.

Smith, Nigel, J. T. Williams, Donald L. Plucknett, and Jennifer P. Talbot. *Tropical Forests and Their Crops*. Ithaca, NY: Cornell University Press, 1992.

Smith, Robert A. "Oil Palms: The West African Industry and Mr. Ormsby Gore's Report." *The Planter* 7, no. 8 (March 1927): 220–23.

Smith, Robert S. *The Lagos Consulate, 1851–1861*. Berkeley: University of California Press, 1979.

Smith, Watson. "Report on Section III of the Manchester Royal Jubilee Exhibition–The Chemical and Allied Industries. *Journal of the Society of Chemical Industry* 6, no. 10 (1887): 624–43.

Smith, William. *A New Voyage to Guinea: Describing the Customs, Manners, Soil, Climate, Habits, Buildings, Education, Manual Arts, Agriculture, Trade, Employments, Languages, Ranks of Distinction, Habitations, Diversions, Marriages, and Whatever Else Is Memorable among the Inhabitants*. . . . 2nd ed. London: J. Nourse, 1745.

Smith, Woodruff D. *The German Colonial Empire*. Chapel Hill, NC: University of North Carolina Press, 1978.

Snodgrass, Katharine. *Margarine as a Butter Substitute*. Stanford: Food Research Institute, 1930.

Soames, James. "The Soap Duty," 1853. https://www.jstor.org/stable/60221493.

Soh, A. C., C. C. Tan, P. S. Chew, S. P. Chong, Y. W. Ho, C. K. Wong, C. N. Choo, H. Nor Azura, and K. Kumar. "Commerical-Scale Propagation and Planting of Elite Oil Palm Clones: Research and Development towards Realization." *Journal of Oil Palm Research* 23 (2011): 935–52.

Soluri, John. *Banana Cultures: Agriculture, Consumption, and Environmental Change in Honduras and the United States*. Austin: University of Texas Press, 2005.

———. "Bananas before Plantations. Smallholders, Shippers, and Colonial Policy in Jamaica, 1870–1910." *Iberoamericana* 6, no. 23 (2006): 143–59.

———. "Campesinos and the Hidden History of Biodiversity." *RCC Perspectives* 7 (2013): 67–72.

Soskin, Selik. *Die Ölpalme: ein Beitrag zu ihrer Kultur*. Berlin: Kolonial-wirtschaftliches Komitee, 1909.

Soumonni, Elisee. "The Compatibility of the Slave and Palm Oil Trades in Dahomey, 1818–1858." In *From Slave Trade to "Legitimate" Commerce: The Commercial Transition in Nineteenth-Century West Africa*, edited by Robin Law. Cambridge: Cambridge University Press, 1995.

———. "Dahomean Economic Policy under Ghezo 1818–1858: A Reconsideration." *Journal of the Historical Society of Nigeria* 10, no. 2 (1980): 1–11.

South, F. W. "The Position of the Coconut and Oil Palm Industries." *The Planter* 15, no. 11 (November 1934): 423–27.

Sowunmi, M. Adebisi. "The Significance of the Oil Palm (*Elaeis guineensis* Jacq.) in the Late Holocene Environments of West and West Central Africa: A Further Consideration." *Vegetation History and Archaeobotany* 8, no. 3 (September 1999): 199–210.

Sparnaaij, L. D. "The Analysis of Bunch Production in the Oil Palm." PhD diss., Landbouwhogeschool te Wageningen, 1959.

Spiekermann, Uwe. "Redefining Food: The Standardization of Products and Production in Europe and the United States, 1880–1914." *History and Technology* 27, no. 1 (March 2011): 11–36.

Spinage, Clive Alfred. *African Ecology: Benchmarks and Historical Perspectives*. Berlin: Springer, 2012.

Spring, F. G. "Cover Crops and Clean Weeding." *Federated Malay States Agricultural Bulletin* 9, no. 3 (1921): 169–74.

Stamler, J. "Diet and Atherosclerotic Disease." *Journal of the American Dietetic Association* 34, no. 7 (October 1958): 701–7.

Staples, Amy L. S. *The Birth of Development: How the World Bank, Food and Agriculture Organization, and World Health Organization Changed the World, 1945–1965*. Kent, OH: Kent State University Press, 2006.

Steinberg, Daniel. *The Cholesterol Wars: The Skeptics vs. the Preponderance of Evidence*. London: Academic, 2007.

Stepina, V., and V. Vesely. *Lubricants and Special Fluids*. Amsterdam: Elsevier, 1992.

Stilliard, N. H. "The Rise and Development of Legitimate Trade in Palm Oil." MA thesis, University of Birmingham, 1938.

Stockdale, Frank. "Report on a Visit to Malaya." *Malayan Agricultural Journal* 27 (September 1939): 371–81.

Stockpile Report to the Congress July–December 1956. Washington, DC: Office of Defense Mobilization, 1956.

Stockpiling—Palm Oil: Hearings before the United States Senate Committee on Government Operations, Permanent Subcommittee on Investigations, Eighty-Third Congress, First Session, on Feb. 25, 1953. Washington, DC: GPO, 1953.

Stoler, Ann. *Along the Archival Grain: Epistemic Anxieties and Colonial Common Sense*. Princeton, NJ: Princeton University Press, 2010.

——. *Capitalism and Confrontation in Sumatra's Plantation Belt, 1870–1979*. 2nd ed. Ann Arbor: University of Michigan Press, 1995.

Stopler, Wolfgang. *Inside Independent Nigeria: Diaries of Wolfgang Stopler, 1960–1962*. Edited by Clive Gray. Aldershot: Ashgate, 2003.

Strickland, C. F. *Report on the Introduction of Co-Operative Societies into Nigeria*. Lagos: Government Printer, 1934.

Strickrodt, Silke. *Afro-European Trade in the Atlantic World: The Western Slave Coast, c.1550–c.1885*. Suffolk: James Currey, 2015.

Stumpf, P. K. "Biosynthesis of Fatty Acids in Higher Plants." In *Oil Crops of the World*, edited by Gerhard Röbbelen, R. Keith Downey, and Amram Ashri, 38–62. New York: McGraw-Hill, 1989.

Suppiah, Ummadevi, and Sivachandralingam Sundara Raja. *The Chettiar Role in Malaysia's Economic History*. Kuala Lumpur: University of Malaya Press, 2016.

Suret-Canale, Jean. *French Colonialism in Tropical Africa 1900–1945*. Translated by Till Gottheiner. New York: Pica Press, 1971.

——. "L'industrie Des Oléagineux En A.O.F." *Cahiers d'outre-Mer* 11, no. 3 (1950): 280–88.

Surre, Christian. *L'Institut de Recherches pour les Huiles et Oléagineux: 1942–1984*. Paris: CIRAD, 1993.

Sutton, Inez. "Labour in Commercial Agriculture in Ghana in the Late Nineteenth and Early Twentieth Centuries." *Journal of African History* 24 (1983): 461–83.

Sutton, Keith. "Agribusiness on a Grand Scale—Felda's Sahabat Complex in East Malaysia." *Singapore Journal of Tropical Geography* 22, no. 1 (2001): 90–105.

Sutton, Keith, and Amriah Buang. "A New Role for Malaysia's FELDA: From Land Settlement Agency to Plantation Company." *Geography* 80, no. 2 (1995): 125–37.

Swank, James Moore. *All about Tinplates: Great Britain's Monopoly in the Manufacture of Tinplates for the American Market*. Philadelphia: American Iron and Steel Association, 1889.

Swansea Chamber of Commerce, and Edwin P. Jones. *Commercial Year Book*. London: Bemrose & Sons, 1915.

Sweet, James Hoke. *Recreating Africa: Culture, Kinship, and Religion in the African-Portuguese World, 1441–1770*. Chapel Hill: University of North Carolina Press, 2003.

Syfert, Dwight N. "A History of the Liberian Coasting Trade, 1821–1900." PhD diss., Indiana University, Bloomington, 1977.

Szczepanski, Kallie. "Land Policy and ADAT Law in Indonesia's Forests." *Pacific Rim Law & Policy Journal* 11, no. 1 (2002): 231–55.

Székely, László. *Tropic Fever: The Adventures of a Planter in Sumatra*. Translated by Marion Saunders. Kuala Lumpur: Oxford University Press, 1979.

Talbot, Frederick Arthur Ambrose. *The Oil Conquest of the World*. Philadelphia: J. B. Lippincott, 1914.

Talbot, Percy Amaury. *In the Shadow of the Bush*. London: Heinemann, 1912.

Talib, Rokiah. *Raja Muhammad Alias: The Architect of Felda*. Bangi: Penerbit Universiti Kebangsaan Malaysia, 2009.

Tandon, Prakash. *Punjabi Saga (1857–1987)*. New Delhi: Viking, 1988.

Tate, D. J. M. *The RGA History of the Plantation Industry in the Malay Peninsula*. New York: Oxford University Press, 1996.

Taussig, Michael. *Palma Africana*. Chicago: University of Chicago Press, 2018.

Teaiwa, Katerina Martina. *Consuming Ocean Island: Stories of People and Phosphate from Banaba*. Bloomington: Indiana University Press, 2014.

Teltscher, Kate. *Palace of Palms: Tropical Dreams and the Making of Kew*. London: Picador, 2020.

Tempel, F. J. *Raw Materials and Pricing*. Rotterdam: Unilever NV, 1966.

Teoh Cheng Hai. "Key Sustainability Issues in the Palm Oil Sector." Discussion paper. Washington, DC: World Bank, 2010. http://siteresources.worldbank.org/INTINDONESIA/Resources/226271-1170911056314/Discussion.Paper_palmoil.pdf.

"The Varieties of the Oil Palm in West Africa. (*Elaeis guineensis*, Jacq.)." *Bulletin of Miscellaneous Information (Royal Botanic Gardens, Kew)* 1909, no. 2 (1909): 33–49.

Thee Kian Wie. "Colonial Extraction in the Indonesian Archipelago: A Long Historical View." In *Colonial Exploitation and Economic Development: The Belgian Congo and the Netherlands Indies Compared*, edited by Ewout Frankema and Frans Buelens, 41–59. London: Routledge, 2013.

———. *Plantation Agriculture and Export Growth: An Economic History of East Sumatra, 1863–1942*. Jakarta: KEKNAS-LIPI, 1977.

Theodoridis, Dimitrios. "The Ecological Footprint of Early-Modern Commodities Coefficients of Land Use per Unit of Product." *Göteborg Papers in Economic History* 21 (2017).

Theodoridis, Dimitrios, Paul Warde, and Astrid Kander. "Trade and Overcoming Land Constraints in British Industrialization: An Empirical Assessment." *Journal of Global History* 13, no. 3 (November 2018): 328–51.

Thomas, Northcote Whitridge. *Anthropological Report on the Edo-Speaking Peoples of Nigeria, Part I: Law and Customs*. London: Harrison and Sons, 1910.

Thompson, George. *The Palm Land; Or, West Africa, Illustrated. Being a History of Missionary Labors and Travels. . . .* Cincinnati: Moore, Wilstach, Keys, 1859.

Thompson, H. N. "The Forests of Southern Nigeria." *Journal of the Royal African Society* 10, no. 38 (1911): 121–45.

Thompson, Peter. “Henry Drax’s Instructions on the Management of a Seventeenth-Century Barbadian Sugar Plantation.” *William and Mary Quarterly* 66, no. 3 (2009): 565–604.
Thompson, Virginia, and Richard Adloff. “French Economic Policy in Tropical Africa.” In *Colonialism in Africa*, edited by Peter Duignan and Lewis H. Gann. Vol. 4. Cambridge: Cambridge University Press, 1975.
——. *French West Africa*. New York: Greenwood Press, 1969.
Tichit, M. “L’amélioration Des Palmeraies Naturelles de La Zone Littorale Du Cameroun.” In *Conférence Franco-Britannique*, 77–90. 14. Paris: Centre Technique d’Agriculture Tropicale and IRHO, 1957.
Tignor, Robert L. *Capitalism and Nationalism at the End of Empire: State and Business in Decolonizing Egypt, Nigeria, and Kenya, 1945–1963*. Princeton, NJ: Princeton University Press, 1998.
Tindall, George Brown. *South Carolina Negroes, 1877–1900*. Charlestown: University of South Carolina Press, 1952.
Tingey, Paul. “An Epitome of the Results of Transporting Oil Palm Fruit from West Africa, and the Storage of Same in England.” In *Oil Palms and Their Fruit*, edited by A. A. L. Rutgers, 18–23. London: Griffith, 1922.
Tinker, Bernard. *My Life’s Adventures*. Bloomington, IN: AuthorHouse, 2017.
Tippetts-Abbett-McCarthy-Stratton, and Hunting Technical Services Ltd. “The Jengka Triangle Report: The Outline Master Plan.” Federal Land Development Authority, 1967. http://lib.perdana.org.my/PLF/Digitisation_MTAR/PART1/Jilid%204/MTAR_PDF_04/The%20Outline%20Master%20Plan.pdf.
“To Thomas Motley.” *London Journal of Arts and Sciences* 4 (May 1822): 183–84.
Tomich, Dale. “Une Petite Guinee: Provision Ground and Plantation in Martinique.” In *The Slaves’ Economy: Independent Production by Slaves in the Americas*, edited by Ira Berlin and Philip D. Morgan, 68–91. London: Frank Cass, 1991.
Torday, E. “Bushongo Mythology.” *Folklore* 22, no. 1 (1911): 41–47.
Torres, G. A., G. A. Sarria, G. Martinez, F. Varon, A. Drenth, and D. I. Guest. “Bud Rot Caused by *Phytophthora palmivora*: A Destructive Emerging Disease of Oil Palm.” *Phytopathology* 106, no. 4 (April 2016): 320–29.
Tovar, Ernesto Vargas. “Décadas 1960 y 1970. La Palma de Aceite: De Fincas a Empresas.” *Palmas* 23, no. 3 (2002): 86–92.
“The Training Schools of Price’s Patent Candle Company.” *Fraser’s Magazine* 46, no. 272 (July 1852): 106–12.
Traugott, Mark. “The Economic Origins of the Kwilu Rebellion.” *Comparative Studies in Society and History* 21, no. 3 (1979): 459–79.
Tsing, Anna. *Friction: An Ethnography of Global Connection*. Princeton, NJ: Princeton University Press, 2005.
——. “Land as Law: Negotiating the Meaning of Property in Indonesia.” In *Land, Property, and the Environment*, edited by John F. Richards, 94–137. Oakland: ICS Press, 2002.
Tucker, S. *Abbeokuta; or, Sunrise within the Tropics: An Outline of the Origin and Progress of the Yoruba Mission*. London: James Nisbet and Co., 1853.

Tullis, Paul. “How the World Got Hooked on Palm Oil.” *The Guardian*, February 19, 2019. https://www.theguardian.com/news/2019/feb/19/palm-oil-ingredient-biscuits-shampoo-environmental.

Tully, John. *The Devil’s Milk: A Social History of Rubber.* New York: New York University Press, 2011.

Turner, Peter D., and R. A. Gillbanks. *Oil Palm Cultivation and Management.* Kuala Lumpur: Incorporated Society of Planters, 1974.

Tyler, Geoff, and Grahame Dixie. “Investing in Agribusiness: A Retrospective View of a Development Bank’s Investments in Agribusiness in Africa and Southeast Asia and the Pacific.” Agriculture and Environmental Services Discussion paper. Washington, DC: World Bank, 2013. https://openknowledge.worldbank.org/bitstream/handle/10986/16660/810830REVISED0otingoinoAgribusiness.pdf?sequence=1.

Tyson, Adam, Helena Varkkey, and Shofwan Al Banna Choiruzzad. “Deconstructing the Palm Oil Industry Narrative in Indonesia: Evidence from Riau Province.” *Contemporary Southeast Asia* 40, no. 3 (December 2018): 422–48.

Uchendu, Victor Chikezie. *Igbo of Southeast Nigeria.* New York: Rinehart and Winston, 1965.

Udo, Reuben. K. *Geographical Regions of Nigeria.* Berkeley: University of California Press, 1970.

———. “Sixty Years of Plantation Agriculture in Southern Nigeria: 1902–1962.” *Economic Geography* 41, no. 4 (October 1965): 356–68.

Udom, D. S. “Nigerian Government Policy: Schemes for Smallholders’ Oil-Palm Planting and Rehabilitation between 1928 and 1981.” *Journal of the Nigerian Institute for Oil Palm Research* 7 (1986): 134–75.

Uekötter, Frank, ed. *Comparing Apples, Oranges, and Cotton: Environmental Histories of the Global Plantation.* Frankurter am Main: Campus, 2014.

Ukegbu, Basil Nnanna. “Production in the Nigerian Oil Palm Industry, 1900–1954.” PhD diss., University of London, 1974.

Uku, Edward Kanu. *Seeds in the Palm of Your Hand.* Edited by David P. Henige. Madison: African Studies Program, University of Wisconsin–Madison, 1993.

“Unguents: Their Chemical and Physical Properties.” *Mechanics’ Magazine* 13 (30 June 1865): 409–10.

Unilever Art, Archives & Records Management. “Loders & Nucoline Ltd.,” 2019. http://unilever-archives.com/Record.aspx?src=CalmView.Catalog&id=GB1752.LNL.

United Africa Company. “The Future of the Nigerian Oil Industry.” *African Affairs* 47, no. 186 (1948): 41–51.

United States Tariff Commission. *Summaries of Tariff Information.* Vol. 1, pt. 4. Washington, DC, 1948.

Unwin, Arthur Harold. *West African Forests and Forestry.* London: T. F. Unwin, 1920.

Ure, Andrew. *A Dictionary of Arts, Manufactures, and Mines: Containing a Clear Exposition of Their Principles and Practice*. London: Longman et al., 1839.
USDA. *Inventory of Seeds and Plants Imported . . . from July 1 to Septemer 30, 1916*. Washington: GPO, 1921.
Usoro, Eno J. *The Nigerian Oil Palm Industry: Government Policy and Export Production, 1906–1965*. Ibadan, Nigeria: Ibadan University Press, 1974.
Uzozie, Levi Chukwuemeka. "Tradition and Change in Igbo Food Crop Production Systems: A Geographical Appraisal." PhD diss., University of London, 1979.
Valencia, Sebastián Ocampo. "Agroindustria y conflicto armado. El caso de la palma de aceite." *Colombia Internacional*, no. 70 (2009): 169–90.
Van Allen, Judith. "'Sitting on a Man': Colonialism and the Lost Political Institutions of Igbo Women." *Canadian Journal of African Studies/La Revue Canadienne Des Études Africaines* 6, no. 2 (1972): 165–81.
Van Der Ven, F. J. M. "Jurgens [Hzn.], Antonius Johannes (1867–1945)." Biografisch Woordenboek van Nederland, 2013. http://resources.huygens.knaw.nl/bwn1880-2000/lemmata/bwn4/jurgens.
Vanderyst, Hyacinth. "Les Palmeraies Dan Leurs Rapports Avec l'Ethnologie." In *Institut Colonial de Marseille. Mémoires et Rapports Sur Les Matières Grasses. T. III. Le Palmier à Huile*, 117–23. Marseille: Institut colonial, 1928.
———. "Origine Des Palmeraies Du Moyen-Kwilu." *Bulletin Agricole Du Congo Belge* 10 (1919): 70–78.
———. "La Récolte Des Produits de l'Elaeis." *Bulletin Agricole Du Congo Belge* 11 (1920): 22–36.
Van Itallie, T. B. "Nutritional Research in Atherosclerosis; a Progress Report." *Journal of the American Dietetic Association* 34, no. 3 (March 1958): 248–53.
Van Pelt, Gaston. *La Culture du Palmier à Huile et La Préparation des Huiles et Amandes de Palme*. Marseille: Institut Colonial, 1930.
———. "La Culture et l'Exploitation moderne du palmier à huile, situation respective des Indes Néerlandaises et de l'Afrique Occidentale Française." In *Mémoires et rapports sur les matières grasses. T. III. Le Palmier à huile*, 157–86. Marseilles: Institut colonial de Marseille, 1928.
———. "The Oil Palm: Report of the Mission to West Africa." *Agricultural Bulletin of the Federated Malay States* 8, no. 1 and 2 (1920): 7–10 and 100–106.
Van Heurn, J. C. "De Bereiding van Palmvet." *Teysmannia* 32 (1921): 522–34.
Vansina, Jan. *Being Colonized: The Kuba Experience in Rural Congo, 1880–1960*. Madison: University of Wisconsin Press, 2010.
Van Stuyvenberg, J. H. *Margarine: An Economic, Social and Scientific History 1896–1969*. Liverpool: Liverpool University Press, 1969.
Varkkey, Helena. *The Haze Problem in Southeast Asia: Palm Oil and Patronage*. London: Routledge, 2015.
Varkkey, Helena, Adam Tyson, and Shofwan Al Banna Choiruzzad. "Palm Oil Intensification and Expansion in Indonesia and Malaysia: Environmental and Socio-Political Factors Influencing Policy." *Forest Policy and Economics* 92 (July 2018): 148–59.

"The Varieties of Oil-Palm in West Africa. (*Elaeis guineensis*, Jacq.)." *Bulletin of Miscellaneous Information (Royal Botanic Gardens, Kew)* 1914, no. 8 (1914): 285–88.

Vaux, Charles Grant de. *The History of Mauritius, or the Isle of France, and the Neighbouring Islands: From Their First Discovery to the Present Time. . . .* London: Bulmer, 1801.

vE, G. "Expanding N. I. Palm Oil Industry Depends on U. S." *FES* 7, no. 13 (June 1938): 153–54.

Veeser, Cyrus. "A Forgotten Instrument of Global Capitalism? International Concessions, 1870–1930." *International History Review* 35, no. 5 (2013): 1136–55.

Vincent, Brian. "Cotton Growing in Southern Nigeria: Missionary, Mercantile, Imperial and Colonial Government Involvement versus African Realities c. 1845 to 1939." PhD diss., Simon Fraser University, Burnaby, Canada, 1977.

Voeks, Robert. "African Medicine and Magic in the Americas." *Geographical Review* 83, no. 1 (1993): 66–78.

Vogt, John L. "The Early São Tomé-Principe Slave Trade with Mina, 1500–1540." *International Journal of African Historical Studies* 6, no. 3 (1973): 453–67.

Voigt, J. O. *Hortus Suburaanus Calcuttensis: A Catalogue of the Plants Which Have Been Cultivated in the East India Company's Botanical Garden, Calcutta. . . .* Calcutta: Bishop's College Press, 1845.

Vos, Jelmer. "Coffee, Cash, and Consumption: Rethinking Commodity Production in the Global South." *Radical History Review* 2018, no. 131 (2018): 183–88.

Vriese, W. H. de. *Tuinbouw-flora van Nederland en zijne overzeesche bezittingen. . . .* Leyden: A. W. Sythoff, 1855.

Waddell, Hope Masterton. *Twenty-Nine Years in the West Indies and Central Africa: A Review of Missionary Work and Adventure. 1829–1858.* London: T. Nelson and Sons, 1863.

Wadström, Carl Bernhard. *An Essay on Colonization, Particularly Applied to the Western Coast of Africa, with Some Free Thoughts on Cultivation and Commerce. . . .* London: Darton and Harvey, 1794.

Warde, Paul. "Trees, Trade and Textiles: Potash Imports and Ecological Dependency in British Industry, c.1550–1770." *Past & Present* 240, no. 1 (2018): 47–82.

Wardlaw, C. W. "Notes on a Visit to the Cameroons, Nigeria and the Belgian Congo, 1948." London: United Africa Company, 1948.

Warner, Melanie. *Pandora's Lunchbox: How Processed Food Took Over the American Meal.* New York: Simon and Schuster, 2013.

Wartena, Dorothea. "Styles of Making a Living and Ecological Change on the Fon and Adja Plateaux in South Benin, ca. 1600–1990." PhD diss., Wageningen University, 2006.

Waterman, H. I. *Hydrogenation of Fatty Oils.* New York: Elsevier, 1951.

Watkins, Case. "African Oil Palms, Colonial Socioecological Transformation and the Making of an Afro-Brazilian Landscape in Bahia, Brazil." *Environment and History* 21, no. 1 (1 February 2015): 13–42.

——. "An Afro-Brazilian Landscape: African Oil Palms and Socioecological Change in Bahia, Brazil." PhD diss., Louisiana State University, Baton Rouge, 2015.

——. "Landscapes and Resistance in the African Diaspora: Five Centuries of Palm Oil on Bahia's Dendê Coast." *Journal of Rural Studies* 61 (2018): 137–54.

——. *Palm Oil Diaspora: Afro-Brazilian Landscapes and Economies on Bahia's Dendê Coast*. Cambridge: Cambridge University Press, 2021.

Watson, John Selby. *The Reasoning Power in Animals*. London: Reeve & Company, 1867.

Wedel, Pailin. "'When We Lost the Forest, We Lost Everything.'" Human Rights Watch, September 22, 2019. https://www.hrw.org/report/2019/09/23/when-we-lost-forest-we-lost-everything/oil-palm-plantations-and-rights-violations.

Weeks, John H. *Among the Primitive Bakongo; a Record of Thirty Years' Close Intercourse with the Bakongo and Other Tribes of Equatorial Africa, with a Description of Their Habits, Customs & Religious Beliefs*. New York: Negro Universities Press, 1969.

Weigert, Stephen L. "Congo/Zaire: The Kwilu Rebellion, 1963–68." In *Traditional Religion and Guerrilla Warfare in Modern Africa*, edited by Stephen L. Weigert, 49–66. London: Palgrave Macmillan, 1996.

Werkö, Lars. "End of the Road for the Diet-Heart Theory?" *Scandinavian Cardiovascular Journal: SCJ* 42, no. 4 (August 2008): 250–55.

West Africa Commission. "Technical Report." London: Leverhulme Trust, 1943.

West Africa: Correspondence Respecting the Grant of Exclusive Rights for the Extraction of Oil from Palm Fruits. Cd. 6561. London: HMSO, 1913.

West Africa. Palm Oil and Palm Kernels. Report of a Committee Appointed by the Secretary of State for the Colonies, September, 1923, to Consider the Best Means of Securing Improved and Increased Production. Colonial no. 10. London: HMSO, 1925.

"The West African Oil Palm. (*Elaeis guineensis*, Jacq.)." *Bulletin of Miscellaneous Information (Royal Botanic Gardens, Kew)* 1918, no. 2/3 (1918): 121–24.

West, Paige. *Dispossession and the Environment: Rhetoric and Inequality in Papua New Guinea*. New York: Columbia University Press, 2016.

Western Africa Regional Office. "Appraisal of Nucleus Estate/Smallholder Oil Palm Project, Rivers State, Nigeria." Washington, DC: World Bank, 1978.

White, Nicholas J. *British Business in Post-Colonial Malaysia, 1957–70: "Neo-Colonialism" or "Disengagement"?* London: RoutledgeCurzon, 2004.

——. *Business, Government, and the End of Empire: Malaya, 1942–1957*. Kuala Lumpur: Oxford University Press, 1996.

——. "Surviving Sukarno: British Business in Post-Colonial Indonesia, 1950–1967." *Modern Asian Studies* 46, no. 5 (September 2012): 1277–315.

——. "'Ungentlemanly Capitalism': John Hay and Malaya, 1904–1964." *Management & Organizational History* 14, no. 1 (2019): 98–122.

Whitford, John. *Trading Life in Western and Central Africa*. Liverpool: The "Porcupine" Office, 1877.

Whitten, Anthony J. "Indonesia's Transmigration Program and Its Role in the Loss of Tropical Rain Forests." *Conservation Biology* 1, no. 3 (1987): 239–46.

Whorton, James C. *The Arsenic Century: How Victorian Britain Was Poisoned at Home, Work, and Play.* Oxford: Oxford University Press, 2010.

Wickens, G. E. "A Preliminary Note on the Quaternary Leaf Impressions from Jebel Marra and the Status of the Oil Palm (*Elaeis guineensis* Jacq.) in the Sudan." *Sudan Notes and Records* 55 (1974): 185–88.

Wickizer, Vernon D. "The Smallholder in Tropical Export Crop Production." *Food Research Institute Studies* 1, no. 1 (1960): 49–99.

Williams, Gavin. "The World Bank in Rural Nigeria, Revisited: A Review of the World Bank's 'Nigeria: Agricultural Sector Review 1987.'" *Review of African Political Economy* 15, no. 43 (November 1988): 42–67.

Williams, W. Mattieu. "The Chemistry of Cookery." *Popular Science Monthly* 24 (November 1, 1883): 98–104.

Wilson, Charles. *History of Unilever.* 2 vols. London: Cassell, 1954.

——. *Unilever 1945–1965: Challenge and Response in the Post-War Industrial Revolution.* London: Cassell, 1968.

Wilson, George F. "On the Manufactures of Price's Patent Candle Company." *Journal of the Society of Arts* 4, no. 166 (1856): 148–56.

——. *On the Stearic Candle Manufacture (Lect.). To Which Are Appended Papers on the Slave Trade and the Soap Tax.* London: Lewis and Son, 1852.

Wilson, Louis E. "The 'Bloodless Conquest' in Southeastern Ghana: The Huza and Territorial Expansion of the Krobo in the 19th Century." *International Journal of African Historical Studies* 23, no. 2 (1990): 269.

——. *The Krobo People of Ghana to 1892: A Political and Social History.* Athens: Ohio University Press, 1991.

Wilson, W. Arthur. "A Slump Soliloquy." *The Planter* 1, no. 6 (January 1921): 19–21.

Winson, Anthony. *The Industrial Diet: The Degradation of Food and the Struggle for Healthy Eating.* New York: New York University Press, 2013.

Winterbottom, Thomas Masterman. *An Account of the Native Africans in the Neighbourhood of Sierra Leone: To Which Is Added an Account of the Present State of Medicine among Them.* London: C. Whittingham, 1803.

Wisniak, Jaime. "Edmund Frémy." *Revista CENIC* 44, no. 1 (2013): 1–11.

Wolf, Joseph. *Injuries and Health Hazards Associated with Malaysia's Oil Palm Industry.* Kuala Lumpur: Incorporated Society of Planters, 1994.

Wolff, Christiane. "Economics of Oil Palm Production in Chiapas, Mexico." MS thesis, Cornell University, Ithaca, NY, 1999.

Wolsak, Anna Saskia. "Of Fishpots, Bonnets, and Wine: The Cultural History of the Bermuda Palmetto." Vancouver: University of British Columbia, 2017.

Woolford, Donald. "Trouble Smoulders under the Oil Palms." *Pacific Islands Monthly* 42, no. 8 (August 1971): 29.

World Bank. "Appraisal of Integrated Agricultural Development Project: Sierra Leone." Washington, DC: World Bank, June 6, 1972.

——. "Cameroon—Rubber and Oil Palm Projects." Washington, DC: World Bank, 1985.

——. "Implementation Completion Report: Nigeria Tree Crops Project." Washington, DC: World Bank, 1996.
——. "Indonesia Impact Evaluation Report: Transmigration I, II, and III." Washington, DC: World Bank, 1994.
——. "The Jengka Triangle Projects in Malaysia." Impact Evaluation Report. Washington, DC: World Bank, 1987.
——. "Liberia: Decoris Oil Palm Project." Project Completion Report. Washington, DC: World Bank, 1991.
——. "Malaysia: Jengka Triangle Resettlement Project." Washington, DC: World Bank, August 1979.
——. "Malaysia—Sabah Land Settlement and Environmental Management Project." Project Completion Report. Washington, DC: World Bank, 1995.
——. "Nucleus Estate and Smallholders II Project." Washington, DC: World Bank, 1978.
——. "Project Completion Report, Nigeria, Imo Oil Palm Project (Loan 1191-UNI), Rivers Oil Palm Project (Loan 1591-UNI)." Washington, DC: World Bank, 1988.
——. "Project Performance Audit Report: Indonesia, Nucleus Estate and Smallholders I, II, and III Projects." Washington, DC: World Bank, 1989.
——. "Sierra Leone Eastern Integrated Agricultural Development Project III." Staff Appraisal. Washington, DC: World Bank, November 26, 1980.
——. "Sierra Leone Integrated Agricultural Development Project." Project Performance Audit Report. Washington, DC: World Bank, May 22, 1978.
——. "Supplementary Annexes to the Nigerian Oil Palm Projects Appraisal Reports." Washington, DC: World Bank, December 18, 1974.
——. "Zaire—Oil Palm Development Project." Washington, DC: World Bank, 1978.
——. "Zaire—Oil Palm Development Project." Washington, DC: World Bank, 1991.
World Rainforest Movement. "FERONIA in the Democratic Republic of the Congo: Harassment, Violence and Oppression." September 21, 2017. https://wrm.org.uy/articles-from-the-wrm-bulletin/section1/feronia-in-the-democratic-republic-of-the-congo-harassment-violence-and-oppression/.
World Resources Institute. "By the Numbers: The Value of Tropical Forests in the Climate Change Equation," October 4, 2018. https://www.wri.org/blog/2018/10/numbers-value-tropical-forests-climate-change-equation.
Wubs, Ben. *International Business and National War Interests: Unilever Between Reich and Empire, 1939–45*. Abingdon, UK: Routledge, 2008.
Wynter. "Candle Making." *Once a Week* 1, no. 4 (July 23, 1859): 78–80.
Yacob, Shakila. "Government, Business and Lobbyists: The Politics of Palm Oil in US–Malaysia Relations." *International History Review* 41, no. 4 (2019): 909–30.
Yacob, Shakila, and Khadijah Md Khalid. "Adapt or Divest? The New Economic Policy and Foreign Businesses in Malaysia (1970–2000)." *Journal of Imperial and Commonwealth History* 40, no. 3 (2012): 459–82.
Yacob, Shakila, and Nicholas J. White. "The 'Unfinished Business' of Malaysia's Decolonisation: The Origins of the Guthrie 'Dawn Raid.'" *Modern Asian Studies* 44, no. 5 (2010): 919–60.

Zallen, Jeremy. *American Lucifers: The Dark History of Artificial Light, 1750–1865*. Chapel Hill: University of North Carolina Press, 2019.

——. "American Lucifers: Makers and Masters of the Means of Light, 1750–1900." PhD diss., Harvard University, Cambridge, MA, 2014.

Zaugg, Roberto, and Craig Koslofsky. "Ship's Surgeon Johann Peter Oettinger: A Hinterlander in the Atlantic Slave Trade, 1682–1696." In *Slavery Hinterland: Transatlantic Slavery and Continental Europe, 1680–1850*, edited by Felix Brahm and Eve Rosenhaft, 25–44. Woodbridge, UK: Boydell, 2016.

Zeide, Anna. *Canned: The Rise and Fall of Consumer Confidence in the American Food Industry*. Berkeley: University of California Press, 2018.

Zen, Zahari, Colin Barlow, Ria Gondowarsito, and John F. McCarthy. "Interventions to Promote Smallholder Oil Palm and Socio-Economic Improvement in Indonesia." In *The Oil Palm Complex: Smallholders, Agribusiness and the State in Indonesia and Malaysia*, edited by R. A. Cramb and John F. McCarthy, 78–108. Singapore: NUS Press, 2016.

Zen, Zahari, and John F. McCarthy. "Agribusiness, Agrarian Change, and the Fate of Oil Palm Smallholders in Jambi." In *The Oil Palm Complex: Smallholders, Agribusiness and the State in Indonesia and Malaysia*, edited by R. A. Cramb and John F. McCarthy, 109–54. Singapore: NUS Press, 2016.

Zeven, A. C. "On the Origin of the Oil Palm (*Elaeis guineensis* Jacq.)." *Grana Palynologica* 5, no. 1 (January 1964): 121–23.

——. "The Partial and Complete Domestication of the Oil Palm (*Elaeis guineensis*)." *Economic Botany* 26, no. 3 (July 1972): 274–79.

——. *The Semi-Wild Oil Palm and Its Industry in Africa*. Agricultural Research Report 689. Wageningen: Centre for Agricultural Publications and Documents, 1967.

Zimmerman, Andrew. "A German Alabama in Africa: The Tuskegee Expedition to German Togo and the Transnational Origins of West African Cotton Growers." *American Historical Review* 110, no. 5 (2005): 1362–98.

——. *Alabama in Africa: Booker T. Washington, the German Empire, and the Globalization of the New South*. Princeton, NJ: Princeton University Press, 2010.

Index